About the Authors

It was my immense privilege and pleasure to have been given the opportunity to recount the amazing and often life-defining travels and memoires of the enigmatic and incomparable Bob McGuire. I was a boyhood friend of his two younger brothers, Ken and John, who were one year my senior, and one year my junior, in that order. It was the Fall of 1959, when my family and I moved to the quiet, suburban neighborhood of East 44th Avenue near Joyce Road in southeast Vancouver, B.C. Our house was just a few doors down from the McGuire homestead.

While I quickly gravitated toward Ken, John and the other boys of similar ages living on the block, the older McGuire brothers, Bob and Mike were of a different generation. Not unlike my own older brother, they were visible to me, but mostly unapproachable. In those days, it was an unwritten rule that one kept to their own peer group and strayed neither too far above nor below it. However, it was always fascinating to surreptitiously observe the behaviors and life-styles of those "Cats," we perceived to be worldly, and who we would one day attempt to emulate.

Without the need for detailed explanation, my home-life as a youth was extremely difficult. It was filled with an over-abundance of negativity and external abuses. Thankfully, I attribute the rejuvenating qualities of friendship with Ken, John and the other boys from the block, as the defining factor that kept me more or less grounded and functional as a developing youth. Unfortunately, even with their positive influences, I still often felt like an outsider. While they all enjoyed childhood activities such as organized sports, musical endeavors, parental affection, family outings and the like, I experienced none of those. I was born first-generation Canadian, from blue-collar immigrant parents. I was trapped in a loveless, dysfunctional family dynamic where daily emotional existence was a struggle. I was not encouraged or even permitted to play sports, my parents were not interested in assimilating into Canadian society, nor did I have the ability to express myself in any way, shape, or form.

Needless to say, as I grew, I developed no measurable athletic ability. I never participated in any organized club or sporting league, I was never given the opportunity to play a musical instrument, and even "Hockey Night in Canada," that iconic Saturday evening television program religiously viewed every week by millions of Canadians, never graced our television set. I didn't even own a bicycle. So, when all the other boys would go 'riding' for the day, I would be stuck at home, sitting on our front steps, waiting for them to eventually return.

When my parents' relationship eventually imploded in the early Fall of 1969 and divorce was their only recourse, I discovered an opportunity to improve my miserable existence. At the time, I was in my final year at Killarney High School and was hoping to graduate from there. The family house sold quickly and my parents were ready to move on. However, I had absolutely no interest in going with either one of them. By a stroke of what I now consider to be divine intervention, the Freeman family, who lived directly across the street from the McGuire's, graciously offered to take me into their home, so that I could complete Grade 12 at Killarney. I believe that this was a turning-point in my life.

This was also a time when we would see more of the elusive Bob McGuire, who when not out globe-trotting, would frequently visit his family's home. His arrival was always an event. Amongst our circle of friends, 'Bobby' had become a near mythical personage. With his shoulder length, flaming red hair, frequent variations of outlandish beards and moustaches, and sporting the latest hippie attire, he cut a distinctive and unmistakable figure. Add the usual cadre of amigos and hangers-on that normally hovered about him, and you completed the visage of a psychedelic Messiah accompanied by his fervent disciples.

One was always guaranteed a good time in the company of Bob McGuire. With his magnetic and uninhibited personality, infectious sense of humor, superior knowledge of current musical trends and world events, plus an all-around welcoming demeanor, he was just the sort of person that a group of long-haired, impressionable, teenaged wannabees would be attracted to. That was us!

I can honestly say that the eight months I spent living semi-independently with the Freemans, attending school and interacting with my many friends on East 44th Avenue, was one of the happiest times of my life. In June of 1970, I graduated from high school and reluctantly moved away from the neighborhood where I had spent my formative years. I continued seeing the 'old gang' from time to time, but gradually those visits became less frequent, and then eventually stopped altogether. Invariably, my destiny now began pulling me in a different direction and it no longer meshed cohesively with my previous existence.

I married younger than most practically-minded people would have recommended, and before my mid-20s, I had fathered the first of my two children, that were to become the foundation of my subsistence. Professionally limited by a lack of post-secondary education and possessing few technical skills or qualifications, nevertheless, I still wanted to better myself and provide my family with the best possible life-style. As a boy, policing had always interested me, and over time, I had read numerous articles and books involving the investigations of Scotland Yard, the FBI, Sherlock Holmes, the Texas Rangers, plus the Royal Canadian Mounted Police. Although not certain if I stood a chance of acceptance, in late 1975, I applied to become an RCMP officer. In October of 1976, I finally received the call I'd been anticipating.

During the ensuing twenty-five years, I served my community and my country as a proud and enthusiastic member of Canada's national police force. The experience left an indelible mark upon me and still causes my chest to swell whenever I think back to my many accomplishments, or whenever I see someone wearing the iconic scarlet tunic, high-brown boots and Stetson hat. The inconsequential, youngest son of immigrant parents with no definitive talent and little hope of fulfilling a bright future, had now become a man of substance. Throughout the course of my career, and with the aid of my colleagues, we investigated and successfully concluded numerous local, regional and national criminal files. Years of diligent endeavor culminated in the year 2000, when I received the coveted 'Police Officer of the Year' award while posted in the diverse and always energized city of Surrey, B.C.

June of 2020 was the 50th anniversary of my graduating class from Killarney High School, and my wife Mary, found a reunion notification on her computer. As she searched the site, her efforts were noticed by my old friend John McGuire and he reached out to me as a result. As they say, the rest was history. Since then, John, Bob and several others from our former social group have come together and rekindled our dormant friendship. After an absence from one another of nearly fifty years, I can't believe my good fortune to have been reacquainted with this fabulous group of now mature men.

Shortly before our first reencounter, I had self-published a paperback novel depicting one of my more interesting criminal investigations during the late 1980s. Several of the 'boys' generously purchased copies, and it became a point of discussion. During one of our visits, and with Bob in attendance, I discovered that during the majority of his earlier travels, he had faithfully maintained journals, and devoutly recorded his daily activities. Who would have guessed that a young man intent on entering the Guinness Book of World Records for bedding every unmarried woman on the European Continent and breaking the all-time 'party-hearty record' for a single human being, would have possessed the forethought to do that? In any event, when it was suggested that I acquire the journals and compose a unified and cohesive document highlighting these excursions, I jumped at the chance.

Few people in this world get to realize their life-long dreams or achieve the pinnacle of their goals. Not so in Bob's case. His aspirations manifest themselves, and his thirst for travel and accumulative knowledge was thoroughly quenched. He also very nearly entered the Guinness Book of World Records for his previously attempted exploits at lust and Bacchanalia, but sadly, just fell short of that accomplishment. However, after all was said and done, he did in fact traverse the globe several times over and positively impacted all those with whom he met. For me, it has been a true joy to chronicle his thoughts, his actions, and his encounters within this following memoir.

H.J.P.

This body of work is dedicated to every inquisitive mind and adventurous soul who ponders, "What are my origins?", "Why am I here?", and "What knowledge, challenges, or opportunities lay before me?" All of the world's hidden treasures await those with the courage to search out and grasp them.

Harold J. Pokorny and
Robert P. McGuire

Naked Soccer on the Beach

An Unabridged Account of The Most Memorable Journeys And Reminisces of Canadian-Born Urban Legend "Bobby" McGuire

Austin Macauley Publishers™

LONDON * CAMBRIDGE * NEW YORK * SHARJAH

Copyright © Harold J. Pokorny and Robert P. McGuire 2023

All rights reserved. No part of this publication may be reproduced, distributed, or transmitted in any form or by any means, including photocopying, recording, or other electronic or mechanical methods, without the prior written permission of the publisher, except in the case of brief quotations embodied in critical reviews and certain other non-commercial uses permitted by copyright law. For permission requests, write to the publisher.

Any person who commits any unauthorized act in relation to this publication may be liable to criminal prosecution and civil claims for damages.

All of the events in this memoir are true to the best of the author's memory. The views expressed in this memoir are solely those of the author.

Ordering Information
Quantity sales: Special discounts are available on quantity purchases by corporations, associations, and others. For details, contact the publisher at the address below.

Publisher's Cataloging-in-Publication data
Pokorny, Harold J. and McGuire, Robert P
Naked Soccer on the Beach

ISBN 9798886932409 (Paperback)
ISBN 9798886932416 (ePub e-book)

Library of Congress Control Number: 2023914889

www.austinmacauley.com/us

First Published 2023
Austin Macauley Publishers LLC
40 Wall Street, 33rd Floor, Suite 3302
New York, NY 10005
USA

mail-usa@austinmacauley.com
+1 (646) 5125767

Having now formally documented my earliest travel experiences and once again relived them through the totality of this composition, my spirit has been completely re-energized. It has also reinforced my former claim that I have lived a fantastic life. After decades of dwelling on this planet, and through my countless eventful experiences, I have fortunately now been able to answer the three primary questions noted on the dedication page of this work. "I have come to know my origins. I now understand why am I here. Plus, I have acquired invaluable knowledge, faced and overcome many challenges, and took advantage of the numerous opportunities that lay before me." In other words, I have been fulfilled.

In life, one does not fully develop or evolve without the influence of others. In my case, I was blessed to have been surrounded by a plethora of amazing people throughout the entirety of my existence. And although they are too numerous to mention, I will attempt to accredit those who made the greatest impact on me.

First and foremost is my immediate family. My parents, Margaret and Larry plus brothers Mike, Ken, and John were my kindred and emotional foundation and will forever remain in my heart and thoughts.

Mr. Peters, my Grade 11 geography teacher and coach of the senior boys' soccer team at Killarney High School, deserves a huge debt of gratitude from me. A great and magnanimous Englishman, if ever there was one. It was him that positioned me as captain of the senior boys' soccer team, and it was his superior tutorial and enthusiasm that initially sparked my interest in foreign cultures and global exploration. Sir, I raise my glass to you.

Herman 'Hermie' Gruhn, is one of those larger-than-life characters that has been a main-stay with me for the majority of my adult life. "Half-a-Case," as we called him then, or "Six Pack," as he would be referred to in today's vernacular, is a handsome, tough, hard-working and thoroughly entertaining gentleman. A pirate of the high seas and a rover on dry land, this King of Cobble Hill holds a special place in my heart forever. Together, we have fished, partied, travelled and battled all those who were foolish enough to challenge us. Hermie always had my back, and it was him that first introduced me to the West Coast Fishing industry. He also continued to ensure that I was always gainfully employed in that occupation. Hermie, I consider you my fourth brother.

Special, heartfelt thanks must include Tom May, the captain of MV Katlin and owner of a large and prosperous salmon fish hatchery located on the Sunshine Coast of British Columbia. After leaving my job of salmon and herring harvesting, it was Tom that offered me a position at his hatchery. He taught me that raising and nurturing fish in the controlled environment of a hatchery, was just as important as catching them in the wild. The final years of my working life were spent in that remote and beautiful slice of paradise. Tom was like a second father to me and I will never forget his kindness, guidance and friendship. Honorable mention also goes out to Tom's beautiful wife, Michelle, for her continual support and for laughing at all of my jokes, even when they weren't funny.

Lastly, I wish to offer my deepest appreciation to Harry Pokorny and my youngest brother John for their involvement in this project. First, to Harry for taking my raw journals and transforming them into a fluid and cohesive story. And next Johnny, for his superior grammatical capabilities and keen editorial eye. Had it not been for these two gentlemen, this enterprise would never have seen the light of day. Mere words cannot fully express my gratitude.

And for all those persons who remain unnamed, yet assisted in making me the man that I became, thank you and God bless.

R.P.M.

Table of Contents

Introduction	13
Chapter 1: The Kid (Last Friday)	15
Chapter 2: The Hook (Saturday After Arrival)	34
Chapter 3: The Beckoning Road Awaits (November 13, 1965)	59
Chapter 4: The Cradle of Civilization	97
Chapter 5: From Disaster to Redemption	124
Chapter 6: Interlude Poseidon's Bounty	176
Chapter 7: It's Better the Second Time Around	193
Chapter 8: Twice Is Nice but Three's a Charm	249
Chapter 9: Rogues and Rastafarians	286
Chapter 10: Bobby Thai's One On	316
Chapter 11: The Eagle Has Flown	344

Introduction

I was extremely blessed to have been born at a time and place where the fates allowed me to achieve many of my lifelong goals. I was able to fulfill most of the dreams and desires that propelled me forward as a youth and continue to do so even now that I have become a wiser and slightly more cautious individual. Had I been born a decade before, or after, I truly believe that my life's story would have been written much differently.

Change for the better occurred in 1957, when my family vacated our modest, older home situated in a lower-income neighborhood of Central Vancouver, B.C., in favor of a brand-new house several miles further to the southeast. We were hardly living 'high on the hog', however, our circumstances were definitely improving. Although my mother, Margaret worked outside of the home and my father, Larry owned and operated Yellow Cab #38, we were still living on a shoe-string budget, and no one left the dinner table until every morsel of food on our plates had been fully consumed. Regardless of our financial status, my three brothers and I were taught respect, manners and that hard work produced dividends. The fact that my parents had just purchased a new home through their own diligent toil, supported that latter theory.

Like all concerned parents, Margaret and Larry wanted the best for their children, and a good education was the surest path toward success. As a result, my school report cards and those of my brothers were always closely scrutinized by Momma and Papa Bear. Moving into a new neighborhood also meant enrolling into a different school. Killarney High School was a short walk from our house and its construction had just recently been completed. This was to be my institution of higher learning for the next six years. I was both nervous yet excited to begin the pending school term and I was hoping to start off on a strong note. Proudly on display in a local shoe store, I had been eyeing a pair of blue suede boots with white piping around the soles. With these boots attached to my feet, I felt certain that I could conquer the world. So, I 'put the bug' in my mother's ear and awaited her response.

I was devastated when she answered, "As long as I'm buying your clothes, you'll bloody well wear what I've bought you!" What I ended up with was a stiff pair of laced, black leather Army issue ankle boots that made me look about as cool as a chimpanzee wearing a Beatle wig. From that day forward, I vowed that I would never again rely on anyone for the things I desired, and I immediately applied for a newspaper delivery route. Within the following week, my application had been accepted and I then commenced the lengthy and arduous task of delivering the city's Province Newspaper to approximately forty residents in our area. At 4:30 every morning, except Sundays, I picked-up my quota of newspapers and then grudgingly delivered them. I was now earning my own money and purchasing all the things I deemed necessary. Feeling a greater sense of purpose, I also began ironing my own clothes, making the occasional sewing repair on a torn garment and helping with additional chores around the house.

Excluding my disdain for Chemistry and Physics, I performed well in high school, both academically and in sports. The defining moment in my life came during Geography class in Grade 11. The teacher for that class was Mr. Peters, and he first kindled and then later stoked the flame that would become my passion for foreign cultures and world travel. He also made me the captain of the senior boys' soccer team, an honor that I would always treasure. Near the end of our final semester in Grade 12, my best friend Jim Bridge and I formed a pact that once we were out of school, we would travel the globe. After graduation, being true to my word, I sold my beloved '55' Chevy Bel Air hardtop and began miserly hoarding every penny I earned. I even stopped going out with friends, just so that I could save money. Unfortunately, Jim became irrevocably attached to his new car and a pair of girlfriends, so he broke our pact. His reversal was difficult to accept but I was still firmly committed toward adventure and global exploits. As you will read, I commenced my journey alone, but it never remained so for long.

Travel is not to escape life, but so that life does not escape you. (Unknown)

Chapter 1
The Kid
(Last Friday)

Nightfall was rapidly descending upon the unassuming but solidly constructed dwelling near the junction of Leonard and Gillies Bay Roads on the west-central quadrant of the island. Imperceptibly, the heavens had been repainted from a dull and apathetic slate gray in the evolving twilight, to now a deep black and blue tint, as though a gigantic unsightly, atmospheric bruise was unabatedly enveloping the entire landscape. As interior lights gradually began revealing themselves at various locations within the structure, the question as to whether or not life existed inside, was now confirmed.

The house itself was an older version of what today might resemble a double-wide mobile home in a typical trailer park, with its flat roof topping the single-story dwelling, but fully clad in local cedar siding rather than aluminum panels normally coating the latter abodes. It boasted seven rooms within, and included a small covered verandah attached to the main entrance. In addition, an impressive stone chimney climbed the west side of the building, terminating 3-feet above the roofline. Being situated in a remote location and surrounded by an abundance of trees, it was not surprising that the main heating source for the house consisted of wood fuel. In addition to the residence itself, the plot also supported a weathered Quonset Hut that was the termination point of the lengthy gravel driveway. It was large enough to accommodate tools, several vehicles and a small workshop. There were also three utility sheds placed indiscriminately throughout the grounds, whose immediate purpose were yet to be determined.

The island, known as Texada or affectionately called "The Rock" by the 1300-souls currently inhabiting it, was approximately 50-kilometers in length and 10-kilometers in width and was nestled within the pristine waters of

Georgia Strait, between the land masses of Vancouver Island and the Province of British Columbia's mainland. Texada was considered the 'jewel' of the Northern Gulf Islands chain and so named after Felix de Tejada, a Spanish rear-admiral exploring the region for King Charles IV of Spain during a 1791 expedition. Archeological studies determined that the isle's original custodians were the Coast Salish and Tla'amin First Nations tribes who could trace-back their history 3300 years.

Once a major mining and logging center in the late 1800s and early 1900s with a significantly larger population than today, Texada's industry gradually began disappearing, as did the majority of its populace. By the 1950s, the once booming land-mass in the Strait had become relegated to a mostly out-of-the-way cottage and camping destination, plus a secluded refuge for individuals wishing to escape the crushing stresses and pressures of the 'big city', that was Vancouver, 100-kilometers to the southeast.

The island was home to a variety of plants and animals native to the coastal region of British Columbia, and was inundated with stands of Douglas Fir, Red Alder, Sitka Spruce and Western White Pine. It was several of these latter species that generously dotted the property surrounding the independent domicile containing its lone occupant, who was now pacing about within and frequently surveying the outdoors through the single-paned window of the great-room. The seemingly inquisitive yet uncharacteristically impatient personage intently scanning the exterior perimeter, had at birth, received from his parents the appellation, Robert Patrick McGuire.

Bob McGuire or 'Bobby' as he was known to his siblings and close friends, was now a tri-quarter centenarian, ex-patriated city-boy and self-professed 'world traveler', who for the last 20-years called the island home. Considering his age, he was still a formidable specimen and could easily pass for someone much younger than his actual years. McGuire stood approximately 5' 11" tall and tipped the scales at a comfortable 215 pounds. His previously trade-marked bright, reddish-orange mane, that frequently flowed beyond his shoulders when younger, had now been substituted by snow-white locks, neatly tied-back in a pony-tail. One physical aspect of the man that had not changed over the many years was the unmistakable, mischievous glint within his pale blue eyes, expressing his true nature. That spark was still as strong as ever. The multitude variations of facial coiffures he frequently sported and interchanged over the years, was now currently an average looking short-cropped, bleach-white beard. However, with his penchant for flair and uniqueness, it probably wouldn't be too long before he would once again revise that growth so that it would resemble some form of lavishly bearded and mustachioed character from the past such as American Civil War Generals Ambrose Burnside or George Armstrong Custer, surrealist artist Salvador Dali or even professional wrestler Hulk Hogan.

McGuire was a passionate, carefree, forthright and infectiously amicable individual, who genuinely cared for humanity regardless of creed, color or ethnicity. One's political or theological stripe meant nothing to him, so long as they wished to live in peace and harmony. His heart was as large as his personality and his affection toward the fairer sex was legendary. He also possessed a love of oratory and spoke so often and so ferociously relentless, that it was often said, "He could talk the ears off of a wooden, cigar store Indian." Had he lived in ancient times, then famous rhetoricians such as Cicero or Demosthenes, would have been hard pressed to compete with McGuire's vocal prowess.

In this manner, Bob McGuire carried himself throughout his life and it served him well. Wherever he went, he was welcomed and whomever he met, he generally befriended. That is not to say that he was saintly or gifted with infinite patience or tolerance, since over the years, a few less congenial persons with whom he interacted, that behaved contrary to his disposition or life-pattern, unfortunately incurred his wrath, and as a result, received a swift and harsh physical rebuke from him. Although wholly justified in inflicting this righteous action upon these callous transgressors, it was also infrequent and only delivered if and when absolutely unavoidable. Overall, McGuire did not relish pugilistic behavior, but neither could he abide disrespect, bullying or cruelty. When one was to describe Bob McGuire, the age-old adage would always come to mind, "What you see is what you get."

Actually, the only two things remotely artificial regarding McGuire or his personae, were the two chromium knee joints that recently replaced his original couplings, and they now began to cause him discomfort. This generally meant a precursor to cold or rain, and judging by the thick and ominously foreboding clouds beginning to hover above the terrain, it probably wouldn't be too long before this aching premonition came to pass. It was now well past the dinner hour and McGuire's stomach was reminding him through a recurring series of groans and gurgles, that it was high-time he paid heed to replenishing the empty space within. Absentmindedly, it had been early morning since he had last eaten. But the constant strain on his brain throughout the day anticipating this evening's appointment, preoccupied him to the point of distraction, and prevented him from focusing on more pressing matters.

As he remained seated on the chair in front of the window facing outward into the now gloomy exterior, he noticed the first droplets of rain beginning to appear on the glass.

Within minutes, those intermittent droplets had become a steady stream of moisture to the point where the window pane was now fully saturated and the once clear view outside had become completely blurred by the multitude of tiny rivers coursing down the glass and striking the sill below.

The reason for McGuire's deviation from his usual congenial attitude, to what was now, an irritated state was as a result of the recent promise he had made to a close friend. If he possessed any noticeable failings, one of them was his inability to deny assistance to anyone who requested it. He constantly derived immense pleasure knowing that his efforts or actions on behalf of another were valued and appreciated. However, now he debated the wisdom of his quick agreement to this latest boon.

Pearl Starchild was one of the first persons he met in his early days on the island and she was beneficial in not only helping him become situated in his current home but was also instrumental in securing him employment in several different ventures within the region. He definitely owed her a debt of gratitude and as a man of integrity and honor, he always paid his bills.

Pearl was an indigenous woman of middle-age who had lived the majority of her life on Texada Island. She was originally born on the island, as was her younger sister Willow, but then later moved to the Lower Mainland as a small child with her sibling and their parents as they searched for greater opportunities.

Unfortunately, the only things the Starchild family found in the 'Big Smoke' was inequality, rampant racism and the destructive complications derived from liquor abuse. Within a few years of their arrival, with very little work available for an indigenous person and even fewer prospects for the future, Pearl's father had become a broken man and quickly devolved into a pattern of alcoholism as so many other men of his ethnicity had done before and were now currently doing.

After staggering out of the Balmoral Hotel on East Hastings Street one Friday night and attempting to cross 4-lanes of traffic in hopes of intercepting the eastbound No. 14 transit bus to Nanaimo Street, Pearl's father was struck and killed by a speeding delivery truck.

Devastated by yet another disastrous turn of events for the family, Pearl, her sister and their mother returned to Texada Island and moved-in with her maternal grandparents, residing near the northern tip of the island at Blubber Bay. As the years unfolded and the sisters matured, Pearl's grandparents and then her mother eventually passed-away.

With her adult sister now living on the mainland and raising a family of her own, Pearl inherited the old but still functional homestead she had known all these many years. Over time, she too married and bore several children to a braggadocios and unproductive dreamer. He eventually left them without warning in the dead of night, to fend for themselves, while he made his way up to the north-west portion of the Province. The last she heard of him was quite some time ago when he was then believed to be a part-time employee at one of the several lumber mills in the port town of Prince Rupert.

No stranger to hardship or disappointment, Pearl took all of these set-backs in stride and still attempted to make a life on the island, as best she could, for her children and herself. This in spite of the fact that very few indigenous persons still actually resided on the island. According to native folk-lore, Texada Island was cursed and had now tipped upside down so that the once livable portion of the land was actually submerged. No one could really state why this was or when it occurred however, it was said that a very few enlightened Tribal Elders possessed that knowledge. With many ancestral peoples still maintaining traditional and superstitious beliefs, it was acceptable to be on the island during daylight hours but it was best to be gone by nightfall.

For years, Pearl commuted daily from her home in Blubber Bay and traveled the approximate 10-kilometers in her 1977, blue and white Ford Bronco, to the Texada Market on Copper Queen Street in Van Anda, where she worked as a cashier. The Bronco had definitely seen better days but even if it broke-down, Pearl was certain that she was still strong enough to be able to traverse on foot whatever distance she might find herself between work or home until she could exact a repair. It was there at the market that she first met Bob McGuire 20-years prior when he stopped briefly to acquire a few essentials. As with many other women in the past, Pearl was immediately smitten with this charismatic stud and nearly sprained an ankle in her haste to first introduce herself and then eventually serve him.

Although never romantically involved with one another, over time, Pearl and Bob became 'fast friends' and socialized regularly. He even came to be known as Uncle Bobby to her three children since they too quickly warmed to his magnetic personality and general good-will toward all. The simple life on the island suited them both and there was very little that occurred in their daily existence that created even the least amount of stress or concern. However, that rhythm was about to skip a beat and falter with the telephone call from the mainland.

It had been nearly a week since Bob McGuire had responded to the rapid and repeated knocking on his front door much earlier than polite company should normally visit. As he swung open the portal, he observed Pearl standing on the porch adopting a submissive posture and with an obvious look of serious consternation about her normally passive facial features. "Can I come in, Bobby?" was Pearl's initial but barely audible request.

"Yah…Yah, sure, come on in," was McGuire's immediate response, curious as to what could be bothersome enough for his friend to attend his residence at this un-godly hour. He quickly stepped aside from the entryway permitting her passage and then politely bid her to take a seat.

"What's up, Pearl, are you in trouble?" was McGuire's first enquiry once she had made herself comfortable on the brown leather settee that comprised but one of the three-piece sectional units within the great room or the living room as some would describe it.

"Oh, Bobby, I'm so sorry for bugging you this early in the morning but I haven't slept a wink all night. I needed to speak with someone that I respect and trust, and you are the first person I could think of that fits that bill, on both counts," was Pearl's shaky response.

Naturally, McGuire was touched by her compliment but he was still no closer to discovering the source of her dilemma. Nodding briefly in acknowledgement, he continued looking at her in silence until she was ready to speak once more. Eventually, Pearl composed herself and began relating her story. "Bobby, you know that my younger sister Willow has been living on the mainland for quite a few years now and that her husband, Jack is on the Tribal Council of the Squamish First Nation in West Vancouver? He is a very important man on the Council and he has a great deal of responsibility. Their only child, Jesse has just turned nineteen and finished high school last year. He is a very bright young man and both Willow and Jack have such high-hopes for him and are expecting him to attend university."

"Unfortunately, this past year, Jesse's attitude has changed and he's been hanging around with a bad crowd on Vancouver's downtown east-side. He used to be such an easy-going and well-behaved boy but now he's withdrawn and quickly becoming a real handful for them. Willow and Jack have both been trying to turn him around but the more they discuss their concerns with him, the further he seems to run in the other direction. They're afraid that if they can't change his behavior soon, he will have already gone too far for recovery."

By now, Pearl's voice had taken on a tone of tearful desperation and she could barely croak-out the next sentence. "Bobby, you know what happened to my dad when I was younger and what continues to happen to so many other men and women of my People stuck in that never-ending cycle of addiction? If Jesse ends-up like that, it would kill Willow and Jack, I just know it! I can't even bear to think of it."

Intently absorbing Pearl's discourse but not knowing exactly how to respond, McGuire merely nodded his head in agreement and mumbled a few generic and impotent retorts such as, "Yah," "I understand," and "That doesn't sound good." But under the circumstances, what else was he expected to say?

Pearl continued, "Willow sounded so upset on the telephone yesterday that I just needed to say or do something to calm her down. I suggested that we get Jesse off the mainland for a while and away from those loser friends of his as soon as possible until we can come up with a more workable plan to get him back on track."

"That's a great idea," exclaimed McGuire, "That'll be an excellent first step!"

"I'm glad you agree," replied Pearl somewhat sheepishly, "I kind of promised Willow that Jesse could come over here to Texada and stay with you for a while."

For a moment, McGuire just stood there looking at Pearl, grinning at her suggestion and nodding his head in agreement. That was until he mentally rewound and then replayed her latter comment in his mind once again and actually processed the gravity of it. Immediately, his nodding stopped and the prior Cheshire Cat grin began quickly drooping downward. "What the hell…Pearl?" declared McGuire. "Why would you tell her something like that? What am I going to do with him?" he questioned incredulously.

By now hot, voluminous tears were visibly flowing down Pearl's cheeks. She had dropped to her knees facing him with clasped hands in a pleading gesture and cried, "Bobby…Bobby, I'm so sorry but I didn't know what else to do. She's my only sister and Jesse is my only nephew and you are the greatest guy I know. You're intelligent and kind and generous and worldly and such a good role model for him…"

"Okay…okay, that's enough," McGuire dejectedly announced, "He can stay with me for a while but he's going to have to earn his keep. I'm not going to babysit him and if he causes me any problems, he's out of here!"

Upon receiving his tentative approval, Pearl immediately sprang to her feet and rushed toward McGuire, grabbing him in a surprisingly strong embrace while commencing a fresh bout of weeping. However, this time it was tears of gratitude rather than sorrow. McGuire began gently patting her on the back while verbally reassuring her that everything was going to be fine. He did this all the while disbelievingly shaking his head and contemplating what he had just gotten himself into?

Once all of the drama had dissipated and the matter was momentarily resolved, McGuire invited Pearl to remain and breakfast with him. Once the meal was consumed, Pearl excused herself and begged his leave. She was now visibly rejuvenated after their intense conversation plus a hearty breakfast and she was confident that her good friend Bobby McGuire was going to be the answer to all of their prayers. Before departing, she informed him that she would be travelling to Vancouver early the following Friday morning by ferry in her old blue and white Bronco. She defiantly announced that, "Come Hell or high water," she would be returning with Jesse that same evening and deliver him to McGuire, preferably in one piece.

She further stated that there was a neighbor in Blubber Bay that would watch her children while she was away, so there would be no encumbrances to impede her attempt at rescuing Jesse and safely transporting him back to Texada with her.

It was now Friday evening and McGuire was sitting alone inside his house, staring out of the front room window, watching the falling rain and wondering whether or not Pearl had actually succeeded in accomplishing her 'mission of mercy'. Hoping to commence the planned intervention on a positive note,

McGuire had engaged in a variety of shopping earlier that day and purchased several items for dinner that he normally prepared well enough to always elicit a plethora of compliments from those fortunate enough to attend his table and partake of the victuals on offer.

McGuire fancied himself a reasonably accomplished cook and he was fully aware that the 'breaking of bread' was not just a means of replenishing the body but could often also be deemed a ritualistic experience. For this evening's fare, he had purchased from the market a variety of local greens for an introductory salad accompanied by an oven-fresh French baguette divided into liberal slices and slathered with herbal butter.

A whole, smaller sized Sockeye salmon with a generous portion of scallops and oysters plus two medium sized Dungeness Crabs constituted the main course that would then be paired with gently steamed, fresh asparagus spears and smoothly blended garlic infused, yellow potatoes. And finally for afters, it was to be a 12-inch Granny Smith apple pie replete with cheddar cheese and vanilla ice-cream. For liquid refreshment to accompany the supper, he would be serving his renowned, home-made wine fermented from Golden Gage Plums. The entire layout cost him a 'pretty penny' and was probably more than they could eat. However, he realized how important this endeavor was to Pearl and her sister so, if this meal became the conduit of what would hopefully culminate into a successful conclusion, then that was a small price to pay.

The clock on the mantle above the fireplace indicated that it was now nearing 9:00 P.M. and still no sign of Pearl and his intended house guest. McGuire was quietly resigning himself to the fact that tonight he would now probably be eating alone and before it got too much later, he decided that he had better get-on with it. As he slowly arose to make his way into the kitchen, he observed the pale white beams of two headlamps coming up the driveway. Breathing a sigh of relief at finally now realizing this much anticipated development, he then quickly altered his prior negative opinion and quietly whispered, "It looks like they've made it after all?"

McGuire did not wish to appear overly anxious at their arrival so, he stood away from the window, backed further into the hallway and waited for them to reach his point. There was a brief interval outside between the closing of the vehicle's two doors and then the assertive tapping of the brass knocker on the front door of the house.

Momentarily hesitating and then thoughtfully adjusting his clothing, McGuire purposefully strode toward the entry while swiftly affixing a welcoming smile on his face. Taking a deep breath, he then pulled open the door toward himself.

Standing in the glare of the single porch light, slightly out of reach of the torrential rain beyond, was Pearl, grinning from ear to ear and looking victoriously elated, although somewhat water-logged. Two steps behind her holding a khaki-colored duffle bag, stood a slightly built youth of medium height with dark brown, straight collar-length hair over his ears. He was wearing a black, full-length slicker over a cream-colored 'Redbone' sweatshirt, camouflage trousers banded at the ankles and a pair of navy blue Doc Martin boots.

His facial features were quite delicate and at certain angles could even be considered somewhat effeminate. It seemed doubtful whether or not a shaving razor had yet ever had an opportunity to scrape his jaw-line and the term 'baby-face' definitely applied to this character. One feature that inescapably dominated the young man's countenance was his eyes. They appeared perfectly round plus slightly larger than average in a deep, hickory-brown shade. They were set evenly spaced above high cheek bones. At the moment, he displayed a blank, unassuming gaze but that could not belie the expressive power of those blinkers nor an undeniable intelligence hidden from within. Without actually staring at him, McGuire could quickly see beyond the initial façade of the youth and was able to discern certain tell-tale mannerisms that informed him that there may be a darker side to this youngster. Beneath the intelligent eyes were predominant dark circles telegraphing probable lack of sleep. He appeared restless and shuffled his feet back and forth while fidgeting with his clothing and frequently scratching his head. McGuire intuitively surmised that these were all possible characteristics of a methamphetamine user.

The visitor stood at the doorway and silently scanned the residential interior. What he could see within was an average looking home with the usual necessary furnishings. They were adequate but not opulent. What was in abundance was a wide variety and array of colorful west-coast First Nations-inspired paintings and related renderings mostly created by the inhabitant of the property and obviously composed with love and care as suggested to by their number and pride of place.

Allowing a few moments to pass and temporarily filing away his observations and opinion of this young man, McGuire was then first to break the silence. "Hey…hey, you made it? I just about gave you two up as a couple of no-shows. Nice to see you both. Come on in out of the rain."

"Thank you, Bobby," responded Pearl as she proceeded into the house with her charge loping slightly behind. "It was quite a trip!"

"Bobby…this is my nephew, Jesse Sdang-Kuug and Jesse this is my good friend, Bob McGuire," remarked Pearl as she introduced the two men to one another. Jesse didn't speak but nodded slightly toward McGuire in reluctant acknowledgement.

McGuire in turn responded, "Pleased to meet you, Jesse. What was your last name again….Sdang-Kuug? That's quite the handle, does it have an English translation?"

Jesse looked directly up at McGuire and answered in a soft-spoken but definitive manner, "It means Eagle Heart in the Haida language. It originates from my father's ancestors and handed-down through the generations."

"That's really cool," admitted McGuire. "Here, I'm going around with a name like Bob," drawing out his name to sound like BAWWB…"and you've got the name Eagle Heart! That's just not fair."

Upon hearing this declaration from her friend, Pearl burst out laughing. And although he attempted his best at stifling any sign of mirth, Jesse too couldn't help but chuckle. Observing this reaction from her nephew, a hopeful emotion then travelled the length of Pearl's frame as she silently reaffirmed her wise decision to involve 'Bobby' in helping her family return Jesse firmly back into their protective and loving embrace. McGuire's charm was already at work and it appeared as though Jesse was also gradually becoming infected by it.

"Okay, guys…let me take your coats and then just make yourselves comfortable while I start preparing dinner," instructed McGuire. "I hope that you both brought along a good appetite? I've got lots of goodies here."

"Don't go out of your way on our account, Bobby, we're not very hungry," Pearl called-out to McGuire but not very convincingly. She was fully aware that he was a great cook and in fact she was famished.

In her determination to bring Jesse over to the island in a timely fashion, she had completely neglected to consider eating or drinking anything for the majority of the day. Now that the most difficult task had been completed, she actually had a moment to consider her own needs.

"No problem whatsoever!" shouted back McGuire from the kitchen into the great room. "I've had the potatoes cooking low and slow for a while now plus most the other ingredients take only a few minutes to prepare. I hope that you both like seafood?"

"Hey, Kid…I've got a pretty good record collection over in the corner. Why don't you pick-out something and throw it on the turntable?" continued McGuire.

Jesse flashed a glare at Pearl when he heard McGuire refer to him as 'Kid' but Pearl raised a forefinger to her lips and gently shook her head from side to side indicating that he should let the remark pass. She knew that it wasn't made through malice or disrespect but rather from affection. That was just Bobby's nature and she quietly whispered this fact to Jesse.

Jesse loved and trusted his aunt Pearl and also knew that she had a 'good head on her shoulders'. If this overaged transplant from the Psychedelic Era was her close friend, then he would accept him as well. As a result, he dismissed the perceived slight to his person, shrugged his shoulders and accepted her explanation.

He then rose to his feet and walked over to the corner of the room where the 5-piece component stereo stood surrounded by neatly placed stacks of 12-inch vinyl record albums, each protected in their individual brightly illustrated or photographed protective cardboard sleeves, situated on either side of the massive dual speakers.

The stereo system appeared to be a vintage set-up from the late 1960s or early 1970s and resembled pictures in magazines promoting the A & B Sound music store on Seymour Street that Jesse had seen in the attic of his parent's home in West Vancouver when he was a small child. As he reached the unit, he observed that the receiver/tuner was a 4-Channel Marantz Model 233b with nearly 200-Watts of power per channel. The separate turntable was a Marantz 6150 fully manual direct-drive model. Next, Jesse could hardly believe that another one of the components was actually a Pioneer Model H-R100 8-Track cassette player.

He had only known of these electronic devices from conversations overheard from older music aficionados but he never dreamed that he would actually ever see one for real. Lastly to complete the set, were the two massive

Wharfedale-Pro floor speakers with 15-inch woofers. All-in-all, Jesse could see that this ensemble in its entirety was ancient.

However, it was definitely well cared for. All of the wooden cabinetry was recently oiled and polished plus the clear, plastic cover of the turntable was completely dust-free. Jesse bent at one knee and methodically began leafing through the stack of albums nearest him. As he had earlier internally predicted, he wasn't recognizing any of the discs passing through his fingers.

'STINK' by McKenna-Mendelson, 'MASS IN F-MINOR' by the Electric Prunes, 'HAPPY TRAILS' by Quicksilver Messenger Service and 'DISRAELI GEARS' by Cream were the first four platters he handled and none resonated with him. His fifth choice was 'ELECTRIC MUSIC FOR THE MIND AND BODY' by Country Joe and the Fish. Jesse had never heard of them either but since they were going to be eating seafood for supper, perhaps the 'Fish Band' with a 'Fruit de Mer' connotation might be a good selection?

Whether or not it was a good decision, he thought that perhaps Aunt Pearl and 'Mr. Bobby' would catch the hint of irony attached. As he dropped the needle on the first track of Side A, the tune 'Flying High' began to play. McGuire heard the opening stanzas of the song from within the kitchen and called-out, "Good choice, Kid! We're going to make a Modern-Day Hippy of you yet. You can turn it up louder if you want."

Jesse bristled at hearing McGuire refer to him as 'Kid' once again and looked toward Aunt Pearl for a potential reaction. She too caught the comment but as before in pantomime fashion, she discouraged him from reacting negatively. Realizing that he was more or less defeated on this point and now suddenly not feeling all that well, Jesse merely shrugged his shoulders, rolled his eyes and privately wondered how he was ever going to get through this weekend.

McGuire was now in his element and operating like a well-oiled machine. With his favorite comical apron bearing a semi-engorged, faux phallus pinned on the front now firmly attached to his torso plus a large glass of wine on the countertop within arm's reach, he busied himself about the stove. Earlier that afternoon he had prepared the salad and was keeping it chilled inside the refrigerator.

The salmon was fully dressed with the head and tail removed and ready to accept the accoutrements with which it would soon be grilled. The shellfish had been kept in a vat of cold water in order to keep them fresh and he could see that they were still looking great. The asparagus had also been prepped, washed and set into the steaming basket awaiting their brief appointment with a pot.

He had peeled the potatoes a few hours prior and not long before the arrival of his guests, had placed them on the stove top at a low heat so, it wouldn't take much longer before they were tender enough to blend. Taking a generous gulp of wine, McGuire began arranging a bed of shallots, parsley, lemon slices and a shallow pool of olive oil on the bottom of a roasting pan.

McGuire then deftly laid the entire salmon into the pan and inserted it into his pre-heated 450-degree oven. Now speaking to himself aloud, he began placing selected pots on various elements of the stove. "Okay, it will take 12 to 17-minutes to grill the salmon. Approximately 10 to 15-minutes are needed to cook the crabs. The oysters and scallops require about 7 to 9-minutes before their shells open and the asparagus should be cooked to perfection within 2 to 4-minutes! Finally, I'll drain the potatoes add the butter, milk and minced garlic and begin whipping them while everything else is cooking." Now with his game-plan in order and his various timings firmly embedded within his brain, he sprang into action.

About the same time as Side A of 'ELECTRIC MUSIC' had completed its revolutions on the Marantz turntable, McGuire called-out, "Okay, you guys…dinner is served!" Without attempting to appear overly anxious but still not wasting any time, Pearl breezed into the kitchen with Jesse in tow and immediately observed the bounty set before them.

The table was not elegantly bedecked with fine china and gilt-embossed cutlery nor was it populated with crystal glasses and decanters. However, what sat atop of it was edibles fit for Royalty. Jesse had not spoken much since arriving and he was still remaining relatively mute even now but McGuire could see by his facial expression and by the darting of his eyes from side to side as they surveyed the breadth of the table, that he seemed impressed with the offerings. Pearl was much less reserved and profusely gushed with praise and thanks to McGuire for his accomplishment. "Oh, Bobby! This food looks absolutely wonderful and the aroma is beyond description. Thank you so much for doing this for us."

Jesse looked over at his aunt Pearl, thinking that she might be laying-on the compliments rather thick but now seeing the glow on her face and the moisture rising to her eyes, he could tell that her gratitude was genuine. For his part, McGuire was about as puffed-up as a male peacock 'on the make' with his face fully flushed and steamy from his efforts and wearing an extravagant smile that stretched the entirety of his bearded mug.

Never standing on formality, McGuire bade his guests be seated and 'tuck-in' while it was still hot. Although Jesse initially just picked at his food, it wasn't long before the true magnitude of the banquet overcame him and within minutes all three participants commenced systematically devouring the treasures of the deep along with their accompaniments, until there was barely a morsel left to sample.

In between mouthfuls, being the excellent host that he was, McGuire continued topping-up their glasses with his cherished wine until by the end of the meal, it was evident from their appearances and occasional belches that all three were fully sated. Jesse in particular appeared as though he was ready to burst and after having consumed numerous glasses of the potent plum wine, he gently swayed on his perch like some drunken parrot, seeming as though he were about ready to drop off. One thing was certain, his restlessness had noticeably decreased.

After several minutes of satisfied groans and musings, McGuire inquired, "Okay, who's ready for dessert?" Both Pearl and Jesse merely lobbed their heads in decline from side to side before she eventually stated, "No thank you, Bobby. I think Jesse and I are both filled to the gills, no pun intended. There's no room for even one bite more. Maybe later?" McGuire looked over at Jesse and burst out laughing. Jesse was not able to even utter a word. He just sat there staring at the empty plates, seemingly on the verge of keeling over and looking like the "cat that just ate the canary."

"All righty then," declared McGuire, "Let's head into the living-room and rest our guts while I throw another licorice pizza on to the turntable. Washing the dishes can wait until tomorrow!"

As they made themselves a bit more comfortable in the great-room and settled down to digest their feast, McGuire singled-out the 1972 debut album from Steven Stills and Manassas. This was one of his all-time favorite records and one that he listened to regularly for its melodious vocal harmonies and catchy lyrics. He dropped the stylus on Track 2 of Side 'A' and cranked-up the volume.

As he plopped himself down in his favorite spot, an Orientale inspired rattan Emperor's chair, Stills and the remaining band broke into a rousing rendition of 'Medley Rock & Roll Crazies/Cuban Bluegrass'. The trio now spoke very minimally, allowing themselves to be transcended by the music plus the gradual relaxing of their distended bellies. They were all in agreement that there was no need for further refreshment tonight.

Their mellow-time together now trickled by like some peacefully meandering, clear-water stream. McGuire had traded the former responsibility of Head Chef to now, that of master DJ and instinctively selected a variety of retro Psychedelic tunes played by an ensemble of now mostly defunct and long since forgotten bands that musically floated and collided about the room.

In addition to the set of spell-casting melodies being methodically expelled by the stereo, was the mesmerizing effect of the Lava Lamp prodigiously perched on a shelf in one corner of the room with its randomly separating shapes of a wax-like substance ponderously rising and falling within its heated, clear-glass vessel. These shapes and colors reflecting on the living room walls uncannily hinted to the early 'light shows' often displayed in 1960s North American Night Clubs and Concert Halls like Bill Graham's Fillmore West in San Francisco.

All of this atmosphere together with the pungently aromatic mini-pyramid of sage incense burning within the belly of a perforated, porcelain Buddha, transported Jesse back to a time in which he had never existed. He had previously seen the Woodstock Music Festival Documentary with his parents plus a couple of older movies such as Easy Rider and Alice's Restaurant, but now he realized that this must have been exactly how the average persons of his age would have spent their Friday and Saturday nights within their parent's basements 'back in the day'. Overall, Jesse found it rather pleasing.

Before they realized the full passing of time, it was now nearly midnight. Pearl stated that she needed to depart in order to relieve the neighbor who had been watching her children. She had been with them since early morning and it wouldn't be fair to impede upon her good-will any longer. Jesse was rather hoping that he could leave with Pearl and spend his time at her residence instead of remaining here where he was. McGuire too, secretly wished that he could be left to himself after this evening's revelries. However, without speaking their thoughts aloud, all three knew exactly what their roles would be in the coming days.

Pearl would leave her nephew with 'Bobby' and allow him to work his magic. McGuire realized that he was expected to perform some type of societal, readjustment exorcism on 'The Kid'. And for his part, Jesse was fully aware of the conspiracy between his parents and aunt Pearl to re-assimilate him back to whatever they considered to be the 'mainstream'.

The question begging an answer at this juncture was, how would they accomplish that feat on this remote island, against his will and with the involvement of a congenial and well-intentioned but otherwise 'Flower-Power' throw-back from the Summer of Love? That was anyone's guess. For the moment, he would reluctantly play along with this interlude until such time as he felt that he'd had enough interference from others in what was clearly his life!

McGuire and Jesse remained at the front door of the residence until the lights from Pearl's Bronco became just a ghostly, dissolving glow through the foliage and then disappeared altogether as she reached the main road. The rain had since ceased but still coated exterior surfaces with a clear, dripping glaze. The shower had also left behind the fresh smelling scent that always prevailed after a proper atmospheric cleansing. McGuire closed and locked the front door. He then picked-up Jesse's duffle bag and began shuffling down the hallway. "Come on. Kid," announced McGuire, "I'll show you to your room."

With Pearl now gone, Jesse no longer felt his need for restraint and voiced his displeasure at his host's continuous usage of what he felt was a derogatory term. "Excuse me…Sir. My name is Jesse, not Kid. Could you please call me Jesse?"

"Yah…yah, no problem K…i…d…Sorry, I mean Jesse. No offence, eh? I don't mean nothin' by it," apologized McGuire. "It's just that to me, anyone under 40-years old is still a kid. And forgive me for saying but you're not even close to 40. And since we're asking favors of one another, there's no need to call me Sir. That's just for Cops or Old Men and I'm neither of those," chuckled McGuire. "Just call me Bob."

Jesse could tell by McGuire's response, demeanor and tone of voice that he was displaying a genuinely neutral and unassuming attitude. He could now clearly see that McGuire was indeed just a non-judgmental, fun-loving person and he actually began believing that he held no malice or disrespect toward him. He felt slightly embarrassed by his childish, defensive outburst but kept that fact to himself.

Jesse then followed behind his host and eventually reached the room where he would be sleeping. It was the first room after the kitchen and like the rest of the house, it was minimally furnished. Within the 12' × 12' bedroom was a twin bed with a wooden headboard, a standard sized, upright chest of drawers plus two end tables situated either side of the bed with each supporting simply designed table lamps. Several oil and acrylic paint art works composed by McGuire hung from the walls.

"Okay, J-E-S-S-E," emphasized McGuire, "This is where you crash. At the foot of the bed is a fresh towel with a bar of soap that you can use and I suspect that you've got your own toothbrush? The bathroom is just at the very end of the hallway to the left. Just so you know right off the bat, there's no lock on the door so, if you're looking for privacy make sure that you flip-over the 'Do Not Disturb' sign that's hanging on the door knob."

McGuire further stated, "Don't worry about me, I'm not shy but I'll try to remember to flip-over the sign in the event that I'm in the tub with a female friend who might be washing my back!"

With this last statement, McGuire threw him a salacious smile and a sly wink of his eye. When Jesse caught the gleam of those pearly whites plus the sass in those eyes, this intuitively informed him that 'Bobby' probably wasn't joking? He probably really was a 'randy' old dog. While attempting to stifle a yawn McGuire then followed with, "Well, that's the extent of the tour and now I'm heading for the sack. I've got some stuff to do tomorrow and I'll probably need your help."

Jesse was curious as to what form or method that assistance might resemble but instead, merely nodded his head in acknowledgement. "Good night, Bob," offered Jesse, somewhat surprised by this nightly send-off that he seldom uttered, even toward his own parents.

"Goodnight, **KID!** See you in the morning," was McGuire's reply as he slowly walked away. Once more Jesse resignedly shook his head upon hearing himself referred to as 'Kid' yet again.

However, this time rather than being irritated by it, he let loose a subdued guffaw and remembered the adage frequently used by his father whenever his mother attempted to change his predictable habits; "You can't teach an old dog new tricks," his father would regularly announce. Just as accurately as that term applied to his father, so did it definitely seem to apply to Mr. Bob McGuire as well.

As Jesse eventually ensconced himself between the bed sheets, still wearing his street clothing, he turned out the light from the end table lamp and was immediately impacted by the eerie silence surrounding him. Missing were all of the sounds he was accustomed to hearing at home. The sound of traffic steadily traveling along the busy street outside of his bedroom window, the whine of sirens from emergency vehicles off in the distance near the downtown core, the occasional low flying aircraft overhead either approaching or departing the Vancouver International Airport and even the frequent and annoying barking of neighborhood dogs. All of these cacophonous but somehow strangely comforting distractions for an 'urban dweller', were now suddenly absent. Jesse lay there within his nocturnal cocoon and stared into the blackness.

With all that was going on in his life at the moment, his indecision and latent fear as to where his future lay plus now this unexpected transversal to Texada Island and what that entailed, he realized that slumber and the fully desired escape it brought, would not be forthcoming any time soon. He needed to settle himself down but lacked the method. As a final thought before drifting off in to a semi-conscious state, he felt a distinctive and gnawing exigency within his brain. Attempting to recognize this sudden need, and then subsequently quench it, he eventually believed to have determined what he was missing. The conclusion was direct and unmistakable. Jesse deduced that his crystal pipe would really come in handy about now!

Chapter 2
The Hook
(Saturday After Arrival)

McGuire arose slightly later than usual the following morning, but still much earlier than the average retired person with little responsibility constricting their daily routines. He wanted to allow his guest some additional time to sleep-off the previous night but not too much time that it would end up wasting the entire day.

The master bedroom was on the opposite side of the house and accessed only from the living room. As McGuire raised the window shade, he could see that the former night's clouds and rain had now been replaced by a relatively clear and promising lofty ceiling. That was very common with island life. The wind was constantly blowing and would regularly usher-in or usher-out both high and low weather systems. One day could be clear and bright while the next would be overcast and inclement.

The old, spring-wound Big Ben alarm clock with the ear-piercing chime sitting on top of the end table indicated that it was 7:00 A.M. After drinking as much wine as he had the previous night, McGuire was surprised that he had not visited the toilet at least once or twice during his shut-eye? He must have slept sounder than usual and the chronic prostate malady that plagued him this past 20-years must have taken the night-off? However, he now realized that his bladder was demanding to be drained and he had better heed that call very soon before he ended up wetting himself.

McGuire normally graced the bed sheets 'au natural' and would often move about the interior of the residence in that state as well, until such time as he felt that there was a need for alteration. However, realizing that there was a guest within the house this morning and wishing not to frighten him by his uninhibited appearance, McGuire quickly donned a pair of boxer shorts and exited his bedroom.

Padding barefoot through the living room, past the kitchen and then into the corridor toward the bathroom, McGuire observed that the door to the spare bedroom was ajar. Peering inside briefly as he passed, he noticed that the bed sheets were disheveled but Jesse was not within. Thinking that he must be inside the bathroom and continuing to that point, he noticed that the 'Do Not Disturb' sign was not turned-over. He rapped on the door several times without receiving a response.

Finally opening the door, he immediately observed that the room was empty. He thought that was odd but at the moment any investigation as to the whereabouts of his guest would have to wait until he first completed his pressing need. Soon McGuire contained less waste fluid than he had moments ago and was now in a better position to attempt to determine what had happened to the young man that had previously occupied his spare bedroom.

Returning to that room, he pushed open the door to its extremity. Jesse's duffle bag was still sitting on the floor with his black rain slicker draped over the top. There was also a silver, Apple laptop computer with the screen raised in the open position, just off to the side of the bag. However, the remainder of Jesse's clothing and Doc Martin boots were nowhere in sight. He couldn't have gone too far, McGuire thought to himself as he then made his way into the living room. Looking out of the feature window, he methodically scanned the front yard area until his eyes located their mark. Jesse was seated on a wooden bench between two large trees with his back to the house and gazing into oblivion. Satisfied that he had not run-off and probably wouldn't any time soon, McGuire returned to his bedroom and dressed himself.

Jesse became aware of his presence but didn't display any interest in McGuire as he approached him. He just continued staring straight ahead and feverishly wringing his hands. "Do you need it that bad?" inquired McGuire. "What are you talking about?" spat back Jesse defensively. "Come on Kid, I've been around far too long and I'm too well educated when it comes to drugs not to notice that you've got a problem. What is it…Meth?"

"It's none of your business what I do or what I don't do!" Jesse rudely shouted back at McGuire. "Who do you think you are questioning or accusing me of anything?" he continued, "I live my life the way I want to and I don't need you or anyone else telling me what to do!"

"Hey, settle down Kid. I'm not judging or telling you what you should or shouldn't be doing," McGuire reassured Jesse. "I just see that you're in a bit of a bad way and as a friend of your aunt Pearl, I'm concerned."

This latter response from McGuire caught Jesse somewhat off-guard since he was actually expecting some additional lecturing or form of admonishment from him. That was normally what he received from his parents during their disagreements, so why should this circumstance be any different?

When he observed that McGuire remained neutral and without condemnation, Jesse began to settle down and although still wary, he had become less defensive. For his part, McGuire kept silent and merely continued looking at Jesse in a kind and patient manner. After a pause of several minutes, Jesse responded in a challenging fashion, "Okay, so I smoke a little Meth once in a while? What's the big deal!"

"It don't mean nothin' to me, Son," replied McGuire. "It's just that I know that stuff is poison and after a while it will either ruin your life or else it will kill you. I've seen it happen to others so, I can speak from experience. You seem like a pretty nice young man and I'd hate to see you end up with the short end of the stick. Not to mention what it would do to your parents and your aunt Pearl."

Jesse absorbed McGuire's statement but had no viable reply. Instead, he just looked away, tightly crossed his arms across his chest and clamped his mouth shut. This muted stand-off remained intact for several more minutes before McGuire once again took the initiative. Looking at Jesse's sweatshirt he asked, "Hey, Kid. What's the story behind the Redbone shirt? Do you know who they really are?" After a brief pause, Jesse broke his silence and responded to the question.

"Of course, I know who they are! This was my dad's sweatshirt when he was still in University and he gave it to me a couple of years ago."

Wishing to keep the momentum going, McGuire continued pressing his young charge, "Well, go on. Is that all there is to it. The shirt belonged to your dad?" Realizing that he probably wasn't going to leave him alone until he related the entire chronical, Jesse decided to enter into a dialogue. "In 1980, my dad was attending the University of British Columbia majoring in economics. You see, his life was already being laid-out for him by his parents and he had nothing to say about it. His family was living in West Vancouver at the time and they had close connections with the Capilano First Nations

Band. Since my dad had always done well in school, it was decided that he would go to University, study finance and economics and would then work for the Band as a financial advisor or something like that, when he graduated."

"As usual he was 'acing' all of his courses and he was quickly becoming the golden child of his family. On the weekends he used to hang-out at Frank Baker's Attic Restaurant near the Park Royal Shopping Centre."

At this point McGuire interjected, "I know that place. Isn't that the smorgasbord restaurant at Southwest Marine Drive and Taylor Way with the James Bond car out in front inside a glass box?"

"Yes, that's the one," acknowledged Jesse. "It was a silver, 1964 Aston-Martin that 007 drove in the Goldfinger movie. Anyway," Jesse continued with an obvious tone of annoyance in his voice after being interrupted, "Like I said, my dad used to hang around there on weekends and over time he became quite 'tight' with the Management."

"One Saturday night he was introduced to an older Afro-American dude from Los Angles, who was supposed to be some famous record producer or band manager or something like that. His name was Bumps Blackwell. I'm assuming that Bumps wasn't his real first name…whatever. Anyway, after talking together for a while and after my dad gives him his life story, this Bumps guy claims to be the manager of a Native Indian rock band called, Redbone. He also said that they had a big hit in 1974 and it went right up to Number 5 on the American Hit Parade. The guy then opens a briefcase that he had with him and shows my dad a black and white publicity photo of the band."

"My dad was still in his early 'teens' in 74 but he tells this Bumps guy that he remembers hearing the song on the radio and reacting to it, since they were an indigenous group. Being the scholarly type even then, my dad did a little research on the band and discovered that one of the members was a Cherokee, another was an Apache plus there was a Yaqui and a Shoshone as well."

Once again McGuire interrupted Jesse's story and added, "I know exactly what song he was talking about. It was called 'Come and Get Your Love'. It was pretty good!"

By now, Jesse was becoming accustomed to McGuire's interruptions so, he just ignored the remark and continued with his monologue, "I guess they talked for quite a while and eventually shut-down the restaurant for the night. Anyway, the dude tells my dad that he was heading back to L.A. the following day but that he was coming back to Vancouver the next week. He then invites my dad to dinner at the restaurant for next Saturday night. Never saying 'no' to a free meal, my dad agrees and calls it a night. Not really expecting the guy to show the following week, Dad goes back to Frank Baker's anyway and waits. Well, to his surprise, Bumps walks in at around 9:00 P.M. and sits down with my dad. They then both decline the smorgasbord but order from the menu, al a carte, instead. The dude insists that they order the most expensive items on the menu plus he also orders a bottle of champagne."

"Well, during their dinner, the dude puts his briefcase on an empty chair, opens it up and pulls-out the same publicity photo of Redbone that he had shown my dad the previous week. But this time, it's autographed in my dad's name by all of the band members. He also pulls out this same sweatshirt that I'm wearing now and gives it to him. Well, my dad was thrilled to receive those gifts plus he thoroughly enjoyed the great dinner that they shared. Over the next few years, my dad and Bumps communicated often by letter and occasional telephone calls and they would always meet whenever he would fly into Vancouver. There was a big age difference between the two of them but I guess they had a lot in common?"

"Unfortunately, the friendship didn't last too long since, my dad told me that Bumps died shortly after he had graduated from UBC in 1985. Dad still has the autographed Redbone picture in our Rec Room. He kept the sweatshirt mostly in a drawer inside his bedroom to keep it in good condition. That's the story," Jesse concluded.

"Great story, Kid!" admitted McGuire, not only for its content but also that Jesse was able to relate it with as much detail and thoroughness as he did. That was the lengthiest and most candid conversation that McGuire had heard Jesse speak since his arrival and he was quite impressed. Despite the obvious irritation caused by the absence of Jesse's crutch, that McGuire considered his Meth habit to be, perhaps he was finally making a small inroad with him after all?

After his dissertation, McGuire could see that Jesse appeared spent. By the disheveled state of his clothing and the haggard look on his face, it was clear that although he may have laid on top of the bed last night, it was evident that he had not enjoyed much sleep, if any? "Hey, no offence Kid, but you look like shit!" exclaimed McGuire. "How about we go into the house and have some breakfast? We have a long day ahead of us."

"I'm not very hungry," replied Jesse, "What I really want is just to be able to get a few hours of sleep."

"Well, that might be the case," responded McGuire, "But I'm not going to take 'no' for answer. I promised your aunt Pearl that I would watch out for you and she'd kill me if you ended-up starving to death while you were here."

McGuire then threw an arm over Jesse's shoulders and physically guided him back toward the house. With any form of resistance now fully drained from within himself, Jesse said nothing more and allowed himself to be led back inside.

"Hey, Kid…why don't you go to the bathroom and get yourself cleaned-up while I fix us something to eat? Nothing like a good breakfast to set the tone for the day," shouted McGuire from within the kitchen. Jesse for his part, passively followed his host's orders and headed for the bathroom, first stopping briefly in his bedroom to retrieve the towel and bar of soap provided for his use plus a few other essentials inside his duffle bag that he normally employed in order to perform his daily ablutions.

McGuire didn't normally eat a large breakfast since a bowl of porridge topped with some fresh fruit and a drizzling of honey or a couple of bran muffins with a cup of tea would often suffice. However, for this weekend he had purchased some extra provisions during his previous shopping trip that he would not normally buy. Good food was often a convincing persuader, and with Jesse, McGuire knew that he had his work cut-out for him. He needed all of the help he could get.

As he entered the kitchen, he was visually assailed by the culinary debris left behind from last night's dinner. McGuire had forgotten that they had not cleaned-up earlier and now that realization confronted him like some unmistakable truth. Swiftly bypassing the neglected eyesore, McGuire went to the cupboard and managed to locate a few items still unused, that would suffice for this morning's meal. Entering the refrigerator and removing several necessary items, McGuire deftly cracked a dozen eggs into a mixing bowl and then adding some finely chopped onions, diced green peppers plus sufficient salt and cracked black pepper to taste. Subsequently, he began whisking the mixture until all items were satisfactorily combined. Preheating a large, cast-iron skillet and momentarily allowing a generous dollop of fresh butter to melt within, he then poured in the contents of the bowl as well. While the eggs slowly began to sputter and congeal within the skillet, McGuire quickly moved about the kitchen preparing the remaining accompaniments. There were still several items left-over from the previous night's dinner so, they could be put to good use. Tomatoes were sectioned, the remainder of the French baguette slices were placed into a toaster oven plus an ample portion of salmon filet was halved and made ready to accent the now rapidly cooking omelet bubbling in the pan. Reaching into the refrigerator, McGuire removed a carton of fresh orange juice and then poured most of its contents into the two, tall glasses sitting atop of the kitchen table.

Observing that the omelet was now nearly cooked to perfection, McGuire carefully and evenly placed the two pieces of salmon into the center of the molten egg mixture and with a metal spatula, folded one half of the nearly completed omelet upon itself. The fish was already cooked so all it needed was a minute or so sandwiched between the loving embrace of the eggs just to give it some warmth. Removing two plates from within the cupboard and placing them on the table while simultaneously dressing them with the tomato and baguette slices, McGuire then returned to the stove and removed the skillet from the heat. Utilizing the spatula to divide the omelet in two, he then lifted each half in turn and set them on either plate.

Satisfied with the results, he then stood back for a moment to admire his handiwork. *Pretty good, if I do say so myself*, McGuire privately thought. As if on-cue, Jesse walked into the kitchen at the same time and greedily eyed the breakfast-for-two now parked on top of the table. He had been smelling the tantalizing aroma all the way from the bathroom and it was drawing him forward like iron filings attracted to an electromagnet. "Hey, just in time," remarked McGuire. "Have a seat and we'll get down to it. I gotta say, you look a bit better now and hopefully after we eat, you'll look and feel even better still."

Missing his usual weekend of hanging-out with friends and smoking Meth on the streets and back alleys of downtown Vancouver's East-Side, left a void in Jesse that needed to be filled. His nerves were still raw and his body ached everywhere but he knew that he had to get something into his system to alleviate the discomfort, even if it was just a temporary reprieve. Two, six-egg omelets were a considerable meal but both men seemed to plow through them with little effort. Within minutes, the once full plates were now barren and the gratified expressions on both faces declared the repast a success. Rather than spending time ruminating on this latest pleasant experience, McGuire realized that time was passing and he had pre-existing plans for the day that were already well behind schedule.

As a result, he piped up and broke their reverie. "Okay, Kid. Let's get a move-on. As they say, time is money. But before we do anything else, we've got to clean up this mess!" motioning with his eyes and extended right hand throughout the entirety of the kitchen, informing Jesse that he was going to be a major player in this effort. Jesse surveyed the room and cringed within. He didn't normally perform any chores at home, and if he did, washing dishes was definitely at the bottom of the list. Accepting his fate and following his host's lead, both men tackled this challenge head-on, and within 30-minutes had completed the task to McGuire's satisfaction.

"Okay, Kid. It's almost 9-o'clock and we've got to hit the road. 'My Girls' are going to be wondering where I am since I always check on them every morning well before now," declared McGuire. "I'm not going to bother taking a shower because of the time so, after I change clothes, we can leave. I've got a change of duds for you as well. You're not going to want to wear those things where we're going," were his final words as he headed back toward his bedroom. On that note, Jesse also returned to his bedroom to forward his mother a message from his laptop computer. Much to his chagrin, Jesse was a bit of a 'Momma's Boy' and he knew it. Before leaving the mainland, he promised his mother that he would message her daily to keep her apprised of his movements while on Texada. Rather than upset her, he intended to keep his promise.

After sending the brief message from his computer, McGuire's words were circulating within his head and he was attempting to make sense of them. Checking on his girls? What was this guy, some kind of Pimp? And what's with the need for him to change his clothes? Without having to ponder these questions too long, McGuire entered Jesse's room with an arm load of clothing plus a pair of sneakers and tossed them toward him. "These are some left-over things from a couple of guys that stayed with me here on the island several years ago when we celebrated my 70th birthday," announced McGuire. "I wasn't living in this house at that time but rather in a place I called 'The Mansion' because of its size and exclusivity. It was a huge place with 3-floors on 5-acres, 6 bedrooms, a 5-car garage and a big swimming pool. It was a fabulous place to party! I think it was one of the largest homes on the island, if not the largest? Anyway, these things should fit."

Jesse was still curious as to where they were going and what they were going to be doing, but seeing that McGuire was in an obvious hurry, he felt that now was probably not a good time to ask. Instead, he merely began disrobing in order to exchange his own clothes for those provided by his host. Observing that his charge was now complying with his direction, McGuire turned on his heels and began walking away. "I'm getting my truck out of the garage and I'll meet you in the driveway. Just pull the back door closed when you leave the house and it will lock automatically," instructed McGuire. Jesse had nearly completed his clothing change when he heard the large roll-up door of the Quonset Hut slam shut followed by the sound of a combustion engine idling in the driveway near the back door of the residence. As he exited the house, he observed McGuire seated behind the steering wheel of a near-new white, Ford F150 pick-up truck.

He was sporting a white, Panama Hat with a black band, a pair of mirror-lensed sunglasses, and a multi-colored Polynesian-style short-sleeved shirt that was so loud, it was visually deafening. Below the waist he wore a pair of shocking-blue, knee-length Lycra shorts, white ankle socks plus a pair of black canvas, Skechers deck shoes. His fashion statement that morning was that of a person informing the world "he had arrived." Jesse on the other hand, in his mismatched outfit, looked like a vagabond. Of his appearance, Jesse's father would have said, "He looked about as cool as the drool dripping off a hound dog's tool." This thought didn't fill him with confidence.

Jesse jumped into the front passenger side of the truck while McGuire tromped down his right foot on the accelerator pedal. He enjoyed driving fast and often exceeded the posted speed limit whether on public or private roads. With time now at a premium this morning, that frequent habit was increased a notch. Zig-zagging down the gravel driveway on their departure, they soon reached Leonard Road and then a quick right-hand turn soon brought them to the intersection where Leonard met Gillies Bay Road. Making a quick southbound turn onto Gillies Bay Road and chirping his tires while doing so, McGuire now lowered the driver's side window of the truck and rested his left forearm on the door frame. Looking over at Jesse, he chuckled internally as he saw that his seat belt was tightly fastened and he was earnestly grasping the support grip positioned near the upper portion of his interior door post. For McGuire, this was normal driving. If this style of motoring was already an issue for Jesse, he wondered how he would be reacting if and when he was really driving with intent?

Jesse seemed to have returned to his quiet-self once again so McGuire decided to avoid all unnecessary conversation and left him alone. Instead of talking, he turned-on the vehicle's AM/FM radio and cranked-up the volume. The F150 continued cruising along for approximately 10-kilometers when McGuire then indicated for a left-hand turn. Accessing an unmarked dirt road just wide enough to accommodate a single vehicle traveling in either direction plus bordered by dense foliage and trees on either side, he continued driving an additional 2 or 3 kilometers. If Jesse assumed that McGuire would significantly decelerate the pick-up on this now diminished roadway, he was sadly mistaken. This was an artery frequently traveled by McGuire and over time he had memorized every bump, rut, twist and turn en route to, and or away from his intended destination. As a result, the pick-up's trajectory had barely altered from highway speed and was now virtually careening forward, to where Jesse could only imagine.

The greenery engulfing them on either side was merely an emerald-colored blur as they streaked past. If he had been bracing himself before, as was the case along the route here, he was now literally holding-on for dear life. McGuire was fully aware of Jesse's terror and he was loving every minute of it! He was completely at ease behind the wheel and confident in his ability to safely navigate this route at any speed. After several minutes passed without voicing his concern, Jesse was just about ready to speak-out and demand that McGuire either slow-down or let him out of the truck.

However, that soon became unnecessary as the once narrow pathway gradually opened-up to a vast and deforested clearing. As the vehicle finally began decelerating, Jesse could see that they were now approaching a large, arched-roof and interlocking aluminum pole-ribbed greenhouse that was polyethylene-encased and fully enclosed within its own 8-foot chain-link fenced compound. The sides of the greenhouse also actually rolled-up in order to provide better airflow for the plants within. Now Jesse's curiosity was definitely peaked as McGuire placed the vehicle in 'park' and shut-off the engine. He had a pretty good idea of what might be inside the greenhouse but he wanted to wait before assuming something that later proved to be incorrect. "Okay, here we are," exclaimed McGuire. "Now I'll introduce you to my girls."

Both men exited the pick-up truck and Jesse followed as McGuire walked toward the locked gate of the compound. Producing a key from within his pocket, McGuire disengaged the padlock from the securing clasp and then pulled open the gate. Beckoning to Jesse, they then both entered the compound and moved toward the main entry door to the greenhouse. Once that barrier was opened, Jesse was immediately assailed by the nearly overpowering scent of that earthy, herbal and woody plant, commonly known as marijuana. His earlier intuition had now been confirmed, but he was still surprised that a 'straightlaced' person like his aunt Pearl would have not only befriended someone involved in the cannabis trade, but also thought so highly of him. Making a closer examination of the greenhouse's interior, Jesse could see what appeared to be at least 20 to 30 mature plants optimally spaced in four separated rows, and positioned in a uniform fashion within the main area of the greenhouse. All were rooted in 10-gallon black, plastic gardening containers, while 16 smaller plants also sitting in similar containers sat off to one side in an empty corner of the structure.

While the mature plants were nearly the height of an average man, the smaller plants were only 2 to 3 feet in size.

"What do you think?" questioned McGuire of Jesse, while adopting a prideful attitude and displaying an obvious look of satisfaction on his face. "These are my girls!"

"What the hell, Bob?" responded Jesse in an incredulous tone. "Are you some kind of a drug dealer?" Without realizing the magnitude of this question, and perhaps not intending to insult his companion, nevertheless, that's exactly what he did.

"Hey, Pal!" spat back McGuire. "I'll have you know that I am a licensed grower of medical marijuana, fully authorized and approved by the Province of British Columbia, and I'll ask you to watch your mouth!"

Usually, Jesse kept his feelings and opinions to himself, but seeing McGuire's extensive 'grow operation', sparked an anger within him that surprised even himself. "That medical marijuana stuff is a load of crap!" hissed Jesse. "It's just an excuse for a bunch of old burned-out Hippies to be allowed to smoke pot legally using some bullshit claim they're not able to take anything else for their illnesses."

"All this time you've been riding me for smoking a little Meth with my friends on the weekends, while secretly you're in the shadows actually producing your own drugs out here in the backwoods! What a hypocrite!"

Jesse's unexpected outburst struck a nerve in McGuire and his former positive countenance quickly dissipated. He could feel his blood begin to boil and that heat had rapidly caused his face to flush. At this moment, it took considerable effort for him not to lose his cool and give this presumptuous young upstart before him a good strong cuff behind the ear. Instead, he took a deep breath and barked, "Okay, JESSE, that's enough! You don't know shit, and you had better quit now while you're still ahead!"

It wasn't lost on Jesse that McGuire had just called him by his actual name rather than 'Kid'. This fact and the dark look on McGuire's face, that seemed it would require the intense beam of a 3-cell halogen flashlight to brighten, returned him to a better frame of mind. He was smart enough to realize that he had just stepped over the imaginary line separating peace from war so, rather than incur any further wrath from his companion, he astutely chose to quickly backpedal.

"I'm sorry, Bob," Jesse earnestly remarked. "That wasn't fair of me to say that. You're right, I don't know anything about medical marijuana or really about any other natural remedies or medicines. I just got upset there for a minute and honestly, without making excuses, I'll still feeling pretty lousy."

As quickly as he had become incensed over Jesse's former comments, so now had McGuire calmed himself once again after receiving the lad's apology. Identifying that there was something else underlying this uncharacteristic diatribe, McGuire now took the higher road. He decided that this would be a good time to impart a portion of his considerable knowledge regarding the cultivation of cannabis. Perhaps he'd assist with Jesse's education on the topic of herbal medicines, including the traditional methods and uses of natural, psychotropic drugs utilized by ancient peoples and cultures.

Leaning against the interior of the chain-link fence near the gate were two folding, canvas deck chairs. McGuire retrieved them both and handed one to Jesse. "Have a seat," he instructed as he simultaneously unfolded the other. After both were seated, he continued further.

"Okay, let's start off by clearing the air," began McGuire. "It's no secret that I have used and have also been affiliated with certain types of drugs throughout the majority of my adult life. Until recently, some of them were considered illegal according to the Laws of the Land. However, I never was nor will I ever be a 'drug dealer', so get that out of your head right now! As you know, the law has now since changed regarding cannabis and its related by-products. What I told you earlier is true. I am a producer of medical marijuana, licensed by the Province, to a maximum of 40 plants. What you see here is all completely legal. And yes, there really is such a thing as marijuana for medical purposes." McGuire paused here momentarily, allowing Jesse to digest his words.

"There are many different variants and potency levels of marijuana but there are 3 main strains. Sativa and Indica are the most common plus the most important of the three. The third and less common strain is named Ruderalis. Most medical marijuana actually contains very little to no THC. It is THC or tetrahydrocannabinol, to use its scientific designation, that causes 'users' to become high when smoked or ingested internally. Contrary to your previous claim and to dispel your notion that only 'a bunch of old burned-out Hippies' use medical marijuana in order to get high, bona fide patients using medical marijuana come from all walks of life and don't actually become 'high' when administering their doses."

"Certain chemicals and properties from within the plant itself assist them with their illnesses without the need for intoxication."

McGuire could see that Jesse was intently listening to his tutorial so, he continued with the lesson.

"Because of its beneficial properties together with no harmful side-effects, medical marijuana is often used for the treatment of Chronic Pain, Insomnia, Depression and Anxiety, PTSD, Multiple Sclerosis, Crohn's Disease, Parkinson's Disease, Migraine Headaches and Cancer Pain just to name a few. I could keep going but then we'd be here all day."

"Now what do you think?" questioned McGuire. "Have you changed your mind?"

Jesse took a moment to consider his choice of words and then responded, "Yah, I'm feeling like a bit of a dick right now. Obviously, I never considered that it was anything more than a recreational drug or an excuse to get high. I didn't realize that medical marijuana contains little or no THC. Thanks for setting me straight and sorry once again for pissing you off."

"Okay, I have to admit that you had me pretty worked-up there for a minute with those comments but I accept your apology. Plus, now that you've agreed that you're clueless, I'm going to continue with your education," st McGuire. He then continued further, "Considering your ethnicity, I'm surprised that you don't have a better understanding of what's going on here?"

"Indigenous peoples all over the world, and especially in North, Central and South America have used natural substances like plants and roots in their medicines and spiritual ceremonies for thousands of years. Many Shamans and Medicine Men of tribes in the Western Hemisphere used marijuana and additional hallucinogenic materials in their sacred ceremonies. Even tobacco was utilized to communicate with 'The Creator'. The leaves would be tied together in bunches and burned openly so that the subsequent resulting smoke would rise to the Heavens, intermingled with the prayers of the faithful. Or the leaves were shredded and stuffed into a pipe. Once lit, the holy man would puff to create smoke but he would not inhale it into his lungs. It was believed that the smoke, considered to be an offering, should not first enter the body of a man but rather travel straight towards heaven."

McGuire was now 'on a roll' and he was talking a mile-a-minute. This was a subject close to his heart and he loved imparting his knowledge on the topic whenever possible. Looking at Jesse, he could see that he was now truly engaged, so before the moment was lost, he quickly continued.

"Mescalero Indians who lived throughout the American southwest frequently used the Peyote Cactus, Mazatec Indians of Oaxaca, Mexico regularly used upward of 10 different varieties of mushroom flora for healing and to reach an altered mind-state. They even ground the seeds of the Morning Glory plant and made a tea from them in order to the calm the nerves of upset persons. Native tribes of Peru used the mescaline-rich Trichocereus-Pachanoi plant and also often served it as herbal tea. They also chewed coca leaves that would relieve hunger and fatigue plus stimulate stomach function, treat Asthma and assist with common colds."

"Bushmen of Dobe in southern Africa would incise their scalps and then rub the onion-like bulbs of the Kwashi plant into the wounds. This would create hallucinations for them."

Although Jesse hadn't interjected with comments or questions during McGuire's dissertation so far, it was obvious from his expression that he still held his attention. He then continued, "Indigenous peoples weren't the only ones to catch-on to the benefits and special properties of natural substances. Some European healers in pre-modern times also took advantage of certain natural materials in their ministrations and treatments. Antropa, Belladonna, Henbane, Mandrake Root and Datura plants were all used, just to name a few."

"Here's a little-known fact that I bet will surprise you. It's common knowledge that many native people world-wide are superstitious and their cultures abound with a plethora of legends and mystical beliefs. Superstition and legends are also common in European culture as well, especially in ancient times. For many years, the old wife's tale of Witches flying through the skies at night on their broomsticks was commonly regarded. Even today at Halloween and on children's cartoon shows, there are images of haggard old crones with long crooked noses, bony fingers, wearing black flowing robes with pointed hats, caressing black cats and soaring on their broomsticks. After much study of these beliefs, historians have come to a different set of conclusions."

"What they now believe is that Witches 'of old' would actually rub hallucinogenic ointments on their broomsticks during ceremonies and then self-administer the drug by sitting astride the handle and rubbing up and down on them. The soft tissue of their vaginas would absorb the drug and it would then produce the desired effect."

"Rather than traveling through the actual skies at night on their broomsticks, what probably occurred was that they traveled across the hallucinatory landscape of their minds, holding on to their broomsticks for dear life!"

With this last anecdote, Jesse burst out laughing and kept on so for quite some time. McGuire also chuckled and allowed the moment to continue until it had naturally passed.

"You just made-up that last story!" cried Jesse as he attempted to regain his composure.

"No way! That's what I read," replied McGuire, attempting to remain straight-faced. "You can either believe me or not."

Waiting a few more minutes until their mirth was now fully expended, McGuire spoke-up again. "Okay, enough fooling around. We've got work to do."

McGuire had risen from his chair and was now walking into the greenhouse. Following his lead, Jesse entered the structure as well. Going to a tool rack near the entrance, McGuire selected a long-handled shovel and a pickaxe. He then handed both to Jesse along with a pair of rough leather gardening gloves. He then spoke. "So Kid, this is your lucky day. You're going to play an important part in a 'ground-breaking' enterprise that I am entering into today." He then let out a short laugh and reiterated, "As I said, for you this day is going to be ground-breaking…literally."

"Do you see those young ladies sitting in the corner over there in those black plastic pots?" After receiving an acknowledging nod of the head from Jesse, he continued. "Well, we're going to plant those babies outside in the ground. I've never tried this before so it will be a new experience for both of us. Unfortunately, I've got a couple of bum knees plus I'm having ongoing issues with my neck and shoulders so, you're going to be playing a pivotal role in this venture. Are you up for it?"

Before Jesse could even answer, McGuire was nudging him out of the greenhouse and leading him to a section of cleared ground, still within the enclosed compound, that had been earmarked for the purpose mentioned earlier.

There was also a large mound of freshly sifted soil off to one side. "Wait here for a minute while I get a tape measure out of my truck. If we're going to do this, we're going to do it properly," remarked McGuire. Until now, Jesse had led a fairly carefree existence. He was barely expected to perform even basic chores at his parent's home, let alone manual labor. He wasn't really relishing the idea of breaking a sweat and potentially getting dirty but he also didn't want Bob to think that he was a wimp. Now the need for a change of clothing made sense to him. McGuire returned shortly afterward with the tape measure and began selectively and uniformly measuring out the places where he felt the marijuana clones should be planted.

This process continued for several minutes before McGuire was satisfied that sufficient locations were identified and that they were all evenly spaced in a grid pattern. Returning to the first measured location, McGuire called Jesse over to him.

"Okay, I've got a trolley inside the greenhouse so, I'm going to be wheeling the 'girls' out one at a time while you start digging the holes. As you can see, I've clearly marked each spot where I would like to see them planted and there are 16 in all. The ground might be a bit hard in some spots so you'll need to use that pickaxe in order to loosen the ground enough to be able to shovel out the dirt."

"Any questions before we start?" asked McGuire with a look on his face that clearly indicated that he wasn't expecting any.

"Just one question," offered Jesse. "Why do you keep referring to those plants as 'girls'?"

McGuire let-out a sustained belly laugh and then eventually responded. "Man, you're really receiving an education today! You're as green as those plants that you're going to be putting into the ground. The reason I call them 'my girls' is because the female marijuana plant is the predominant of the species. They produce the most potent buds compared to the male plants. In fact, we limit the number of male plants in any grow operation since they can potentially contaminate a crop of female plants by continuously fertilizing them. We don't want that to happen. Does that answer your question?"

With an obvious expression of embarrassment, Jesse quietly responded. "Yes, it does and it makes sense now. I should have been able to figure out that one for myself. I guess it was just too obvious."

"Okay, before I start bringing out the girls and the nutrients, the holes need to be about 3-feet in diameter plus 3-feet deep. You might as well get on with it and I'll be back in a few minutes," instructed McGuire as he returned to the greenhouse. Internally, Jesse cringed at the thought of having to dig holes that deep. He wasn't expecting that. Excavations that large were about half the size of a human grave and he had 16 to dig! This was going to take a while. Resigning himself to his fate, Jesse donned the gardening gloves and grabbed the shovel. As he attempted to thrust the blade into the ground, it barely penetrated the surface. His depressive state increased exponentially as he now bent over to retrieve the pickaxe, beginning to employ it with only minimal success.

By the time McGuire returned approximately 20-minutes later with the first plant and accompanying nutrients, Jesse's first hole was only about 25% completed. McGuire was disappointed with the lack of progress but he didn't say anything. He realized that this kid was as soft as a marshmallow but if allowed to, McGuire was going to toughen him up. Not wishing to crowd him too much, he then stepped back several feet and seated himself on one of the folding deck chairs. As time elapsed, Jesse finally finished digging his first hole and plopped himself on to the ground awaiting inspection. McGuire viewed the youth's handiwork, quickly utilized his tape measure again to check-out the dimensions and then to Jesse's relief, deemed the crater satisfactory.

"Okay Kid, you can take a break from the digging but I want you to carefully watch what I'm going to be doing next. Who knows, one day it might come in handy for you?"

Jesse moved back slightly from the hole while McGuire wheeled his cart closer. As he now commenced his portion of the planting process, he apprised Jesse of his movements. With the practiced hand of an expert, he began potting the clone. "So, first-off we're going to take some of the earth you just removed from the hole and combine it with some of the newer, sifted soil from that pile over there and return it into the hole. Then we're going to start adding some of the nutrients. I've got quite an assortment of things here but trust me, they're all necessary to produce a strong and healthy plant. Along with my 'baby girl' and working in specific increments, I'm going to be adding worm castings, a homemade fertilizer mix from a buddy of mine, called Flower Power, coconut fiber, Green Sand, bone meal, blood meal, a measured amount of chicken and sheep manure, Dolomite Mix and then another portion of the old and new soil combination. And lastly, a healthy dousing of water."

As McGuire's instructional tutorial came to an end, the first clone was successfully planted in the ground and now on the road to eventual maturity. Jesse had no previous idea that the process of growing marijuana, or anything else for that matter, was so involved and that there was an actual science behind it. He admired McGuire's passion and knowledge on the subject and quickly developed a new-found respect for him. Without additional prodding required, Jesse now picked-up his tools and moved to where the second hole had been selected. McGuire nodded his head in approval and called back to Jesse as he headed in the direction of the greenhouse.

"I've got some work to do with the mature plants now so, I'll be gone for a while. Just keep digging the other holes like you did the first and then we'll get the other girls into them lickety-split. If you run into any problems, come and find me in the greenhouse." On that note, he returned to the greenhouse while Jesse continued his labor.

By the end of the day, only 5 holes had been dug with their intended herbage safely tucked within. The amount was less than McGuire had anticipated but overall, he was pleased with Jesse's efforts considering his lack of work experience. Perhaps now the remainder could be completed during their next visit to the site, since somewhat of a rhythm and methodology between the pair had been established? Jesse appeared visibly relieved when McGuire instructed him to return his tools to the greenhouse and prepare himself for departure.

After the greenhouse and fenced compound had been re-secured, the two men boarded the Ford F150 and buckled themselves to their respective seats. As McGuire fired-up the pick-up's engine he turned to Jesse with a serious expression and said, "Now that we've got some of the girls out of the greenhouse and into the ground outside, it's extremely important that the gate to the compound is constantly kept closed. I don't want Linda coming by and eating them. I've worked far too hard and too long getting them to this stage so, I don't want her ruining all of that!"

It had been a long, difficult day for Jesse and his muscles were aching where before he didn't even know he had muscles. In addition, in spite of the thick gardening gloves, he also sustained several blisters on the palms of both hands. The last thing he needed now was another riddle. Without speaking, but simply flashing McGuire such a helpless and dumbfounded look, Bob couldn't help but burst-out laughing. Feeling merciful toward his charge, he decided to explain.

"Okay, Kid you can close your jaw now, I'll fill you in. Linda is a pregnant deer that has been coming around here and some other nearby sites looking for a free meal. She loves to eat the soft, new growth and buds of young plants. She recently visited a nearby site belonging to a friend of mine who thought it would be funny to name the deer after a very special woman currently in my life. The deer not only ate numerous plants but also trampled a considerable portion of his crop because he had not secured it properly. I don't want that happening to me. So, if you ever see Linda snooping around here, try to chase her away and make damn sure that the gate is always closed!"

Relieved that he didn't have to hurt is head deciphering this latest perplexing ambiguity from McGuire, Jesse merely nodded his agreement and grunted somewhat of an understanding. On that note, McGuire delivered another one of his trademark laughs, dropped the transmission lever of the truck into the drive position and thrust his right foot down hard on the accelerator pedal. Here we go again, thought Jesse as the pick-up shot forward like a rocket ship. He immediately braced himself and instinctively reached for the support handle on the passenger-side door post.

They quickly reached Gillies Bay Road and retraced their movements now back toward the residence on Leonard Road. McGuire then spoke up. "You did great work today, Kid. I'm proud of you! I don't feel like cooking when we get home so how about I take you out to dinner as a bit of a reward for helping me out? What do you think?"

"Yah, that sounds good," voiced Jesse agreeably. "I'm actually quite hungry. I guess digging those 5 holes burned up whatever calories I had stored up inside of me."

"Great!" responded McGuire. "We'll head home first to get cleaned-up and changed, and then we'll drive into Van Anda. The Texada Island Inn serves up a pretty good spread so, we'll go there." With McGuire's, Formula 1 Gran Prix style of driving, the pair were home in record time and both took turns refreshing themselves. Approximately 45-minutes later, they were back in the F150 and on the road driving in a north-easterly direction toward the tiny hamlet of Van Anda. Since everywhere on the island was within close proximity to everything else due to its diminutive size, it wasn't long before the pick-up was pulling into the parking lot of the Texada Island Inn at 1108 Gillies Bay Road.

The lot was only sparsely populated since full commencement of the tourist season was still several weeks away. As a result, McGuire was able to park right up near the front door of the establishment.

Entering the building, they sauntered into the 58-seat restaurant, walked through its interior and then went outside on to the open patio. The weather that day had been fabulous and even though the major warmth of the sun had since dissipated, it was still a very pleasant late Spring evening. There were no other patrons on the patio at the time so, the pair had their choice of tables. Seating themselves to afford the best view of the Malaspina Strait and Coastal Mountains beyond, they then patiently waited to be served.

Within minutes, the server had taken their identical meal orders of deluxe double cheese burgers with fries and gravy and was just now returning to their table with two ice cold bottles of Sleeman's Honey Brown Ale. It didn't take long for their meals to appear and again not too much longer before McGuire and Jesse fully consumed them, right down to the last French fry and garnishments of pickle and tomato slices. The server eventually returned to collect their empty plates and asked the men if they had any additional room for dessert. Remembering that there was still an untouched apple pie from last night back at the house, McGuire declined that offer on behalf of them both.

"I've got an idea," McGuire directed toward Jesse. Without waiting for a response, he continued, "It's not quite 7 o'clock yet and I'm pretty certain that your aunt Pearl is working at the Texada Market until 8:00 P.M. It's literally just a couple of minutes down the road from here and I'll bet she would love to hear how you spent your day? What do you think?"

Obviously, Jesse had nothing pressing on his time at the moment and nowhere else to go so, he agreed. Taking their bill to the cash counter and fulfilling the reckoning, the pair re-boarded the white, Ford F150 pick-up truck and drove the short distance to the Texada Market a few blocks away. Entering the store, they found Pearl alone inside. Watching two of her most favorite people in the world entering the store, Pearl lit-up like a votive candle and promptly greeted them both with a heartfelt embrace.

The men visited with Pearl for an extended period and when pressed, Jesse described to her his activities from earlier that day. Pearl was thrilled to hear his rendition of events and discreetly threw McGuire an appreciative glance. She also informed Jesse that he must be very special indeed to have been introduced to McGuire's 'girls' so soon since, this was an honor only reserved for people that he truly liked and trusted.

Knowing now that the time was drawing near for Pearl to commence cashing-out her till and closing the store, McGuire and Jesse excused themselves and prepared to leave.

"Jesse!" Pearl called-out as they reached to door. He turned to face her. "When do you want me to take you back to the mainland?" Having now spoken this question aloud, she held her breath for the response.

"I don't exactly know just yet, Aunt Pearl?" was Jesse's response. "I still need to help Bob with a bunch of work and it's pretty slow going. I'll let you know."

Her heart skipped a beat since this was exactly the response she was hoping for. Not wishing to outwardly display the sense of excitement she now felt inside, Pearl remained expressionless and merely returned him a nonchalant, "Okay, that's fine. Just say when."

Once back on the road, McGuire and Jesse were back home within minutes. Jesse sat on the couch while McGuire went into the kitchen to cut the apple pie and grab them both a couple of drinks. Usually, life on the island was barely one step above living inside a monastery with practically nothing to do socially, especially at night. Aside from the one or two restaurants and drinking establishments throughout the entire island, which all closed relatively early, residents were mostly left to their own devices for ensuring that they found a 'good time'. There was a satellite dish attached to the back of McGuire's house that fed a Personal Video Recorder or PVR, along with his flat screen television set inside the living room. With the wide-variety of channels and programs, he could receive plus the ability to record them at will, over time he had accumulated a considerable library of visual entertainment.

McGuire divided the pie into four quarters and then removed two sections that he placed on to individual plates. He then filled two large juice glasses with his Golden Gage Plum wine and settled all four items on to a flat, wicker carrying tray. Lifting the tray and walking back into the living room, he found Jesse fast asleep on the couch. He released a muffled snigger at the sight of the reclining youth and deposited the tray on to the coffee table. Giving Jesse a gentle nudge in order to awaken him, he then escorted the exhausted waif to his bedroom. Lowering him on to the mattress fully dressed and then throwing a comforter over top, Jesse was back asleep seconds after hitting the pillow. Closing the bedroom door behind him, McGuire then returned to the living room and turned-on the television set. There was apple pie to eat plus wine to drink so he might as well be entertained while these items were consumed.

Grabbing the remote control, he selected the pre-recorded rerun of the 2002 FIFA Men's World Cup soccer final between Germany and Brazil. Having previously watched the game he already knew how the match ended but what the heck. He wasn't yet quite ready for bed and after all, it was a great game.

McGuire rose at 7:00 A.M. the following morning and began preparing himself for the day. Peering briefly into Jesse's room, he observed him still asleep and seeming as though he hadn't moved an inch throughout the entire night.

Breakfast this morning was going to be light and there were still a few things for McGuire to rustle-up for today's lunch at the greenhouse. Once these initial tasks were completed, he returned to Jesse's bedroom and found him sitting on the edge of the bed. He had overheard the commotion in the kitchen and realized that it was time to rouse himself.

"Good morning," offered McGuire. "You were out like a light!"

"Yah, I was super tired. I actually slept through the whole night!" Jesse excitedly exclaimed. "I haven't done that in a long time. It must be this countrified air and slugging it out with that pickaxe and shovel." After making that statement, he commiserated momentarily and appeared to be lost in thought. It was only a very brief pause but appeared to have been meaningful.

"Whatever it was, I hope that you're ready for a big day. We've got lots of work ahead of us. Hurry up and wash your face and then come into the kitchen for breakfast. We're heading out as soon as you're finished," McGuire instructed.

Shortly after 8:00 A.M. they were back at the greenhouse and ready to settle into their assigned routines. McGuire had his mature plants to tend and Jesse had his holes to dig. With only minimal words spoken between them, both men then commenced their work. Without actually feeling the need to approach him, McGuire would occasionally peek his head out of the greenhouse door to check on Jesse's progress. Each time he checked he was surprised but gratified to see that Jesse was taking his job seriously and although the holes weren't being dug any quicker than yesterday, at least he had not yet given-up. At the noon-hour, he felt that they had both deserved a break, so, he instructed Jesse to temporarily rest.

He then beckoned him over to where the deck chairs were placed. An overturned wooden crate was employed to act as their table. Walking to his truck, he retrieved their lunches and then returned to their makeshift, outdoor cafeteria.

As both men ate their lunches, conversation gravitated toward personal likes and dislikes. The discussion was casual and disarming so, gradually, and with all the stealth and cunning of a hunting predatory beast, McGuire began carefully probing into Jesse's personal life. It didn't take long to discover that Jesse's life at home was a loving one, if not a bit over-protective from both parents. He had been an honor roll student at Delbrook High School in North Vancouver, but in his senior year, he had become disillusioned with the state of the world and its many injustices.

Personally, he had not experienced any real bigotry or racism from others while growing-up, but he was fully aware that it existed, especially toward First Nations peoples. Even in their own backyard!

This entire country had once belonged to them generations ago, and now it was almost completely lost. Jesse had befriended another indigenous male student while he was in Grade 12, and it was this person that was beginning to alter his attitude. He had introduced him to drugs and was encouraging him to engage in activities that his parents were vehemently opposed to. This young man was also quite militant and was attempting to form some type of anti-establishment group that he hoped would then confront the Government. At the end of the day, it appeared as if Jesse was being torn between a lifestyle he had always known, and overall quite satisfied with, versus that of an angry and vengeful Indian. His friend believed that their culture and birthright was stolen by Western Europeans, and he was hell-bent on having those returned to his People. If that were not possible, then he would take any action or means necessary to disrupt the Status Quo.

It was clear that Jesse was in a state of internal turmoil and was desperately attempting to choose his path. Which path it would be was not clear at the moment, but if McGuire could help him decide, then he would do whatever he could. True to his nature, he was beginning to develop a certain affection for this young man and as they say, "it takes a community to raise a child." As McGuire listened to Jesse's plight he thought back to his own youth. Although his upbringing and social circumstances were completely different to those of Jesse's, he too spent many a sleepless night wondering where his future lay.

He understood what interested him most, and what did not appeal to him. However, those facts didn't culminate into a definitive answer. He proffered this revelation to Jesse in an attempt to have him realize that he wasn't the only person in the world that ever suffered this dilemma, nor would he be the last.

"What finally did it for you, Bob?" enquired Jesse. "What was it that got you heading in the right direction?"

"Well, what got me started believe it or not, was taking a Social Studies class in Grade 11 with a really great teacher! By the time he got finished with me, all I wanted to do was travel and see the world. Even now at my age, I still just want to grab my backpack and hit the road to some place I've never been before."

"Going to school is one form of education, but traveling to foreign countries and meeting different people and then learning about their cultures, is also another form of education. That's what turned my crank then, and it still does now. I remember my first big trip like it was yesterday. I took a cruise ship out of New York City and sailed to the Mediterranean Sea."

Jesse absorbed McGuire's words and for several minutes, just sat quietly. He then said, "Hey Bob, is there any chance you can relate a few of your experiences to me? I'd really like to hear them."

Well, asking Bob McGuire if he is interested in speaking, is like asking a 'streetwalker' if she is interested in attending the Vancouver Dockyards, in order to greet the sailors of Canada's Pacific Fleet, having just returned to port after a six-month 'tour of duty' from the expansive and lonely seas. It is more than likely that the response would be a resounding, "HELL YES!"

"Yah, I don't mind," replied McGuire matter-of-factly. However, he was actually extremely pumped at being able to re-visit times and places that were instrumental in defining the person he had now become. "Okay, I might as well start from the beginning. It just so happens that I've been re-reading some of my old travel journals. I've got my first one inside the truck right now. Give me a minute and I'll go get it."

McGuire retrieved the compact and battered, red journal from his pick-up and reseated himself. He took a moment to get comfortable on the deck chair and then leaned back slightly. Closing his eyes and clearing his mind of 'the here and now', he sent his thoughts back to the time when it all began.

Chapter 3
The Beckoning Road Awaits
(November 13, 1965)

"I remember that it was very early in the morning when I awoke the day I left for my maiden voyage into the Great Unknown," commenced McGuire with a bit of melodramatic flair that he so often liked to use when relating something meaningful. "It must have been near 4:00 A.M. and I wanted to do some last-minute checking to make certain that I had everything in order. I was going to be gone for several months and the last thing I needed was to be in some foreign country somewhere only to discover that I forgotten to pack my baby powder or something important like that." He then looked over at Jesse for a reaction. Naturally, Jesse grinned as a result of the last remark. McGuire then continued.

"My family was still asleep but they would also soon be rising since, they were all coming along to the Greyhound Bus Station to see me off. Time passed quickly and it was then time to leave. Well, as you can imagine the mood was quite somber, especially for my parents who still weren't certain if this trip was a smart move on my part. They weren't unsupportive, they were just a bit worried. I tried to lighten things up a bit by referring to my family with the 'pet names' I had given them. My parents were Mama and Papa Bear while my younger brothers, Kenny and Johnny, were Brother Bears. My older brother Mike was already out of the house and living on his own, beginning to climb the 'corporate ladder' with the Canada Safeway grocery chain."

"I was amused to see my brothers struggling to carry my jam-packed duffle bag and backpack outside to my dad's car which was Yellow Cab #38. Yes, my dad owned his own taxi and with that he was able to earn enough money to build the new house in the Collingwood District of Vancouver where we now lived. Previously, we lived in Vancouver's Mount Pleasant District so, this was definitely a step-up."

"We all piled into the car and drove to the Greyhound Bus Station downtown. My bus was leaving for Kelowna at 7:00 A.M. but we made the journey with time to spare. I was surprised to see that my old soccer buddy Ken McKenzie, from Main Street, had also arrived in his own car to bid me farewell. While we waited the short time for me to board the bus, Mama Bear kept reminding me that I should visit a barber regularly during my trip."

"I thought this was a bit strange since, she didn't seem too concerned that I might be robbed or kidnapped or killed during my travels but making sure that I cut my hair on a regular basis was a really big deal. Naturally, I reassured her that I would follow her advice but at the moment, that was the furthest thing from my mind."

At this point, Jesse interjected with a question. "Bob, excuse me but I'm a bit confused? You said that you took a cruise ship out of New York City but then why would you be taking the bus to Kelowna? That's not exactly a direct route."

McGuire responded with a question of his own, "Okay, Kid. Let's see if you're smart enough to answer your own question? Since time in memorial, what three things have motivated young men?" For a moment, Jesse thought long and hard. Then the expression on his face informed McGuire that an imaginary lightbulb inside his head had just been lit. With a gigantic grin on his face, he offered, "I'm pretty certain that I know the answer…SEX, DRUGS, & ROCK and ROLL!"

"You got it Kid, that's exactly what it is. Well done!" responded McGuire approvingly. "There was a fresh, ripe young 'tomato' living in Kelowna that I had met previously in Vancouver and she said that if I was ever in town, I should look her up. Since I was leaving anyway, why not make a bit of a detour and pay her a little friendly visit? I was actually hoping to start-off my Mediterranean Cruise with a 'bang', if you know what I mean?"

There was that look again on McGuire's face that Jesse had noticed previously, whenever he spoke of women. The salacious smirk on his lips plus the mischievous glint in his eye could only mean one thing. The mating season for Bob McGuire appeared to last 365-days of the year.

Well, 7-o'clock had rolled around pretty quick and I was on the bus with my gear safely stowed away inside the cargo hold. The bus was full of passengers as we began to pull out of the Bus Depot. The Family Bears were all waving to me and then my little brother, Johnny, stuck-out his tongue at me. I thought to myself, *You little bugger* so, I threw him the middle finger. Mama and Papa Bear did not appear impressed but Kenny, Johnny, and me, were all laughing our heads off.

"Well, I was finally on my way and who knew what fantastic adventures lay before me? My only regret was that my best friend Jimmy Bridge wasn't with me. When we were still both students at Killarney High School in southeast Vancouver, Jimmy and I had decided to travel the world soon after graduating, and after we had accumulated a few bucks."

"After high school, I kept my part of the bargain by selling my cherished 1955 Chevy Bel Air and saving every nickel I earned at whatever job I could find. I became the human version of the Walt Disney cartoon character, Scrooge McDuck and had become so miserly, that I wouldn't even leave the house, for fear of spending money. Unfortunately, Jimmy's resolve wasn't quite as strong, and before long, he had acquired two new girlfriends plus purchased a new car. Needless to say, that was a kick in the gut for me, but I was committed, and if Jimmy wasn't coming along, then I would go myself."

"Here I was finally on my way! I had $1400.00 in American Express traveler's cheques in one pocket of my jeans plus my one-way ticket for the cruise ship in the other. As I tilted back my chair and closed my eyes in order to catch a bit of a snooze, I let the soothing and hypnotic sound of rubber tires caressing the asphalt road lull me to sleep. The bus trip to Kelowna was interminably long and we were constantly stopping to either drop-off or pick-up passengers. Naturally, each time the bus would stop, I would awake so, if I told you that I was experiencing a restful sleep, then I would be lying. It was 5:30 P.M. when we finally arrived in Kelowna. Although, I had barely raised myself from my seat during the entire trip, I was exhausted. I eventually checked into a cheap motel for $4.00 per night and just crashed on to the bed like a giant Sequoia tree being felled by Paul Bunyan."

"The next day I met up with Shirley, who was the reason for my detour but she was with other friends so, unfortunately, I had to share her company. Finally, the next day Shirley and I found some time alone and we truly consummated our friendship." There was that look again, thought Jesse.

"Well, now that my mission was accomplished, it was time to hit the road again. In any event, Kelowna was deader than the Mountain View Cemetery at East 41st Avenue and Fraser Street in South-East Vancouver. I left Kelowna at noon on November 16th and managed to snag a ride with a drunken firefighter all the way to Vernon. Unfortunately, I got stuck in Vernon for a few hours on the side of the highway trying to thumb a ride the remaining distance to Kamloops. It was wickedly cold outside, and I damn near froze my balls off! I finally caught a ride in the late afternoon and made it into Kamloops at 6:00 P.M."

Jesse now piped in again. "Weren't you nervous riding with strangers? Especially that drunk guy? You could have been in an accident or they could have robbed you…"

"Nah, we didn't worry about that kind of stuff in those days. It was almost an unwritten rule of the road that if you saw someone hitch-hiking and they didn't look like an axe murderer, then you picked them up. Guy or girl, no matter. You just did it and they knew that they were going to be safe to jump in," McGuire explained.

"Anyway, since the Canadian National passenger train for Toronto was leaving at 2:00 A.M., there was no use grabbing a room for the night. I just ate dinner at a local restaurant in town, and then caught a ride to the station at 500 Lorne Street. Upon arrival, I purchased my one-way train ticket to Toronto and then just waited around the station for the few hours, until it was time to board. Being early morning, it was pitch black outside and nothing to see. As I was getting stowed away, another guy about my age entered the compartment. Apparently, he was going to be my bunk-mate. I can't even remember his name right now, but I just ended-up folding down my bunk and then 'hitting the sack' for forty-winks. The compartment was designed to accommodate two people, but it was smaller than the average jail cell. The dimensions were 7-feet/2-inches by 4-feet/11-inches. Crammed into this tiny space were two folding bunks, a private toilet, a mirror over a miniature sink attached to the wall, plus a luggage storage rack. Just to give you an idea about the mattress size, it was 71-inches by 31-inches. Under those circumstances, I wasn't going to be hosting too many orgies inside that ridiculously small space any time soon!"

"We were awoken later that morning by a knock on the compartment door. I opened it wearing only my boxer shorts, hoping that there was a gorgeous chick on the other side. Unfortunately, it turned-out to be a male steward wearing a smart-looking charcoal gray CN Rail uniform and holding a tray that I assumed was our complimentary continental breakfast. The trolley cart beside him supported additional trays with identical offerings. I accepted the tray from the steward and examined its contents. Breakfast consisted of two Peak-Frean shortbread cookies, a banana plus a tepid cup of milk-tea, sans sucre. What a treat!"

"By the time I cleaned-up myself in the mini-sink and got dressed into the same clothes that I wore the previous day, the train was already half-way through Alberta, on its east-bound journey towards Ontario. I began exploring all of the coaches on the train, but that didn't take very long. After that, I decided to sit down and do a little sightseeing. However, if you've ever been to the Prairies, you'll understand that there's nothing to see. It's completely barren and as flat as Olive Oyl's chest (using cartoon character, Popeye's girlfriend as a comparison)."

"Without seeking his attention, my bunk-mate never-the-less followed me around like a lost puppy. I wasn't rude to him but I didn't pay a lot of attention to him either. He had just been fired from a job at a lumber camp in Northern B.C. and was heading to Toronto to seek his fortune. On the 18th, we stopped in Winnipeg, Manitoba to pick-up more passengers and I observed a younger, red-haired and bearded male board the train. He could have passed as my brother."

"The name of the new arrival was Daryl and we hit it off pretty good. Anyway, we spent most of the time thereafter sitting together in the Lounge Coach sipping beer at the exorbitant price of 53-cents per bottle. (You've got to remember that a glass of draught beer in the Kingsway Hotel near where I lived was only 20-cents). It was irking me to no-end since, I was now spending money that I was actually trying to save for my trip. Possibly wanting to impress me and Daryl as we were nearing arrival to Toronto the next day, my bunk-mate confesses that he had approximately 50-feet of detonation cord plus several sticks of dynamite, that he had previously stolen from his lumber camp job, inside his luggage and both were actually on the train! I was incensed that this idiot would endanger our lives plus that of the other passengers with this hazardous material onboard and I gave him a good piece of my mind. I actually wanted to beat the crap out of him and I still don't know why I didn't? Instead, Daryl and I just left him there in the lounge and sat by ourselves for the remainder of the ride. No wonder that jackass got fired!"

"It was dinner time when we arrived in Toronto on the 19th and just as I was about to disembark, Carl, the bartender from the train asked if I needed a place to stay? Never turning down anything 'free', especially when I was now trying to save every penny I could, I took him up on his offer. Carl had his own apartment downtown and he spared no expense taking me around and showing me the sights of T.O. One of the highlights was my first subway ride. I always thought that Vancouver was a big city but Toronto put it to shame. The nightlife on Yonge Street was fantastic and one night, I forgot myself and blew 15-bucks trying to impress a couple of 'birds'. Needless to say, later that night I wanted to kick myself for being so stupid! And in case you're wondering, 'No', I didn't score with the chicks!"

"Well, I was having a fine time at Carl's place eating from his refrigerator at will, watching T.V. and basically just having a run of the place. On two occasions, he bought me a Chinese dinner and overall, he was very generous. Unfortunately, as they say 'nothing in life is free,' and that was the case in this instance as well."

"The night before I was going to leave Toronto, I was at Carl's place in bed sleeping when I suddenly felt someone crawl in beside me. It was around 3:00 A.M. and at first, I thought I was dreaming. Unfortunately, I wasn't dreaming, and to make matters worse, the pair of bare legs that were attempting to entwine themselves around me, had more hair on them than mine. Well, Bobby doesn't play that game, and I pushed the guy out of the bed real hard. I was also going to beat the crap out of him, but then decided against it. I was going to be leaving town in a few hours anyway, so what was the point? Anyways, when I got up a few hours later that morning, it was still dark outside and everyone else inside the apartment should still have been sleeping. However, there was not a sole in sight. Carl, Reg, Marcel, Murray, Joe and whatever other guys had been hanging around the apartment the last couple of days, had all flown the coup. I guess someone had a guilty conscience and they probably thought that I was going to kill somebody? In any event, I never found out which one of those jokers tried to 'poke' me that night, but by now, I didn't care anymore. I was on the move once again."

Jesse now broke into the conversation with a few additional comments of his own. "Holy cow, Bob! You haven't even actually started your real trip yet and already all of these things have happened to you?"

"Never mind, Kid. It gets better!" McGuire responded.

"After leaving Carl's apartment, I made my way to Toronto's main Greyhound Bus Depot and caught the 8:00 A.M. bus to New York City. The one-way ticket cost me $18.00. From Toronto all the way to Buffalo NY, I was sitting beside a Protestant Minister so, naturally, I was on my best behavior. I'm not overly religious, but I still didn't want to tempt fate by aggravating him. I finally arrived in New York City at 7:30 P.M., and was now in a bit of a quandary. I was now by myself in the 'New Amsterdam' on a cold, damp night and completely lost. I had absolutely no idea where I was supposed to go. Trying to remain calm and focused, I grabbed my bags and started walking toward the brightest lights that I could see on the skyline. As luck would have it, I eventually found myself on Broadway and was then able to ask for directions toward my hotel."

"When I originally booked my trip with the Travel Agent in Vancouver, she made all of the additional arrangements I required while awaiting to board the ship for the Mediterranean. She pre-registered me at the Mayfair Hotel at 242 West 49th Street. The cost of the room was only $4.00 per night, so I wasn't expecting anything as grand as the name of the hotel suggested since, Mayfair was a very fashionable district in London, England. Well Kid, let me tell you, I'm not overly fussy but that place was an absolute 'dump'. I'm certain that the interior of the jail cells at San Quentin Penitentiary were more comfortable than this place. I couldn't believe that this 'hole' was only half a block away from Broadway, where there were all kinds of swanky joints plus flashy clubs and restaurants."

McGuire took a moment to catch his breath and then continued the rendition. The recollection was still strained even though its occurrence was all those many years ago. "It was the most austere room I had ever seen. Inside the place was a single, metal-framed bed complete with squeaky coil springs and a wafer-thin mattress covered by a soiled, woolen blanket. There was an old beat-up chest of drawers in one corner and a single, naked light bulb suspended from the ceiling by a threadbare electrical cord. The wallpaper was peeling in serval different locations throughout the room, it smelled of lingering sweat, stale cigarette smoke and the musky odor of countless sex-acts that had been performed within those depressing four walls over the years, and had never been properly exhausted from its interior. So, now that reek permeated the entire space."

Jesse wrinkled his nose and grimaced upon hearing McGuire's description of his room and just shook his head in disbelief. The story then continued. "Needless to say, I didn't spend a lot of time there and essentially slept on top of the blanket while wearing all of my clothes. I didn't trust getting in between the sheets!" McGuire quickly glanced inside his journal at the applicable date, read a few lines and then continued his dissertation.

"December 4th was my first full day in the 'Big Apple' so, I decided to leave my hovel of a hotel room as soon as possible and began exploring. As I said earlier, the hotel was very close to Broadway and I headed there first. It was amazing to see all of those famous theaters that I'd always heard about sitting and watching television back home, with the marquees announcing all of the popular plays of the time. Musical shows such as 'The Sound of Music' with Julie Andrews and 'Funny Girl' with Barbra Streisand were playing, plus all of the other variety shows like The Ed Sullivan Show and the Merv Griffin Show with all the signs and displays lit-up in neon lights. It was really impressive!"

"You could also find anything you wanted to eat within a few blocks radius in any direction. Every 100-feet you would bump into a hot dog stand, or someone selling burgers or pretzels on the sidewalk and then there were all of the restaurant choices from every country in the world that you could imagine. One of my favorite spots was Katz's Deli on the Lower East Side of Manhattan on Houston Street. They made a 'killer' corned beef sandwich on rye bread served with a giant, Kosher dill pickle."

"In addition to all of the eateries in the city, the other locales I noticed the most were the Sex Shops. Stuck in between every second or third storefront would be a Triple-X store or a magazine stand or something else in combination of the two. I surmised that these Yanks were obsessed with sex! I don't want to shock you Kid, but I loved it! That was right up my alley." Jesse wasn't surprised upon hearing this senior storyteller's admission since, in the short time he'd know him he had already formulated that McGuire was a 'sex-maniac'.

McGuire now took another moment's pause and gulped a drink of water. Speaking for him was not hard work but it was definitely thirsty work. Before long he was back into his reminisce. "One thing that really caught me off guard was the disparity between the rich and the poor. On one street, you'd be surrounded by luxury hotels or apartments and fancy clothing stores and then two blocks further on you'd be in a ghetto."

"It was so bad that it made Pigeon Square on East Hastings Street in downtown Vancouver look like the British Properties. There was garbage everywhere and most of the buildings looked as though they were bombed-out. Scores of homeless people were shuffling around on the sidewalks like a bunch of zombies and it really made me feel out of place."

"I also couldn't get over the amount of people who didn't look anything like me or the people I knew back home. Trust me Kid, I'm not a prejudiced person but I had never seen so many darker-skinned and mixed-race people in my life! To be honest, it made me feel a little uncomfortable. Whenever I watched American movies or shows and commercials on the T.V. almost everyone was affluent and usually 'white'."

"Yah, you'd see the odd colored or Asian person on the screen working as a butler or doing some other menial type of job, but beyond that we never really saw how it actually was in their cities. It didn't worry me but it was a real shock to my system. Trying not to spend any money if I could help it, I just walked around all day doing my version of a 'poor man's' sightseeing tour. Luckily, I was used to walking so it didn't bother me too much plus the weather was nice. However, I probably wore half an inch off the soles of my shoes! I returned to my hotel well after dark and just lay on top of the bed for about 6-hours without falling asleep. It was either my grumbling stomach from lack of food or else the decrepit state of my room that kept me awake. Or maybe it was a combination of the two."

"The next day, I started out with the same routine as before but after a while I was getting bored with just walking around. I was also actually becoming a bit worn-out. Forced to break into my cash-stash once again, I decided to splurge and headed for the Empire State Building. At the time, it was still the tallest building in the world. I reluctantly shelled-out the $1.50 and rode the elevator all the way up to the Observation Deck. Wow, what a sight! It was clear that day so, I was able to see 25-miles out across the entire city. Cars and trucks below on the street looked like Dinky Toys. And the people on the sidewalks looked like ants…not uncles," interjected McGuire while releasing a forced chuckle.

Jesse forgave him for this weak attempt at humor and then forced a smile of his own in acknowledgement.

"Speaking of the Empire State Building," McGuire suddenly announced, "I've got some cool trivia that would probably interest you? It's common knowledge that New York City is the skyscraper capital of the world but did you know that many of the tallest buildings were all built with the contribution from indigenous construction workers?"

Hearing this, Jesse raised his eyebrows and turned his head from side to side indicating that he was unaware of that fact.

"I didn't think so," concluded McGuire. "It was never really publicized or even acknowledged that generations of specialist construction workers living on the Canadian side of the border in specific Mohawk Indian Reserves such as Akwesasne and Kahnawake in Ontario and Quebec, would actually regularly commute to New York City from Canada to work on these buildings. These workers were fearless when it involved extreme heights and many of these skyscrapers might not have been completed without their valuable skills. The predominantly 'white' architects and executive types probably wanted to highlight their own achievements and not those of peoples who were considered to be below their classes."

Jesse was justifiably impressed and proud of these facts imparted by McGuire plus he was appreciative of him since, he considered that this portion of the story was specifically imparted for his benefit, being an indigenous person himself. Slowly but surely, Jesse's respect and sense of esteem for this elderly, new-found mentor continued to grow. Allowing Jesse time to digest those previous details, McGuire briefly referred to his journal and then continued.

"I spent about 30 minutes on the top of the Empire State Building and took about a half-dozen pictures with my camera. It was windy as hell up there so, I decided it was time to plant my feet back on solid ground. I ended up walking to Greenwich Village and checked-out the 'open air' markets. They were selling everything under the sun! After spending a couple of hours there, I decided to walk down to the dockyard to acquaint myself with the layout. There wasn't much to see there so, I started walking back toward the Mayfair Hotel. I didn't realize how far I had actually traveled from there."

"I was now becoming a bit paranoid with spending money and I needed to think of something that would prevent me from purchasing anything but the barest essentials like food and drink. On my way back, I stopped in a book store and bought a copy of the Guinness Book of World Records. It was pretty thick and chock-full of information so, I figured that it would take me quite a while to read through it. Well, my premonition was correct because I spent the next few days penned-up in my stinky hotel room reading the book and saving my money."

"On December 10th, I finally ventured out of my room for a while and did a last little bit of sightseeing in the Naked City. I also went to a laundromat and washed all of my dirty clothes. Wanting to please my mom, I also got a haircut and shaved the sides of my beard, leaving only a goatee, or Van Dyke as the snooty people called it. Afterward, looking in the mirror, I had to admit that I was a pretty handsome devil!" With this last statement, McGuire stopped speaking and made eye contact with Jesse. Within seconds, both suddenly burst-out laughing.

McGuire now rose from his chair in order to stretch his legs and ease the pressure on his lower back but then resumed speaking as he paced about in a small circle. He was now into the heart of the story and nothing short of a natural disaster could have stopped him from continuing.

"So, the big day finally arrived! It was December 11th and I was going to be heading out onto the briny deep. I checked-out of my so-called hotel around noon and gradually made my way down to the docks. I arrived in good time and headed for Pier 90. As I passed muster at the ticket office and was confirmed on the ship's passenger manifest, I spritely made my way toward the wharf. There she was! The S.S. Independence, all decked-out in majestic splendor and festooned with all manner of colorful streamers, ribbons and balloons. There were hundreds of people on and around the ship all drinking and laughing and creating a real party atmosphere. There was even a 'live band' playing on the pier, although it wasn't my kind of music."

"The ship itself looked very impressive with a creamy-white hull and superstructure plus twin gold-colored smoke stacks capped with black, circumference stripes on each. The ship wasn't as large as some of the other cruise ships plying their trade throughout the world. However, it had a very storied past and already transported notable celebrities such as President Harry Truman, Walt Disney and Alfred Hitchcock to name a few. Even Grace Kelly, who in 1956 was the world's most famous female actress, sailed on the S.S. Independence on her way to Monaco to marry Rainier Grimaldi, the Prince Regent of that tiny European Principality. She later became Princess Grace as a result. And now little old Bobby McGuire from Vancouver, B.C., Canada was also going to be added to that list of famous people." With that last sentence, he once again let go a snicker with Jesse following suit.

"After, eventually managing to squeeze myself through the crowd of revelers and onto the ship, I made my way through the maze of decks and corridors and finally located my cabin. A few minutes later a young American, also named Bob, entered the cabin with his gear. He was en route to Genoa in Italy. It was a 4-bunk cabin but it appeared as though it was going to be just the two of us for this ride. That was fine with me since, this meant that there was room enough for a couple of chicks to share the cabin with us and we didn't waste any time getting back up to the main deck to start hunting."

"It took quite a while for the crew and a legion of security officers to get all of the drunks and non-passengers off of the ship but they finally managed and at 4:15 P.M. we pulled up anchor and with the aid of several tugboats pulling us away from the wharf, we gradually started chugging out of the harbor and toward the open sea with 720 passengers, including me and Bob number two. Total cost to me for this 14-day excursion of a life-time was the unbelievable price tag of $280.00."

Upon hearing the total cost of the cruise, Jesse's jaw dropped and he let-go a long and expressive whistle. "You've gotta be kidding me!" Jesse incredulously exclaimed. "That's practically the same price as a return trip for two persons and their vehicle on a B.C. Ferry from Horseshoe Bay to Nanaimo! WOW!"

McGuire then continued, "We were only out of port for a couple of hours when dinner was already served. After not eating very often in New York in order to conserve money, I now ate two plates of roast lamb with all of the trimmings. That sure 'hit the spot'! We shared our table with two other guys named Tony and Arnie. The latter were also from Vancouver. Now, you need to remember that the ship's passengers were divided into three separate groups. There were the 1st Class passengers, the 2nd or Cabin Class passengers and then there was us, 3rd or Tourist Class passengers. Each group had their own areas of the ship exclusive to them with all of the privileges, or lack of them, that came with the territory. Naturally, me and the rest of the 3rd Class passengers were confined to certain areas and were not permitted to 'rub elbows' with the Upper Classes within their areas. Bob number two and I attempted to 'crash' the 1st Class lounge and dining room a few times but on each occasion, we were discovered by Security and told to move along. After a while, it started to become a bit of a game for us."

McGuire paused here momentarily and smiled to himself as he became lost in thought. Jesse surmised that this must have been a happy time for Bob and he remained quiet as he waited for him to begin again.

Within moments, McGuire seemed to awaken from his dream-state and then resumed his recollection. "Like I said, we got rousted each time we attempted to breach the imaginary 'gilded curtain' but we didn't really care. There were a lot of other things for us to do such as ping-pong, checkers, shuffle board and of course, drinking. And if we tired of those things, there was even our own library that we could access plus nightly showings of near, first-run movies. During our first night we watched 'Cat Ballou', with Lee Marvin. The sea was quite rough that night and some of the passengers were on deck puking over the railing. However, I felt pretty good. Later, I let the waves rock me to sleep like a baby in its cradle."

"December 12th was our first full day at sea and we got-up quite early in order to make the most of our time on the ship. When we reached the main deck, the reality of where we were actually set-in. All we could see was the open ocean with not a speck of land in sight. There were already quite a few people on deck, playing shuffle board and other games or just reclining on the deck chairs reading books or taking a snooze. Bob number two and I started looking for girls. We hadn't seen many in Tourist Class but it seems we hit the jackpot with the better-heeled clientele."

"Before long we spotted a group of girls singing and playing guitars. A couple of them weren't bad looking. It turns out that they're private school girls from Mexico City en route to Spain. Well, we barely started to chat them up when a couple of Catholic nuns came barreling toward us with looks that could kill. They start speaking to the girls in Spanish, all the while looking over their shoulders at us. Before you know it, the girls are packing-up their guitars and walking away. Like the old adage goes, 'You win some, you lose some.' No big deal, we were just getting started."

"As the days pass the weather is gradually warming and Bob number two and I are settling into a routine of eating, drinking, laying around and introducing ourselves to whatever girls catch our eye. We've now spent time with quite a few of them but nothing serious or entertaining has developed. One night when I was feeling no pain, I managed to get into the Oak Room and danced with one of the Mexican girls. Man, it felt good having a female in my arms but it was also very frustrating since afterward, I knew that I'd be heading back to my cabin alone."

"By December 15th, we're now 1600 miles from New York and I'm really getting used to this 'easy life'. Meal times aboard the ship was 8:00 A.M., 12:00 P.M. and 6:00 P.M. To announce these events, a steward walks all around the ship banging a brass gong. I never miss a meal, and If I'm not careful, I'm going to get fat! We continue to bump into the Mexican school girls during our travels around the ship and each time we stop to chat. Even though I realize that nothing will ever develop with these chicks, I really enjoy speaking with them, and they are the nicest and sweetest girls I've ever met."

"On December 17th, we reach our first port since leaving New York. It is Funchal in the Madeira Islands and it belongs to Portugal. The water temperature here is 69-degrees Fahrenheit. Quite a bit warmer than the December water temperature of where we started-out. We stopped here for a few hours and I made my way ashore. I've learned to say 'Bom Dia' to everyone I meet. It means good morning in Portuguese. There were quite a few kids on the streets wearing rags and begging for money. It was very sad to see. I took a tour into the mountains and saw sugar cane and banana plants. It was also a great view of the island from the top of that mountain!"

Now Jesse jumped-in with a question. "Did you go by yourself or did Bob number two join you?"

"I can't remember for certain, but I do remember that I was drinking quite a bit during the afternoon. That also continued into the night. During the trip, we befriended some of the ship's kitchen staff and that night it was Mario's 44th birthday. He and another Italian guy from the kitchen named Giovanni, came to my cabin and we had a bit of a party. They were great guys and there were a few times that I'd go down to the galley at 4 or 5 in the morning, and watch them prepare the meals for the coming day. They always let me eat whatever I wanted, chicken, bacon, sausages, cookies, cakes, literally anything and everything!"

"I've now also started hanging around with Tony, who was one of my table-mates in the dining room. He's quite a bit older than I am, but we get along like a house on fire. On December 18th, we arrive in Casablanca, Morocco and disembark. Tony knows his way around here and we hire a private guide to take us around to some of the non-touristy places in the city. Everywhere we go the locals are looking at Tony and me as if we were

Martians. Traveling around the city was like being in an episode of National Geographic."

"All the men were wearing long, nightgown style clothing with slippers or sandals on their feet, and wore either Fes hats or cloth wrappings on their heads. The adult women's bodies were completely covered from head to foot, including their faces that were covered by veils. Little kids, both boys and girls seemed to be exempt from this style of dress and mostly wore what you see other kids wear all over the world, except that they looked a little more ragged."

"The guide took us to the Old Medina and we saw all kinds of historic buildings, little shops and kiosks selling everything from soup to nuts plus farmers and other merchants flogging their wares in the market place. There were camels and their handlers plus carts being pulled by oxen or donkeys and a real lack of anything that resembled modern. They were even required to go to the Mosque 5-times a day to pray and they were forbidden from drinking alcohol. It was all very strange and definitely somewhere I could never live. At one point, our guide stopped at a roadside stand and bought us each a small glass of mint tea. It was way too sweet for me. Later, on a bus while returning to the dock, a little boy about 7-years old reached right into Tony's shirt pocket and stole his wallet. Before we knew what had happened, he was off like a shot. Tony had $60.00 in the wallet and it was now long gone. The nerve of that little bugger!"

"Just before boarding the ship, I bought a gold metal women's watch for $6.00 from a local guy hawking jewelry and I later gave it to Irma. She was a young American filly on board the ship that I had been trying to impress for the last couple of days and I was hoping that this gift would possibly break-down her remaining defenses. Well, that back-fired on me since she saw straight through that ploy. Oh yah, she thanked me very politely and gave me a big kiss on the cheek as she took the watch but she was still wearing her knickers as she sauntered away from me."

Again, Jesse was amazed by McGuire's relentless fascination and pursuit of the opposite sex. So, he slowly gazed downward and gently shook his head in disbelief. McGuire ignored this gesture from his audience of one, took another lengthy pull from his water bottle and began once again.

"After nearly the entire first week of the voyage out on the open ocean, we were now landing at different ports daily. Today was December 19th and we were approaching Gibraltar."

"Somehow, I was expecting the 'Rock' to be much larger than it actually was. I guess that's what happens when you only see photographs of something. It distorts one's perception. It was 8:00 A.M. and we were all itching to disembark. However, since the harbor was not deep enough to accommodate the size of our ship, we had to wait for launches from the dock to come out one by one and then ferry us back to shore. Gibraltar was a very compact city with one centralized main street that traveled the length of the entire town and held most of its important buildings, shops and restaurants. Strangely enough it was called Main Street."

With this last statement, McGuire adopted a goofy look on his face until Jesse commenced laughing.

"The prices of goods in these shops are unreal. Everything is way below retail, even expensive jewelry, perfume and custom leather goods. I bought several bottles of Johnnie Walker Black Label Whiskey at $1.95 per bottle. Unfortunately, we were only in Gibraltar for a few hours and by 1:00 P.M. we were already on our way again, now heading for Naples, Italy. A lot of the passengers on board are Italian so, I assumed some of them might be getting-off the ship."

"While we're steaming ahead the crew are beginning to hang Christmas decorations all over the ship. It's looking very festive since they aren't skimping on the variety or amount. It just seems a bit weird because the weather is now constantly sunny and warm rather than cold and snowy like Christmas weather is supposed to be."

"I'm starting to spend a considerable time now with Irma and I'm becoming more hopeful of my prospects with her. What a great girl she is and what a body! I'd crawl on my hands and knees for a mile over broken glass just to have 5-minutes with her alone in my cabin. We've been together during the days and at night we either go dancing or see a movie. There's always something to do on board and if you're not able to have a great time then there's something wrong with you. Unfortunately, I've kind of 'shot myself in the foot' while on board, since over the past week I've spoken to so many girls that now when they see me, they want to stop and talk. Irma doesn't say much but I can see that she's not impressed. At a time like this, I wished that my old buddy Jimmy Bridge was here with me in order to relieve me of these excess women!"

By now, Jesse realizes that this last comment from McGuire is not just idle bragging but probably very truthful. Having known him now for a couple of days he can understand why people, especially women, are attracted to him.

"On December 20th, I'm on my own since Irma had something else going on. I received a Christmas card in the mail from Momma Bear and it momentarily sends my thoughts back home but not for very long, I'm ashamed to say. To be honest, I've been having such a great time since I left home that it's the last place I think of. I'm too busy thinking about what's coming next! Shortly after lunch, I spot the Second Mate on deck and we start talking. He asks if I would like a tour of the Bridge and naturally, I jump at the chance! Not everyone has the privilege of observing the inner workings of this floating community so, I considered myself very fortunate. The tour turned-out to be quite interesting and the Mate explained everyone's job on the Bridge plus the workings of all the gauges and technical instruments. However, I thought it prudent not to ask him if I could take the ship's wheel for a minute."

Once again, Jesse was amazed by McGuire's amicable personality and sense of humor. He really possessed the 'gift of the gab' and could probably sell refrigerators to the Inuit living on the ice-flows of Nunavut if he put his mind to it.

"On December 21st we finally reached land again at Naples, Italy and I'm hanging-out with my kitchen buddies Mario and Giovanni. We had a fabulous time in the town eating spaghetti and fried squid all day plus drinking lots of local wine and beer. The only problem with Naples as far as I could see was that there were too many Italians living there."

Jesse outwardly groaned at McGuire's stale and predictable joke but Bob didn't seem to notice. He just laughed himself into insensibility. Eventually, after wiping the tears of mirth from his cheeks and gradually composing himself, he was ready to continue. He had now reseated himself and sat perched on the edge of the folding deck chair as though he was ready to spring forward at the drop of a hat. Although McGuire had now seemed to have lost all sense of time and space during his reminiscent recital, Jesse realized that it was now well into the afternoon and he had barely advanced his designated task. At this rate, it would take him a remainder of the week to finish digging all 16-holes. However, it was not for him to interrupt this senior orator seated before him. After all, it was Jesse who asked for the story in the first place.

As stated, McGuire was now fully immersed in his past and the details were rolling off of his tongue quicker than a 'hooker' rolling-off a 'John', only to prepare herself to instantly engage another.

"After Naples we visited Genoa. This is where my bunk-mate Bob the second and table-mate Tony, left the cruise. They were on their way to the Main Train Station where they would then take the express train to Vienna and Innsbruck, Austria respectively. I'll miss them. They were both 'good heads'. On December 23rd, we reach Cannes and Monaco. I've met this rich kid named Jeff during the cruise and he and his parents invited me to join them on shore in order to see the sights. Jeff's mom was a real looker and had been a model in New York. She had long blond hair and wore a full-length mink coat. Her leg kept bumping against mine in the back of the taxi on the way into town and I was starting to get a 'chubbie'. I don't know if she noticed or not?"

Jesse was now beside himself. Not only was McGuire chasing young girls around everywhere he went but now he was even after his friend's mother. This guy needs to take a cold shower!

"Jeff's father was an older guy with a bald head and a fat gut, wearing a suit and tie even in this warm weather. He's got a fancy camera and he's snapping photos of everything he sees. Looking at him and then at Jeff's mother it seems to me that she must have married him for his money. He definitely wasn't the Cary Grant type."

"I'm absolutely taken aback by the beauty of Monaco and its surrounding countryside. Had I not seen it with my own eyes, I wouldn't believe that any place could be so gorgeous! We checked-out the most notable sights like the Palace and Grand Casino and then watched the 'Changing of the Guards'. Afterward, Jeff's parents wanted to do some shopping so we visited a few designer stores. I wanted to faint just looking at the prices of some of the items on sale. At one exclusive perfume shop, Jeff's mother purchased $160.00 worth of tax-free perfume. Man, those guys were rich!"

"We left Monaco that evening and set sail for Barcelona, Spain arriving on Christmas Eve. I went into town with Darlene, Sonja and Anne who were traveling around for the past 6-months seeing the world. We all did the usual walking tour around the city checking out the sights and ended up eating dinner at a famous restaurant called Caracoles. In Spanish, caracoles translated to

English as snails. We ordered fried squid with mushrooms and of course, snails. It was all washed down with lots of Spanish wine. When we returned to the ship later that night, we were all 'feeling no pain' and I managed to peel away from the group with Anne from New Jersey. Well, just as we started getting real friendly with one another and I had my hand down her pants, the ship's air horn suddenly blew."

"Since we were on deck in the shadows at the time, the noise almost blasted us out of our shoes. Needless to say, that interruption cooled-off the moment and we ended up parting company near midnight. Inside my cabin, I sat on top of my bunk and waited for the 12:00 A.M. bell to strike. I just wanted to ring-in Christmas Day before going to sleep."

"On Christmas Day, I rise at 11:00 A.M. and then leave the ship for the final time. It's hard to believe that 2-weeks could have flown by so quickly but it has. My cruise is over. I head back into Barcelona with my bags and walk around for about 3-hours. I check-out my finances and decide that it's time to revert back to miser-mode. I'm spending too much money. I've only got $1225.00 remaining from my original $1400.00. I found a cheap room in a Pension for 60-piasters per night complete with breakfast. That worked out to be about $1.00 USD. That's not too bad! For dinner, I went back to the Snail Restaurant and ordered half a roasted chicken plus a bottle of wine. It was very delicious and only set me back $1.50."

"As I'm heading back to my room, I realize that this is the worst Christmas I have ever spent. Yah, I'm in Europe free as a bird but I'm by myself. After 2-weeks on the cruise ship, constantly surrounded by people, I find myself very lonely. I don't know how long I'm going to remain in Barcelona but first thing tomorrow, I'm going to buy a Spanish/English dictionary so that I can start communicating with the locals. I'm also seriously considering purchasing a cheap motorcycle. While on the ship, an American bird that I had befriended gave me the name and address of a guy she knew living in Barcelona. He was a musician from Montreal and his name was Peter. He was my age but apparently had been traveling since he was 15-years old. I made contact with him on the 27th and he bought me a lunch of steak and chips inside his hotel. Peter said that he was going to take me to the Youth Hostel the next day. That night for dinner I ate 3-navel oranges and a loaf of bread for the unbelievable price of 8-cents."

By now, McGuire noticed that the sun was beginning to shine into his eyes and he suddenly realized that he had been speaking for hours. He suggested that they temporarily adjourn their conversation and return to it another time. He then instructed Jesse to dig one more hole while he completed his chores inside the greenhouse. An hour later they were back in the F150 pick-up truck and returning to the house on Leonard Road. McGuire prepared a light dinner for them and with drinks in hand, they later settled down to watch a couple of programs on the T.V.

Heading for their respective bedrooms relatively early, Jesse began massaging his upper body and decided that he was feeling better than he had yesterday. His palms were still raw and sore but he seemed to be adjusting to the discomfort. After spending some time on his laptop computer, he then turned-in and gratifyingly, slept through the night.

The following morning found them both at the greenhouse performing the identical actions as the previous day and during their lunch break, McGuire was once again prepared to re-commence his commentary. "Okay, so where did I leave-off yesterday?" as if he didn't know. He was fully aware of this fact but was simply feigning only mild interest for Jesse's benefit.

"I've been hanging around with Peter for a few days and he takes me around on his 600CC motorcycle. He took me downtown to the nightclub where he sings and plays guitar only to discover that the Fascist Government headed by the dictator Francisco Franco, shut it down. Peter was really upset. Since, he was now essentially unemployed, Peter decided to travel to Switzerland to visit his girlfriend. I'm now staying at the Youth Hostel and it only costs 2 escudo per night. That's literally a few pennies. I've now made a few friends at the hostel and we go around town checking out the sights and the female talent. I don't get a chance to shower everyday but when I do, I feel like a million bucks! I bought a bottle of Smirnoff Vodka to celebrate New Year's Eve with my new found friends Phil, Terry and Derek. Needless to say, I got completely gassed that night and can't even remember ringing-in the New Year. I slept until 2:30 the next afternoon. It was hard to believe that 1965 was now history. I was hopeful for 1966 and I even wrote a letter to my parents wishing them a Happy New Year."

"On January 3rd, my three buddies and I decide that we're going to hitch-hike further south where it's warmer. We walked into town and I went into a bank to cash a traveler's cheque. When I returned, my fair-weather friends were nowhere in sight. The bastards dumped me! I was super-pissed and I thought if I ever saw those suckers again, they're dead meat! I didn't want to leave town on my own so, I grabbed a bus back to the hostel. The next few days ran into one another with my same daily routine but the weather had gone for a bit of a dump. It wasn't raining but it was very cloudy. I experienced a terrible shock when I went into my knapsack on the morning of January 6th. During the night someone had gone into my bag and stole a 1000 escudo note."

"That was the equivalent of $16.00. It doesn't sound like much in today's money but back then that would have lasted me a week. I went ballistic and couldn't stop swearing for several minutes. I was really disappointed that a fellow traveler would do something like that to me. I guess I was quite naïve, since, I should have realized that there are bad people wherever you may go. In the mean-time, I've met a few more guys at the hostel and we all plan on heading south in the morning. Me and another guy named Phil from Winnipeg decide we're going to hitch-hike while Henry, Jock and Bob number three are going to take a train. Phil and I take a bus to the highway and exit. Within half an hour, we strike gold when two Moroccan dudes in a VW bus pick us up and take us 300 miles south before they head in their own direction. The same thing happens the following day and within a couple more days we reach our destination of Alicante on the Mediterranean coast. Hitch-hiking was a great success and Phil and I are really stoked!"

"The beach was covered with hippies from all over the world and we started to ingratiate ourselves into several different groups. By now, my eyes were quickly becoming opened to some of the criminal activity that was always secretly surrounding us. While we were on the beach, we even heard of an incident involving a young girl from town. Apparently, she had long blond hair to her waist and had it cut short, so that she could sell it. The hair was worth about $300.00 and she kept it in her room. While she was out one afternoon, someone broke into her room and stole the hair. Strange but true. I've now become warier of where I am at any given time, and who I interact with. I'm also keeping a close eye on my bags."

"I was also a bit concerned with Spanish police conduct. One day, while we were watching a neighborhood soccer game in a local park, an argument broke out regarding the fouling of a player on one of the teams. Anyway, it got a little heated and someone ended up calling the police. The police arrived and before you know it, they're swinging their batons around like fly-swatters and bashing people on their heads. I thought that was a bit extreme over a sporting disagreement."

"As much as we were enjoying ourselves in Alicante, it wasn't long before we started getting 'itchy feet' again. So, on January 12th Phil and I hit the road and started heading for Almeria, another southeastern city on Spain's Mediterranean coast. This time, hitch-hiking wasn't quite so successful. Phil and I decided to split-up and then try to hook-up again at a later time, hopefully in Gibraltar."

"I'm alone now, and rather than standing in the same spot for hours on end, I decided to start walking down the highway. Each subsequent kilometer becomes a marathon and I'm wearing thin. My bags are feeling as though they are filled with concrete and the highway's asphalt has so effectively absorbed the brutal heat of the sun's rays and spitting it back in my direction, that it seems I'm being pulverized by a blast furnace. The soles of my aching feet feel as though they were walking barefoot over hot coals and I'm dying of thirst. Thankfully, around 6:00 P.M. I see a car come flying around the corner and the driver picks me up. He then took me the remaining 120 kilometers to Almeria. On the outskirts of town, this Good Samaritan bought me a beer and Calamari dinner at a small roadside diner. He left me at the restaurant and after eating, I started making my way back toward the highway."

"Well, lo and behold, who do I end up bumping into but Bob number three. Rather than continue hitch-hiking any more that day and realizing that it would soon be dark, we walked back into the village and booked a cheap Pension for the night. Before we knew it, we were up and at 'em once again at 8:00 A.M. the next morning. But this time rides were few and far between. After nearly 6-hours, we received one measly ride, and that was only for a short distance. I was beyond pissed-off. I was swearing at every vehicle that drove past us and if I had a gun, I probably would have shot someone. Once again, I had to employ my painfully abused feet and it was many kilometers later before I could allow them to rest."

"Bob number three and I bummed around the south coast of Spain a while longer and visited additional cities and towns. As time passed, we decided that we had seen enough of Spain for the moment, and that it was now time to move on. Don't get me wrong, I loved the country and its people with its fantastic cuisine, historic sites and beautiful scenery. However, I only had limited funds and I wanted to see as much of the world as I could with that cash, rather than spending all of my time in one place."

"On January 16th, we had our sights set on Gibraltar and reached the highway at 8:00 A.M. In order to better our chances at catching a ride, we split-up once again and I walked about a kilometer further down the road. Within approximately 10-minutes, a brand-new Mercedes Sports Car driven by a Spanish woman pulls over and I jump in. She was about 40-years old with long blond hair and a great body for her age. She was also wearing the biggest diamond ring I have ever seen in my life. I'm certain that I could have stayed in Europe for an additional 2-years at least, with what that bauble would have been worth."

"The car was a real beauty and drove incredibly smooth. One thing that caught my eye was that there was a record player fastened underneath the dashboard. It was called a 'Highway Hi-Fi' and she kept reaching down to either change the tune or the platter. This friendly, cultured and gorgeous woman took me as far as Marbella and then dropped me off as she headed in a different direction."

Waiting for McGuire to catch his breath, Jesse had a question that required a response. "Hey Bob, if you were ahead of Bob number three on the highway and the Mercedes passed him first, how come he wasn't picked-up but you were?"

McGuire flashed Jesse a huge smile and gestured with both hands traveling down the length of his body and then back upward. "Come on Kid. Just look at this masterpiece. What red-blooded woman anywhere in the world, rich or poor, could even begin to resist?"

Jesse once again shook his head in disbelief at the audacity of this response, but by now he shouldn't expect anything else. McGuire was comfortable with who he was and confident in his ability to charm anyone at will. This was one of his greatest character traits and he constantly utilized it to its fullest extent. Jesse was certain that McGuire could even deal with the Devil and best him at his own game.

McGuire then completed Jesse's enquiry with a shrug, "It didn't matter that I caught the ride in the Mercedes and Bob number three didn't since, I ended-up meeting him again the next day in Gibraltar. That's just how it goes sometimes. You roll the dice and sometimes they come up as a lucky number seven, while other times, it's snake-eyes."

Being somewhat more cognizant of their time constraints this afternoon, McGuire took a moment to glance upward toward the sun as if to ascertain the time of day. Seemingly satisfied that there was still plenty of time to impart a few additional 'war stories' before re-commencing their work, he quickly perused his journal in order to corroborate his facts and began once again.

"As much as I preferred company while hitch-hiking, it seemed that I had better luck when I was alone. Once again, the Highway Gods appeared to be smiling upon me and I was catching rides almost at will. The morning of January 17th found me in Gibraltar, and within hours of arrival I hooked-up once again with Phil and Bob number three, at the TOC-H Hostel. This was a place inhabited mostly by English, American and Canadian travelers, so it was an obvious place to go."

"And before you have the brainwave of asking me what TOC-H stands for…don't bother. I have absolutely no idea. Gibraltar is a relatively small place with not much going on so, Phil and I decided to hike-up to the top of the 'Rock' with all of the other tourists. Like I said earlier, it's not as big as it seems in photos or on the Prudential Insurance of America commercials on T.V., but it's still a challenging climb. All along the way and especially at the summit you can see scores of Barbary Macaque monkeys running around and begging for food. Phil and I are giving them a hard time when one of the little buggers then bites me on the finger. It hurt like hell! The weather was beginning to deteriorate and with not much else happening, Phil and I decided to leave for Tangier the next day. Bob number three has decided to stay behind. I'm thinking this is pretty rank, since Tangier is in Morocco, North Africa. Tomorrow I'll be on another continent!"

"Early the next day we took a launch over to Algeciras near Gibraltar and then the ferry to Ceuta situated on the very northern tip of Africa. It was a pretty short sail before we finally arrived on the African continent. However, now our thumbs need to come-out once again since we still have to make our way to Tangier which is over 78-kilometers away. The weather hasn't changed much since yesterday but our surroundings definitely have. First, the flow of traffic on the highway is much less frequent plus the people and their surrounding circumstances appear much poorer. There is a marked difference between the continent of Europe and the continent of Africa even though they're only separated by 14.5-kilometers of water."

McGuire took a moment to reposition himself on his folding chair and to mentally catalogue his memoires. So, Jesse took that opportunity to once again ask a few questions. There appeared to be somewhat of a game-plan when McGuire originally booked the Mediterranean cruise, the side-trip to Kelowna to visit a girl and then the sojourn in Toronto while en route by bus to New York City. But to Jesse, it now seemed as though his reminiscing tour guide was merely ad-libbing in his movements without any real planning or organization. In response, McGuire reassured him that there was nothing further from the truth. Upon loose examination, this might indeed seem to be the case but in actuality, these locales had been dreamt of for a considerable time-frame and although the current agenda may seem a bit haphazard, it's only because the itinerary was not set hard and fast.

McGuire further responded that he wanted to keep these excursions fluid and decisions were being made as opportunities arose, or who might be in his company at the time plus other similar factors and variables. On the surface, this global trek might appear disjointed but in the mind of its enthusiastic traveler, all was going according to plan.

McGuire took a deep breath and began his recital once again, "It was a real slog for Phil and me to reach Tangier but we finally arrived in the early hours of January 19th. I don't like knocking down any person or the place where they may live but Tangier was a major 'dump'. I was really disappointed. There were piles of garbage on almost every street corner, most of the buildings were completely dilapidated and quite a few of the streets were unpaved. The few that were paved boasted more ruts and potholes than the runways of the Tempelhof Airport in Berlin after they were bombed by the Allies during World War II. Even the people we saw looked half-starved and were dressed in ragged clothes. Like I said before, the difference between Europe and North Africa was like black and white. One good thing was that the prices of goods and services were even a little cheaper here than they had been in Spain."

"We managed to quickly find a cheap room for the night and dropped our bags. We then headed into the Old Medina on foot for something to eat. We soon found a place that served us soup, salad, bread and a potato-meat stew all for the equivalent of 25-cents. Some of the local passers-by were giving us dirty looks and we couldn't understand why. We later discovered that this was the time of a religious celebration for them called Ramadan and for 1-month they were not allowed to eat anything between the hours of 5:30 A.M. to 5:30

P.M. No wonder they all gave us the evil eye, I'd be pissed-off too if I saw someone eating and I couldn't. Luckily, their fasting was only required for a few more days and then they could get back to normal."

"That afternoon we met an American dude named Jack in the Old Medina, drinking tea in a small café. We started a conversation with him and discovered that he and his girlfriend have been in Tangier for several months and had rented a large house outside of the city for 75-dirhams per month. It even came with a maid plus a housekeeper for that price. I did some quick figuring in my head and determined that 75-dirhams was about $15.00 per month in our money. What a deal! Jack's girlfriend wasn't with him at the moment, since she had recently overdosed on sleeping pills and was at their house recuperating. Jack tells us that he smokes a lot of pot and asks if we're interested in buying some for ourselves."

"Well, I'm young and I'm fearless plus I'm on the trip of my lifetime. I want to experience everything I possibly can so, Phil and I don't even contemplate his offer, but rather immediately nod our heads in agreement. We then follow Jack further into the Medina where he makes us wait near the Youth Hostel while he leaves on his own. After a few minutes, he returns and hands Phil a small plastic bag about the size of a walnut. Jack says it contains 'keef' which is finely chopped marijuana. I can't remember how much it cost but Phil gave Jack very little money." Phil stowed the bundle in his pants pocket and we temporarily put it out of our minds. Jack knows his way around the Medina, so he begins showing us the sights. After a while, I decide that Tangier's Medina is no different than Casablanca's. All the shops and stalls are similar and they all appear to be selling the same "kitsch," geared toward foreign tourists. One good thing is that most of the shopkeepers speak English. By early evening, we're already pretty tired from our ordeal of getting here, so we make an early night of it. We headed back to our room while Jack went his own way. Before leaving, he said that he would meet us again tomorrow.

"We slept a bit longer into the next morning but eventually met up again with Jack near the bazaar in the afternoon. Today, he was accompanied by a 9-year-old boy he called 'Charlie', and said that for a few dirhams, Charlie would act as our personal guide. I'm thinking that Charlie is helping to support his family with the 'personal guide gig', so what the hell, I figure we can spring for a few coins. It turned out that Charlie was a pretty good head, and he really gave us our money's worth."

"Later that night we met-up again with Jack and he took us to a real sleazy night club. It's so dark and smoky inside that Phil and I are tripping over ourselves as we attempt to follow Jack to a table. The interior smells of dirty socks and unflushed toilets. While we're sitting at our table drinking, a wicked looking chick in a long, thin dress sashays to the middle of the dance floor and starts gyrating. I'm scratching my head, when I see Jack turn his chair toward the dance floor, and takes off his shoes and socks. As if on cue, the chick starts floating in our direction and starts giving Jack his own private dance. Well, I just about fell out of my chair when I see Jack take one of his bare feet and lifts it underneath the chick's dress. Now, she really starts gyrating while Jack's whole leg is moving up and down. This continues for a couple of minutes and then the dancer leans over our table and grabs one of the bills that Jack has sitting in front of him. As she leaves our table, Phil and I just stare at each other and then burst out laughing!"

"Jack appears to be getting-off from our reactions, and the general good humor at the table, so he's making sure that the drinks keep coming. By the time we leave the club, I'm 'gunned' and as far as I'm concerned, they can 'send-in the clowns'. I'm not sure who it was that purchased the wine afterward, but now each of us also has a bottle of it in our hands, and that too is beginning to spill down our gullets."

McGuire now appears to be building-up to some momentous declaration since he suddenly rises from his chair, removes his Panama Hat and runs one hand through his snowy locks. After replacing his topper, he re-adjusts his clothing and commences pacing in a small circle with eyes closed in concentration. Jesse is now uncertain what's coming next, but he privately bets that it's going to be interesting. As if reaching some undisclosed decision, McGuire stops pacing and opens his eyes. He re-seats himself and earnestly focusses on Jesse.

"So, Kid. As you've heard so far, I've now encountered a few curious experiences since leaving home in November of 1965, and they were either good, bad or both. But what happened to me next forever altered my life. I'm not saying that it was the greatest thing in the world, but it affected me to such a degree, that it continues to affect me even today. As I said, Jack, Phil and me, were all bombed and aimlessly stumbling through the dark and dusty alleys of Algiers. Almost as an afterthought, Phil put his hand inside his pants pocket and pulls-out the 'keef' Jack had sold him the previous day. We all stop walking and stupidly stared at the small plastic bag within his palm, as though we were expecting it to start speaking."

"Jack then reached into his own pocket and withdrew an ornately cast brass pipe with a pearl stem, about the length of his thumb. Without first asking Phil's permission, he snatched the bundle from him, untied the loose knot securing it and using his forefinger and thumb began taking small pinches of 'keef' and then stuffing it into the bole of the pipe. Once completed, he withdrew a shiny chrome Zippo lighter from another pocket and lit the contents of the pipe. Taking a deep pull from the pipe and holding the smoke in his lungs, he then extended the pipe to Phil who then duplicated Jack's actions. Phil then passed the pipe over to me and I took a drag. Being an obvious novice with this activity, I naturally commenced sputtering and coughing, to the obvious delight of my two friends. Being small, the contents of the pipe was quickly consumed and then refilled by Jack several times."

"Eventually, after taking just as many drags from the pipe as Jack and Phil but not feeling any different, I decided that 'keef' was a huge disappointment, and not something I would continue using. It was better that I stuck with alcohol. I didn't really like smoking anyway since it would probably interfere with my running-game while playing soccer."

"As I further contemplated this letdown and swallowed another mouthful of wine, I began noticing an unusual sensation commencing at the soles of my feet and gradually creeping upward through my limbs. The feeling was warm and calming with a slight tingling as it reached my outer extremities. The world around me suddenly became muted and as the feeling entered my chest, I stumbled backward a half-step and my breathing became slightly labored. However, this did not seem to alarm me. I realized that my eyelids were flickering like a strip of celluloid film that had slipped a movie projector's feeding sprocket, and what sounded like pounding surf engulfed my eardrums. I had the impression that I was now smiling but wasn't certain since, I couldn't really feel my facial muscles. As I slowly turned toward Jack and Phil, I could see that they were both laughing at me but I didn't know why. What was so funny? I could see that their lips were moving, as they pointed in my direction, while continuing to laugh. I couldn't hear them properly and I wished I knew what was so comical. I felt myself beginning to physically and emotionally drift away ever so slightly, until I no longer felt anything."

"The next time I could trust my eyes to inform me of where I was and what was happening around me, was the following afternoon. I discovered that I was lying on top of the bed in my rented room and that suddenly, I was feeling sick. Rising too quickly increased my sense of nausea and I barely made it to the sink before I puked. As I attempted to wash the disgusting expulsion down the drain, I realized that I had now plugged it solid with the undigested chunks of last night's supper. What a mess! No sooner had I managed to crawl back into bed, when Jack and Phil came crashing through the door and instructed me to get-up. Phil and I were going to be leaving the next day for Marrakesh in Morocco, and we needed to acquire a Visa from the Consulate Office. Before I could pass-out again, they picked me up and dragged me down the hall where for the price of 1-dirham, I could take a cold shower. Unfortunately, the shower didn't help much, and to further my discomfort, we couldn't find the Consulate Office either. We eventually arrived back at our rented room after I first attempted to down a bowl of thin but spicy soup at a food stand on the way. I eventually crashed for the rest of the day, and never did find-out who unplugged my vomit from the sink's drain."

McGuire now took a moment to capsulize for Jesse what he had just imparted. "The last part of that story was kind of gross what with the puke and all, but it was that initial experience smoking 'keef' on the night of January 20th, 1966 that triggered something inside of me. Without the need to bleed on my wedding night like some maiden bride, I was now no longer a herbal substance virgin thanks to my journeyman buddies, Jack and Phil. I promised myself that I would never mix 'pot' and alcohol again, as a result of the severe hang-over I had just overcome, and for a little while, I kept that promise. However, for some reason from that day forward, I always felt that marijuana would remain an important part of my life. And here I am 55-years later, still involved with it."

Jesse now also wanted to clear the air with a reflection of his own. "Please don't take offence Bob but until recently, marijuana was illegal. That meant all those years before it became legal, you were breaking the law. Didn't that bother you?"

McGuire had heard this question thrust upon him from others many times in the past, and each time he always responded with a patent answer. "Yah, I knew it was illegal. We all knew it was illegal, but I didn't consider myself a criminal. I felt that the Law was wrong in this case. I always considered marijuana and its bi-products as a natural, harmless product. I never imported or exported it, nor did I ever deal or traffic in it for profit. For guys and girls like me that used 'pot and hash' regularly, it was a social tool. We formed and then bonded friendships with it, and we reaffirmed solidarity to our group that was not so much anti-establishment, but rather a group of peaceful people searching for a new world order. It was strictly meant for hospitable purposes, and once in a while, just for kicks."

"Go to any household in Great Britain and the first thing a host will offer you is a cup of tea. That is their social tool. For me and people like me, it wasn't tea but rather marijuana. When you think about it, it's the same principle, only a different mechanism."

Jesse seriously considered McGuire's response and analyzed it. Surprisingly, he couldn't completely disagree with it, but actually found certain points held credible logic. Once again, he was confounded by the person before him. From one angle, he appeared to be an overgrown teenager on a never-ending quest for self-gratification, while from a different angle, he was a very sensitive and multi-faceted man.

"After a much-needed good night's rest, I felt a bit better the next morning and Phil and I checked-out of our room early. We had said our goodbyes to Jack the night before. We never did get a chance to meet his girlfriend. I hoped that she was okay. We caught a bus to the outskirts of the city and exited once we reached the highway. We commenced hitch-hiking and within minutes were picked-up by a younger man driving a large Mercedes-Benz sedan."

"He was a Swedish chemical engineer on vacation and spoke English well. As we climbed into the car, we could see large groups of local citizens heading toward Tangier with all manner of transport, such as mopeds bearing entire families, rickety bicycles, donkey or ox carts and of course, persons merely traversing on foot. All were carrying bundles of personal goods plus their proverbial sheepskin prayer mats. Ramadan was ending soon, and all of the faithful were migrating toward mosques in order to pray and to celebrate the conclusion of their religiously imposed fasting."

"Our Swedish chauffeur is making good time and is attempting to keep the speedometer of his Mercedes as close to 120 km/hr. as possible. As we continue moving eastbound, we begin ascending in elevation. The Swede informs us that this is because we are near the base of the Atlas Mountains. In this arid, desert-like landscape going on seemingly forever, I never considered that North Africa might have mountains. Swede says that the mountains extend almost 2,500 kilometers across Morocco, Algeria and Tunisia. After driving for a couple of hours, it seems as if we were dropped on to a different planet. We are now suddenly entering an unbelievably beautiful grove of date palm trees. The place is massive and I think we must have come to an oasis or something. Our Swedish chauffeur now becomes Swedish professor and he begins explaining that this place is called Skoura, and is over 30,000 acres in size, with over 150,000 palm trees. He tells me that I was pretty close to being correct when I assumed that this area was some type of oasis. In fact, there is an abundance of ground water flowing directly below Skoura near the surface, and the trees draw the water directly to their roots from that source. Our driver takes us far as Marrakesh. Once there, our first course of action is to score a room. We find one in the Old Medina for 3-dirhams per night. We drop off our bags and head-out right away looking for action. It's the last night of Ramadan so, most shops are still closed. We head into the main square and it's filled with thousands of people. Everyone is extremely excited that they will soon be able to eat again, so there's a real festive atmosphere."

"There were acrobats, sword-swallowers and tumblers performing. An Arabic version of clowns wearing colorful costumes staggered around on tall stilts. Religious leaders incessantly preached to the multitude over loudspeakers. We even saw a real snake-charmer handling two sinister looking cobras, spitting out their tongues every few seconds in a menacing fashion."

As he's describing events and locations, McGuire can tell that Jesse is attempting to paint a visual picture of these things within his mind. To assist him with that process, he's attempting to be as descriptive as possible. But he can only go so far. "You know Kid, I can explain to you what I saw and what I did in all of those different places through my words, or by showing you some of my photos taken at the time. But what I can't describe to you are the smells. For that, you had to be there. Just imagine, you're in the Old Medina and it's scorching hot. There is no breeze to help with circulation and the stale air is clinging to you like plastic wrap. You can't take a deep breath because it's too hot and too dry, plus it's filled with dust. Around you there are hundreds of unwashed bodies wearing outfits that probably haven't been changed in a week. They don't use cologne or deodorant, so the smell of their persistent sweat is almost sickening. Occasionally, you walk past a spice stall or food stand that provides your nose a much-needed reprieve, if only temporary. But seconds later, you're once again assaulted with the stench of animal dung or the human sewage that's flowing in the gutter just a few feet away from where you're walking. That's something they can't duplicate for you in a travel poster or a television show."

Jesse had never really considered that aspect of McGuire's story before, but after hearing this additional informative explanation, it definitely helped him to better understand the full extent of Bob's experiences, rather than just a one-dimensional version. Again, he could now see that McGuire was not just re-telling his travels, he was actually re-living them.

McGuire allows his last revelation to register with Jesse and then he continues. "I'm regularly checking my finances and I can see that I'm running out of cash. I've got enough for another night's stay at our hotel and for a couple more meals but after that, I'll be skinned. Phil and I leave the square and start looking for somewhere to cash a traveler's cheque. Everywhere we go is either closed or they refuse to cash a cheque. By the time we head back to our room, I still haven't fattened my purse. Even the hotel staff won't help me."

"The next day, it's the same story, and now I'm beginning to worry. The longer we wait here, the worse things are going to get, so on January 24th, Phil and I leave Marrakesh and head for the highway. My change purse is now empty. We're leaving town on a sour note, but I don't want to 'bitch' too much, since today is Phil's 24th birthday!"

"Today is the lowest point of my entire trip so far, and I'm really depressed. We've been on the highway for 8-hours in the blistering heat and we've barely made any headway. We have no food, no water, no ready cash and we're both knackered. We have walked for kilometers due to a lack of traffic and the unrelenting sun has sucked-out of us, every bit of strength and energy remaining. By now, my shoes are thrashed and they're barely holding together on my feet. I don't know what I'll do if they finally decide to fall apart. Before nightfall, we luckily make it to a small community where we can spend the night. I felt really bad for Phil, since this was a very miserable way to have spent his birthday. He deserved better."

"Our luck improves a bit the next day when between six rides, we managed to travel 250-kilometers, mostly through the kindness of local truck drivers. I still don't have any cash and I'm starving. At one point in between rides, we passed an orange grove. I'm not proud of it, but I stole eleven oranges from the trees as we walked past. I couldn't avoid it. Lack of food or drink, now going on two days, drove me to do it. Near the end of that day on our final push to Fes, we were picked-up by a woman driving a French-made, rear-engine Simca, and she took us almost the entire distance."

"On the morning of the 26th, Phil and I are limping into the Central Moroccan city of Fes. The very first thing I do, is find a place to cash a traveler's cheque, and then we eat. After that, I bought a brand-new pair of sturdy leather ankle boots, and said a fond farewell to my previous 'kicks', that were now destined for the footwear cemetery. We remained in Fes only overnight, and the very next day we're back on the highway, with our thumbs held high. Things are looking-up when we catch our first ride within minutes. Two young Moroccan dudes were in the car that stopped and they're also heading toward the Algerian border. BINGO! En route, they even bought us a great lunch consisting of salad, hors d'oeuvres, a cold meat platter, bread, mineral water and finally a plate of steak and chips. What a feast!"

"As we're back on the road again with our Moroccan land pilots, the landscape has become more inhospitable, and now it's nearly a desert. All we can see is sand and sand dunes. Without saying it aloud, I'm grateful that Phil and I aren't stuck out there on foot. I honestly don't think we would have lasted too long."

"We arrive in the border town of Oujda that afternoon and the Moroccans drop us off at the Algerian Consulate Office, before going their own way. We have the necessary document photo taken at a neighboring shop and then enter the Consulate. For the additional price of $3.00 and with the friendly assistance of a cute, female Consulate employee, we are now the proud owners of an Algerian Visa. Without a hitch, we then cross over the frontier and enter the former French colony of Algeria. We made our way into the town of Tiemcen and immediately checked into the local Youth Hostel. We meet some guys and girls from an International Aid Agency near the hostel and we have a lot of fun with them. At the hostel, we're re-charging our batteries by getting a lot of rest, reading, playing crib as well as several rousing games of ping-pong. We're meeting a lot of real nice French-speaking people and it's inspired me to take a French course when I return home to Canada."

"On January 31st, we reluctantly leave Tiemcen and head toward the coastal city of Algiers. When Phil and I arrive in Algiers soon afterward, one of the first things we do is head to the world-famous Casbah. This is a place of legend and the source of many books and movies. There's not much left of it now but then it was still quite a vibrant and colorful place. The Mediterranean Sea is visible from the docks of Algiers and it's calling Phil and me to come back across to the European side. While we're in Algiers, we met a couple of Americans who owned a Jeep Scout. They're traveling around North Africa and agreed to take us as far as the city of Constantine. On February 2nd, we said goodbye to Algiers with our two new mates, and hit the open road. Within a few hours, we arrive in Constantine and decide to spend the night. The two Americans were last seen riding-off into the sunset in their Jeep Scout. Our time spent in Constantine was pretty much the same as time spent in the previous towns. It was nothing out of the ordinary."

"On February 4th, Phil and I are back on the highway en route to Tunisia, when we ran into a serious winter storm between Constantine and El Bouni. There we were stuck on the highway in the middle of nowhere and totally exposed to the elements. The rain was pounding down on us like some blacksmith's hammer on an anvil, while 60-km/hr. winds were whipping us in every direction. To add insult to injury, no one would stop to pick us up. We couldn't believe it."

"Where before we were at least getting rides from truck drivers or farmers with flatbed trailers, now even they weren't stopping. It was almost as if the passing motorists enjoyed seeing us in our misery. We didn't know whether we should shit or go blind."

"Before long, we knew that we needed to get out of the elements or else we were going to be in some serious trouble. We started walking down the highway looking for an overhang on the side of the rock face or maybe a small cave or depression cut within."

"As we're hunched over and desperately plodding toward some type of shelter, a little blue sports car stops about 50-feet ahead of us and waits. Not knowing what to make of this we stop as well and just stare at the car. Seconds later, the driver's door opens and a woman calls-out to us, 'Are you going to stand there all day? Get in!' Quickly regaining our senses, we boogie up to the car and climb-in. I'm jammed in the back seat with our bags while Phil sat in the front passenger seat beside the driver. She's a very friendly middle-aged Dutch woman and she introduced herself as Madia. Kind of like me, she likes to talk and before long she's giving us her life's history. She looks very rich and sounds well educated. She tells us that she lives in Amsterdam but won't say why she's in North Africa. She drove here by herself all the way from Europe. She claimed to have been in the Dutch Underground Movement during the Second World War and that she was trained to handle a knife and a gun. For some reason, I believed her. She just had that look about her."

"Madia said that she was also heading to Tunis and agreed to take us all the way there. She divulged that she had quite a bit of Algerian money to spend before she crossed the border into Tunisia and invited us to share a meal with her. We couldn't believe our good luck after nearly expiring in the rain storm just a short while ago. True to her word, we stopped in El Tarf just before the Algeria/Tunisia border and she blew her wad on a lavish spread that included steaks, vegetables, wine and all the trimmings. We all enjoyed it immensely! Madia then drove us the rest of the way into Tunis. Before leaving, she gave us a brief orientation of the city so that we could maneuver around a bit easier. What a great woman! I'll never forget her. We quickly booked lodgings for the night and changed out of our damp clothes. Tunis was much cleaner and more affluent than the previous cities in North Africa that we had visited. During our stay, we took a train out of the city and visited the Roman ruins where the city-state of Carthage had once been. That was a really interesting trip and the ruins were actually quite impressive."

"On February 7th we checked-out of our lodgings and made our way to the Port of Tunis. From there we booked passage to Palermo, Sicily and as the blood-red sun was setting over the Carthaginian ruins later that same evening, Phil and I bid North Africa a bittersweet farewell."

"We'd had some great times there but we also faced a lot of hardship. Especially on the highways. I'll be honest with you Kid. There were a few times out on the highway in the burning sun with no food or water or even prospects of catching a ride to the next town, when I just wanted to pack it in. I was that spent and that dejected. But something inside kept telling me to keep going. I'll never regret my time spent in that part of the world and we met some exceptional people. However, I didn't think I'd be rushing back there any time soon."

Jesse now took that opportunity to voice a few observations of his own. "Man, I can't believe that so many people were willing to help you guys out the way they did. You and Phil were essentially strangers, yet they allowed you into their vehicles, drove you endless miles for free, and in some instances, even bought you some pretty nice meals."

McGuire digested Jesse's words for a moment and began to respond. As he did, his voice broke and he had to stop himself. After a momentary wait, he started again. "You're right Kid. We lucked-out quite a few times and I will never forget what all of those people did for me. I can never thank them enough. All I can say is that it was a different vibe then. Throughout history, the world has never been an easy place in which to exist, but back then, it seemed to be a little bit more empathetic. I think when those people saw us standing on the side of the road with our thumbs out, or walking in the gloom of night searching for a place to sleep, they might have seen themselves, or one of their family members in that same situation. If the tables were turned, wouldn't they want someone to extend a helping hand to them? Maybe that's why they did what they did."

At this point, Jesse could see that McGuire was becoming quite emotional and seemed to become lost in some private thought. He also appeared to be struggling with adding any additional comments to what had just been expressed. Jesse decided to keep any further questions to himself. There would be other times when he could ask them. Finally, McGuire shook-off his reverie and piped-up again, "Okay, Kid. That's enough for today. Let's get back to work!"

As a result of the abbreviated recital from McGuire this day, both men actually managed to accomplish some work in the later afternoon, and by the time the white, Ford F150 was back on Gillies Bay Road traveling toward home, an additional five holes had been dug, and now accommodated another five leafy clones.

McGuire had remained uncharacteristically quiet during the remainder of the afternoon, and even now wasn't speaking very much other than to inform Jesse that they would be stopping briefly at the Texada Market before heading home, in order to purchase additional groceries. Jesse realized that McGuire's most recent reminisce today had left him feeling melancholy, so he was intelligent enough to just leave him alone. He was also confident that words of wisdom and corny jokes would soon be rolling off the tongue of his associate once again. Pearl was not working at the market that evening so, the shopping side-trip was completed in relatively short order.

Once back at the house and after both men had performed their respective ablutions, McGuire created the pair, a tasty, but simple meal. The clean-up process afterwards was completed efficiently, but somewhat subdued. Later, rather than watching T.V., McGuire turned-on the stereo and commenced playing a compilation of tunes from various artists and bands. Jesse could see that Bob's mood began improving with each subsequent song played and surmised that music had the ability of altering his emotional state. This in addition to a few glasses of Golden Gage Plum wine, that had now also been consumed. Not knowing him well, but gradually beginning to understand him, Jesse felt that this syncopated activity was therapy for his new-found friend, and what was taking place now, was probably a common occurrence whenever he felt depressed.

The remaining evening passed quickly and by the time each man parted company for the sanctuary of their own bedrooms, McGuire's usual temperament had returned to him. Jesse was pleased by this turn of event, and felt a sense of relief. After all, it was him that encouraged McGuire to impart the details of past travels in the first place, and he would have felt guilty knowing that he had unintentionally contributed to any bad thoughts or memories that might have been conjured-up by those recollections.

Determining that he wasn't quite ready for bed after reaching his bedroom, Jesse retrieved his laptop computer from the floor, switched it on and sat upright on the bed with it.

There were a few things that he wished to check before turning-in. He had also now come to the conclusion that he needed to confront his militant friend on the mainland, as to what their future association might be. Jesse had never been comfortable with his friend's aggressive attitude and suggestions as to what should be done regarding the plight of 'their People'.

He also wasn't certain whether his continued drug use and 'banging around' the lower east-side of Vancouver on the weekends was all that attractive to him anymore. He had a lot to consider. He did know one thing, after being away from his usual haunts and activities these past few days, he was actually feeling a bit better. He was now sleeping through most of the nights, plus when he touched his biceps, he actually felt a slight hardness where before that had been absent. The palms of his hands were still blistered and sore, but they too were slowly beginning to adjust. And after listening to McGuire's travel stories to date, he was beginning to formulate a better understanding and appreciation of his host, the better natures of persons toward others in need, and just things in general overall.

It was close to midnight when Jesse finally shut-off, and closed his laptop computer. He had completed his desired 'site searches', plus he had drafted a distinctly blunt email to his buddy on the mainland. He wasn't certain how his message might be perceived, but now that it was sent, he would have to wait and see what type of a response, if any, was going to result. In addition, he had also sent his mother a brief communication letting her know that he was fine. As he lay back on the mattress and reached over to the end table to turn-out the light, he winced from a sharp pain radiating between his shoulder blades. Rather than being annoyed or concerned by this, Jesse actually smiled to himself as he realized that the discomfort was derived from honest labor. He considered that was something to be proud of. Within minutes of cancelling the light, Jesse was fast asleep and beginning to dream of exotic locales.

Chapter 4
The Cradle of Civilization

Once again, Jesse was awoken by McGuire the following morning at first light. By now, a routine had been adopted. Without uttering a word, he rose, made his way down the hallway to the bathroom and performed the needful. After a brief refresh, he sauntered into the kitchen to find his breakfast already prepared and waiting for him on the table, while McGuire was at the preparation counter constructing their afternoon lunches. The fact that he was quietly singing to himself, albeit somewhat off-key, reassured Jesse that yesterday's emotional episode, was a thing of the past.

Before long, they were back at the greenhouse performing their assigned tasks. Thankfully, the weather was still cooperating with them making the outside work more tolerable. The former pain between Jesse's shoulder blades seemed to have disappeared overnight. In addition, McGuire had given him some ointment for the blisters on his palms, that was infused with CBD oil. As a result, his abused extremities, were now definitely beginning to feel better.

Jesse had just completed digging his second hole of the day when McGuire exited the greenhouse and beckoned him over. It appeared that the shortened recitation from yesterday did not sit well with him so, their lunch break was going to be earlier than usual today, so that he could make-up for lost time. After seating themselves at their make-shift cafeteria, and upon opening their lunch sacks, McGuire adopted his story-telling guise and once again launched into his memoirs.

"Like I said yesterday, we had left Tunis by ship and were steaming toward Sicily, the birthplace of the Mafia. Phil and I wondered if Sicily had a 'Dodge City' mentality, or if that was just hype for writers and movie makers. During the voyage we bumped into Bob number three again. What a coincidence! It was an overnight trip and the passage was quite rough at times. Just like on the

S.S. Independence, there were people throwing-up all over the place. However, I felt fine. I wondered if that was because I had some Viking blood in me. It was common knowledge that the Norsemen were accomplished sailors, and that they frequently sailed great distances over the open seas and invaded coastal towns in England, Scotland, and Ireland during medieval times."

"The fact that I have red hair probably attests to the fact that one of my descendants must have copulated, at one time or another, with a Viking. Whether or not that union was voluntary or forced, begs answering."

"Phil, Bob, and I watched the ship's approach and docking at Palermo, while taking the last few bites of our continental breakfast. After passing through Customs and converting a traveler's cheque into Italian Lira, we begin walking toward downtown Palermo. There are now six in our group, since the three of us have attracted a few groupies. While en route, we all stopped at an ice cream parlor and each bought a cardboard dish of gelato."

"Have you ever had ice cream for breakfast, Kid?" taunted McGuire. "I'll bet you haven't. These are the kinds of things that you can do when you're free as a bird, as we were then. Anyways, we're walking into the city when we see a large group of people marching down the main street toward us. They were shouting some slogans in Italian, blowing whistles and carrying large, colorful banners. As they reach us, some guy in the crowd is waving for us to join them. We've got nothing else planned, so what the hell, we dive into the crowd. The guy hands out whistles to us and motions that we should blow on them. That's exactly what we do. The high-pitched screeching sounds of hundreds of whistles is almost deafening. Next, we see this guy with a professional looking movie camera, and he focusses in on us. Someone in the crowd informs us they are filming a movie about some famous labor movement, and we are all 'extras' in a 'strike' scene."

"Unreal, we've only been in Sicily for 10-minutes and we're already movie stars!" McGuire then looks directly at Jesse and asks if he wants his autograph. Both men then begin laughing whole-heartedly at the absurdity of it all. "Within a few minutes the Director yells 'cut' in Italian and the scene has been shot. After many firm handshakes all around, we leave that group and continue on our way. No one mentioned whether or not we were going to be paid for our cinematic debut, but I guess they were dropping the cheque in the mail. By the way Kid, I'm still waiting for that cheque." After this last statement both men begin laughing anew and continued doing so for an extended period.

"True to form, we asked for directions to the Youth Hostel and made a beeline for there. After dropping our bags and freshening-up a bit, we started walking again in order to take in some of the sights. Palermo is a beautiful old city, with a very diverse history. Over the centuries it was invaded and occupied by the ancient Greeks and Romans, the French Normans, the Moors and finally modern-day Italians."

"Each one of these groups left behind a portion of their influence and culture, that still remains there today. Of course, now the island is officially considered to be part of Italy, but regardless of these incursions made by others over the millennia, the people of Sicily are still fiercely independent. We weren't going to be here long, so, we wanted to make the most of our stay. Within a few hours, we visited a couple of churches and some historic monuments. A few people we met recommended that we visit the ancient catacombs. They indicated that the tour was quite unusual, and I was always up for something different."

"We arrived at our destination in style, being driven by a horse drawn carriage. We were greeted at the frontal façade by an actual Monk wearing full-length vestments. After paying the nominal entry fee, the Monk escorted us through the façade, and down a lengthy flight of stairs. The further we descended, the cooler and mustier the air became. As we reached the bottom of the stairs, the floorplan opened out into several large, electrically lit rooms. There displayed before us, were thousands of human skeletons. It was obvious that these corpses were very old by the varying degree of decomposition to each skeleton. Some were close to becoming dust while others were still in reasonably good condition. There were men, women and even children from all over Sicily positioned throughout the rooms. Some were standing, while others were laying inside coffins, and many of them were dressed in the clothes of former time-periods. I gotta' say that I was kind of creeped-out by it all, but the Monk was acting right casual and verbalizing as though he was introducing us to his extended family. It didn't take long to complete the tour and as we were following our guide back toward the land of the living, I spotted a human tooth lying on the floor. Obviously, it had fallen out of one of the skulls. Checking to see that no one was watching, I picked-up the tooth and stuck it in my pocket. I thought, *Who knows, maybe it will bring me good luck?*"

"After a busy day of sight-seeing we sat down to a typical Italian spaghetti dinner at a small restaurant near the hostel, and then hit the sack early. The next morning, Bob, Phil and I said goodbye to Palermo, as well as the groupies we had acquired the day before. Not long afterward, we were back on the highway exposing our thumbs. It didn't take long to figure-out that three guys and their packs weren't going to be getting a ride anytime soon, so Bob headed-off on his own, while Phil and I stuck together. The first ride in a large garbage truck only took us about 50-kilometers, but the second brought us all the way to Messina. The driver was a good head, but he couldn't drive worth a crap. I don't know if he was trying to impress us, but he was continuously speeding, grinding gears, driving into on-coming traffic, as well as passing around blind corners."

"What a jackass! Phil and I were completely stressed the entire trip. By the end of it, we were thankful that we weren't killed." Upon hearing this last statement, Jesse just couldn't prevent himself from laughing out loud. When asked by McGuire what was so funny, Jesse responded that he knew just how that felt, since that was exactly what he is experiencing every time he gets into the F150 with McGuire, and his style of driving. He further stated that whether he realized it or not, perhaps that was one habit he picked-up while traveling in Italy. McGuire responded by sticking-out his tongue and flashing Jesse the middle finger of his right hand. Jesse just continued laughing and slowly shaking his head.

"Now if you don't mind," questioned McGuire, "I'd like to get on with the story."

"From Messina to Reggio Calabria, Phil and I caught a small ferry that separated the two communities, for the reasonable price of 1,900 Lira. That was just about $1.50 in our money. As soon as we arrived in Reggio Calabria, we booked a cheap room at a Pension, and ate another spaghetti dinner. Eating spaghetti in Italy is like French-fries for us over here. You just can't escape it. I don't know if it was the Roman air or what, however, that evening was another early night for us both. No excitement. While leaving town the next morning, we soon received a ride from another maniac driver. These Italian guys must all go to the same driving school, 'Lunatics-R-Us'. After being dropped-off 140-kilometers later, my nails were chewed down to the quick. By late afternoon it was becoming more difficult catching rides so, Phil and I split-up again. Phil lost the coin toss, so it was his turn to start walking ahead, while I stayed where I was."

"Before long, this young guy in a Simca 1500 with a 4-speed transmission, stopped and picked me up. Seeing Phil 'hitching' up ahead, he picked him up as well. This fella also drove fast, but unlike the other Bozos we rode with previously, this guy knew how to control the car perfectly. Up and down the mountainous roads and through every hairpin turn, he held the road confidently, shifted gears precisely, and allowed the car to feel it's way over the asphalt, rather than forcing it to perform contrary to how it should. It was now a true pleasure riding with a competent driver, and I was totally relaxed, as we traveled ever northward. Looking out of the windows, I watched the unrelenting surf crash upon the beaches well below the highway. They were the biggest waves I had ever seen in my life, and I became excited for what was to come. Four-hundred kilometers later, we reached the outskirts of Salerno, and me and Phil were sad to leave the confines of our friendly chauffeur's Simca 1500."

"Regardless of the country and treacherous roads, it was one of the best rides that we had experienced during our entire trip, and we truly appreciated that fact. However, rather than continue traveling with that driver all the way to Naples, Phil and I got out nearer to Salerno, so that we could make our way into the magical, but tragic city of Pompeii. This was one place that I had really wanted to visit for years, and I was already getting pumped, just thinking about being there soon."

"Phil and I gradually made our way into Salerno and attempted to find a cheap room for the night. Unfortunately, every vacant space cost more than we wanted to spend. Being so close to a major tourist attraction, I guess they could charge whatever price they wanted. Not knowing what to do, but definitely not wanting to sleep on the street, we kept walking through the city and eventually found a 4-storey apartment building under construction. We easily passed by the securing barrier and entered the building. Climbing up to the second floor, we first spread out our ground sheets, and then laid our sleeping bags on top of them. We didn't have the most comfortable of sleeps that night, but at least we were safe, and out of the elements. We had heard rumors that the large Italian cities were dangerous at night, and we weren't in the mood to test that claim."

"We slipped out of the building early the next morning before the contractors arrived and immediately started 'hitching' toward Pompeii. We didn't even stop for breakfast. Luckily, we made pretty good time and arrived in Pompeii before lunch-time. It was a very eerie feeling to walk through the town, that before August 24, 79 A.D., had been a vibrant and bustling community. Many of the excavated ruins had been well preserved, and you could even see many solidified bodies of Pompeii's lava-encrusted former citizens, laying in the places where they died that day. Although all were coated in pumice, most died from the intense heat of the lava flow, the resulting fires, and the clouds of poisonous gasses being spewed forth by the volcano. Unfortunately, our tour was cut short, since it started to rain and continued raining for the remainder of the day. Even with the rain shortening our stay, it was still a very interesting day, and well worth the visit."

"I guess the tooth that I picked-up at the catacombs in Palermo was indeed bringing me good luck. In short order, we caught a ride out of Pompeii with a young geologist who was heading to Rome. On the way there, we stopped at a roadside diner, and the driver bought us all a turkey dinner. What a great score!"

"Making good time again after leaving the restaurant, our driver continued northbound on the highway, arriving near the outskirts of Rome at 5:00 P.M. After saying our good-byes, me and Phil started trekking toward the city center, but stopped short of our goal, in order to check into the Pension Suquet. It was a reasonably priced lodging, and we figured rooms would be more expensive, the further we entered the city. I guess what they say about turkey is true since Phil and I were suddenly quite tired. We decided to make an early night of it. Tomorrow we would be entering the Eternal City. Although we wouldn't be leading any victory parade into the Forum like a pair of conquering Roman Generals, for us, just being there nearly amounted to the same thing."

Jesse blew a sustained whistle and shook his head. When McGuire enquired as to why, he replied that it was almost unfathomable to imagine that a person could travel the world and visit all of those magical places with very limited funds. And at times, not even spending a dime. Meals, transport, lodging and other benefits were offered them without hesitation, knowing that they would never be repaid. Jesse couldn't believe the kindness and generosity from strangers that McGuire and his peers encountered everywhere they went. Jesse doubted that this could ever be duplicated in today's world.

"It's like I said, Kid. It was a different vibe then," was McGuire's retort.

By 8:00 A.M. the following morning, McGuire and Phil were awake and preparing to head into the city proper. Consuming the light breakfast offered by the proprietress, they first polished their shoes and then hit the cobble stones. Jesse thought McGuire was pulling his leg when he mentioned "polishing their shoes," but Bob assured him, that it was perfectly true. His father, as many other fathers of that generation, had been either ex-military or of a disciplinary disposition so, polishing shoes, short haircuts, etcetera, was an expectation. Also, in those days, most footwear was constructed of leather, not the materials utilized today. And leather always looked better when it was polished.

En route to the city center and the ancient treasures there on display, the pair first stopped at the main American Express office to cash some traveler's cheques. They also wanted to see if they had received any packages or mail. Jesse questioned this latter comment and was informed that it was commonplace for travelers on the move to have messages and other forms of mail forwarded to various American Express offices throughout the world, when they knew that a visit to those locations was imminent.

It was almost like some modern-day Pony Express postal system. Jesse was visibly impressed upon learning this little known, but important detail, and nodded his head in acknowledgement. After making his financial transaction, McGuire was extremely pleased to discover that he had indeed received mail from home, and to his amazement the clerk handed him six letters. There was a letter from Momma and Papa Bear, individual letters from Bear Brothers Kenny and Johnny, plus three more from friends Pat Ireland, Judy Elrick and Rick Witting. Although it was liberating to be independently traversing the globe at will, it was also comfortably reassuring to know that he still possessed a fan-base and support system to rely upon back home. It would take some time to read all of this mail, and at the moment he was commencing what he hoped to be an exciting day. As a result, McGuire placed the letters into his pack and decided that they would keep until the evening when he would then be able to afford them the time they deserved.

Living up to expectations, the first full day in Rome for McGuire and Phil was extremely rewarding. Taking in as many sights as possible, they visited the Forum, the Coliseum, the Palace of the Caesars and the Vatican. Before calling it a day, they went to the world famous Trevi Fountain, where McGuire tossed a coin into the water and made a wish. The wish was for his friend Judy back home, but he would not divulge the nature of that wish to Jesse. This must have been a very private moment for him, since up until that point, he had been disclosing anything and everything to Jesse. Realizing that they had barely scraped the surface of what Rome offered her visitors, McGuire decided to remain there indefinitely, with no exit date yet determined.

During his entire trip, the daylight hours were always spent sight-seeing or relaxing, while the evenings were spent carousing. It wasn't long before McGuire and Phil located a 'watering-hole' popular to backpackers and foreign tourists. They decided to celebrate a belated birthday fete for Phil and quickly befriended a group of likeminded youths, male and female, all intent on having a good time. As the liquor flowed, the inhibitions dissolved, and it was not long before the entire group was 'plastered'. Their significant level of inebriation was aided by the multiple drinking games being played, as well as the Bacchus spirit that inhabited all Roman drinking establishments. By the time they closed down the taverna and staggered back to their lodgings, there was no further time, or ability, to properly read the coveted mail tucked away safely in McGuire's pack. That would be the first thing on tomorrow's agenda, he thought as he dropped down hard on to his bed fully clothed, and quickly drifted off into oblivion.

Later that morning McGuire was awoken by the brilliant Roman sun. A laser-thin beam was streaming through a crack in the thread-bare curtains covering the room's singular window and directly into his eyes, similar to that of an interrogator's intense lamp-light blinding the sight of his prisoner. The interior of his mouth was dry and chalky, and his skull felt twice its normal size, as his temples throbbed with the rhythm of his heartbeat. Phil was still 'out like a light' and probably would be for some time yet to come as a result of last night's revelries. McGuire gingerly rose from his bed and plodded over to the sink in the corner of the room. Placing his head within, he turned on the faucet and allowed the cool water to wash over its entirety, and slowly begin to clear away the proverbial cobwebs. This continued for several minutes until he felt somewhat rejuvenated. Replacing his head, for an open mouth directly under the faucet, he then repeated this procedure until he felt his gob sufficiently flushed. Toweling off and returning to his bed, he removed from his pack an orange and partially eaten bun from the previous day. While beginning to consume these items, he also removed the six letters from home, and in no specific order, opened and read them one by one. By the time he finished reading the last letter, he was now fully awake and feeling somewhat human. He noticed that Phil was also beginning to stir, so hopefully, they would soon be back out on the streets.

The next several days were spent in similar fashion, with sight-seeing during the days, and then partying throughout the nights with a cavalcade of various characters they would meet during their escapades. McGuire was enthralled with the majesty of the city and reveled in all things ancient. He even repeated visiting several sites simply because he enjoyed them so much, and there always seemed to be something new to see, that he missed the time before. During one day and feeling more energetic than usual, the friends visited all seven hills which Rome was built upon. Whether the Aventine, Esquiline or Capitoline, each district was different than the other, with its own unique citizenry and flavor. That same day they ventured out of the city to the Appian Way. This was the original cobblestone road that led into and out of ancient Rome, and was the pre-cursor of today's modern highways.

Being of an artistic nature himself, McGuire was particularly intrigued with the paintings and sculptures of Michelangelo. By sheer coincidence, the blockbuster movie, 'The Agony and the Ecstasy' starring Diane Cilento, Charlton Heston and Rex Harrison, was playing in the theaters during his visit, and he took that opportunity to watch it one night.

The movie was based on the life of Michelangelo and his tumultuous relationship with Pope Julius II, while painting the ceiling of the Sistine Chapel between the years 1508 to 1512. Seeing the movie and then simultaneously, viewing in person, the masterpieces attributed to this genius, for McGuire was just the icing on the cake.

The weather had been fantastic during their Roman Holiday but McGuire was in somewhat of a sour mood on February 17th. First, he had given his soiled clothing to be laundered by the proprietress of their Pension. The clothes were returned in a timely fashion, but at a cost of $5.00 for the service. McGuire thought this to be excessive, and wondered if the laundry soap used to perform the task contained gold flakes. Next, he had previously taken several rolls of expended film into a shop for development, and when they were retrieved, the cost was nearly $10.00. McGuire was seriously 'choked' and immediately decided that these two cash draining functions would not be repeated. Whether or not he was still upset later that evening at the taverna is debatable, but before the night concluded, and while both distinctly 'in their cups', the two traveling companions became embroiled in a heated argument.

The disagreement concerned an effeminate-looking, black man that was also drinking inside the bar that night. McGuire was persistently teasing Phil to try and pick-up the outwardly gay man. However, Phil became increasingly more upset by this unwanted annoyance, to the point where he eventually exploded. To make matters worse, Russ, a newly befriended backpacker, sided with Phil, leaving McGuire feeling both foolish and misunderstood. Phil ended the night by announcing that he would no longer travel with McGuire. What was supposed to have been a nice night out, became the exact opposite.

The remainder of the night naturally became subdued, and the two friends refused to speak to one another. The next day the pair were still together, but remained aloof. They sat on the treads of the Spanish Steps for the majority of the day, essentially people-watching and reading. As time passed, the frigidity between them slowly began to thaw, and they gradually adopted former attitudes. On Saturday, February 19th, Phil headed out on his own to a different part of the city, in order to visit with family friends that had come to vacation in Rome. McGuire and Russ decided to go to Mario's Restaurant. This eatery had become one of their hang-outs in the city center, and today Mario was presenting an Italian version of 'happy hour' with kegs of local beer. McGuire and Russ were on a mission to ensure that they made a good accounting of themselves and as the day evolved, they proceeded to become 'hammered'.

Once again, McGuire was forced to stagger back to his room at the end of the night, and as had occurred so many times before, was basically rendered unconscious once his head hit the pillow. Sometime during the night, Phil returned to the room to find McGuire 'three sheets to the wind'. The next day was Sunday and they had already planned to attend the massive Flea Market being held weekly in the city. It was promoted as being the best market in Europe, where everything and anything could be found and purchased. He hoped that his 'potted' associate now snoring a full-scale symphony, would be able to awake in time. This was their opportunity to potentially pick-up some personal necessities or perhaps something special for someone back home, but it only convened on a Sunday.

It was initially a bit of a struggle, but Phil finally managed to arouse McGuire early the following morning, and eventually motivated him out the door. By the time they reached the Flea Market, McGuire was already lucid and ready to barter. One thing you had to admit concerning McGuire. No matter how hard he worked or how hard he partied, he was always able to achieve functional competency relatively soon afterward. And then with positive cognition and a happy demeanor. Just as advertised, vendors at the market were purveying everything under the sun. The stalls seemed to go on forever. At one location McGuire and Phil attempted to purchase switchblade knives, but unfortunately, were not able to sufficiently lower the asking price. Being steadily on the road and often in suspect locations, the pair felt the need for protection that a knife might provide. In the end, perhaps it was best for the two young men that those weapons were not acquired.

After spending several hours at the Flea Market very little was purchased by the two friends, but both thoroughly enjoyed the time spent browsing the myriad of stalls and kiosks. Before leaving, they came upon a man extolling a game of chance. McGuire immediately recognized the game as a shady, speculative means by the Promoter to relieve some unsuspecting 'mark' of their money. However, Phil insisted that he could best this Shylock and beat him at his own game. For a nominal bet, the Dealer would drop a small square of paper on a table top, and place one of three round discs overtop the square. The positions of all three discs would then be altered on the table top several times by the Dealer, and then halted. It was then up to the challenger, to determine under which disc, the paper square was located. If he guessed correctly, then the 'accumulated pot' would be his. If incorrect, then he would forfeit his bet.

Against McGuire's sage advice, Phil played the game numerous times, and on every occasion, he guessed incorrectly. Well, $7.50 later, Phil decided to cut his losses and admitted defeat. As they left the market, with McGuire roaring with laughter, Phil dejectedly announced, "Kick me in the ass!"

On the morning of February 21st, McGuire and Phil checked-out of their Pension. They found Russ at one of the usual backpacker haunts and said their farewells. They then slowly began making their way out of the city. As far as McGuire was concerned Rome had been a very enlightening and enjoyable city to visit. He found that it was more expensive than desired, and had spent more money here than anticipated. But overall, it was an experience to remember the rest of his life. Once out of the city proper and back on the highway, that persistent dilemma of static inactivity resurfaced. It had been hours, and they had barely traveled any distance at all. Once again, the decision was made to separate for practical reasons, and the pair parted company. Once on his own, it wasn't long before McGuire's lucky Sicilian tooth once again worked its magic in the form of a ride. However, this ride was the harbinger of something darker yet to come.

The driver was a friendly sort with limited English. Unfortunately, he was only traveling an additional 20-kilometers further down the highway. Realizing that this would be disappointing to someone wishing to go much further, he pulled-over the vehicle at the point of his deviation, and then flagged-down the next car coming in his direction. McGuire noticed that the next vehicle was a bit of a 'beater', and contained four occupants. His driver spoke with them briefly and then returned. As best he could, he explained to McGuire that he needed to take a different route, but the occupants of the other car were heading to the coast, and they were willing to take him along. Understanding the gist of the explanation, McGuire exited the one vehicle and boarded the second.

Immediately upon entering, he felt uncomfortable. This new ride contained two unsavory-looking males in the front seat. Both were in their early to mid-twenties and looked about as slippery as a pair of eels coated in Vaseline. In the backseat with him was a pimple-faced, teenaged male and an old, female crone in her late sixties. All four were staring at him in a way that a vulture would stare at a potential meal. It made McGuire's skin crawl. As soon as they commenced driving, the front passenger immediately turned around fully, and demanded cigarettes.

He then threw a barrage of questions at McGuire, all the while pointing at his bags. He kept asking for food and money while pointing over to the old lady, indicating that McGuire should give her something. By now, Bob was becoming increasingly alarmed, and was expecting the worst. He pretended that he couldn't understand what his interrogator was saying, and tried to explain that there was only clothing inside his bags. He also now wished that he had purchased that switchblade knife at the Roman Flea Market, regardless of the price, because he was certain that shortly, he was going to be robbed, or worse. The mood inside the car was now extremely tense and all four Italians were becoming visibly agitated. Not knowing what else to do, but wishing to perform some type of bluff, McGuire placed one hand inside his pack and adopted a hostile appearance. This action appeared to work, since the front passenger now hesitated and appeared to become concerned. He barked something in Italian to the driver and the car then immediately pulled over to the shoulder. Once stopped, McGuire was rudely told to get out of the car.

Doing so gladly, he promptly exited and slammed the door shut. Spitting gravel as it sped off, McGuire fired-off his own barrage of expletives at the departing vehicle, and if given the opportunity, he would have in turn, soundly thrashed each one of them.

Now safely out of that viper's nest, he was nevertheless, stuck on the side of the road once again. However, as quickly as disaster was averted, deliverance was now at hand. On this occasion, it was a large transport truck and trailer, en route to the east coast of the country. For safety and productivity purposes, the truck contained a driver as well as a passenger, so that they could spell-off one another during long hauls. It was now the current driver's turn to rest while the passenger assumed driving responsibilities. The former driver crawled into the tractor's sleeping compartment, allowing McGuire to assume the large, comfortable and now empty passenger's seat.

McGuire was intent on reaching the east coast of Italy however possible, in order to then proceed toward Greece. That was to be his next major destination and the plan was to meet Phil in Brindisi, within the next day or two. It was from Brindisi that they would then sail to Greece. From Rome to Pescara was slightly more than 200-kilometers and essentially due east. You could imagine McGuire's joy, when the relief driver informed him, that Pescara was their intended destination, and that he was welcome to accompany them the entire way. Protectively placing his hand over the lucky tooth tucked within the pocket of his jeans, McGuire gave it an appreciative pat.

At 7:00 P.M., they arrived in Pescara and McGuire jumped out of the truck. He had achieved his goal well above expectations and he was now poised on Italy's Adriatic Coast. Finding a small taverna nearby, he entered and enjoyed a simple but hearty pasta dinner as well as a couple of drinks. Being congenial as ever, he then struck-up a conversation with two local men seated at the bar.

One of them spoke reasonably good English, so McGuire launched into an enthusiastic rendition of his travels to date. After a few laughs and a few more drinks, the men agreed to transport their Canadian visitor to a nearby campsite, since there were no available rooms anywhere in the vicinity. It had been another gloriously sunny, warm day and this translated into a comfortably mild night. McGuire spread his groundsheet and sleeping bag on a bare patch of grass and then slipped inside the zippered bag. Being removed from the bright lights of the city, the night sky was a mass of twinkling stars and constellations. As McGuire lay on his back marveling at this celestial lightshow and hearing the persistent but relaxed chirping of cicadas, he couldn't help but smile. Here he was, an unremarkable and inconsequential young man from southeast Vancouver, British Columbia, thousands of miles away from home and living without a care in the world. He currently occupied a patch of ground in the once most powerful nation on earth. The influence and impact ancient Rome had upon much of the European continent, the Mediterranean and North African regions, and even the British Isles, was profound and unmistakable. Perhaps even now he might be laying in a place where Roman Legions once tread, where politicians and philosophers orated or debated and where indentured slaves from conquered lands tilled the soil. With this last thought in his mind and the sight of a shooting star spiriting across the heavens, he gently drifted off toward the deep and perfect sleep of the innocents.

It was this latter memory that seemed to place McGuire into another state of melancholy and he appeared to want to savor the moment. Jesse had predicted correctly when McGuire suddenly announced, "Okay, what do you say Kid. Should we put this business on hold and get back to work? I'll bet that you're probably bored to tears by now anyways, and it looks like we've got a couple more holes to fill."

Jesse reassured McGuire that nothing was further from the truth. He was actually immensely enjoying the tales from McGuire's younger years, and he told him so. However, he did agree that it would be a good idea to put two more plants into the ground, since the holes were already dug. He rose from his seat and walked toward the planting area.

McGuire entered the greenhouse and soon returned with the cart bearing two of his young ladies and sufficient nutrients to satisfy both. Jesse watched as McGuire planted the clones with loving care utilizing the identical method and ingredients for both. Jesse had been intently observing this process on twelve separate occasions, and now felt confident that he possessed sufficient knowledge and ability to perform the task himself.

There were still four holes to dig and four plants to fill them. It appeared as though there was insufficient time to accomplish that task on this day, but Jesse decided he would ask McGuire if he could perform a complete transplant with at least one of the clones, during their next visit to the greenhouse.

As McGuire brushed the dirt from his knees and removed his gloves, he questioned Jesse. "Hey, Kid. What do you say that we call it a day and lock-up? Then we'll head up to the Texada Island Inn in Van Anda and grab a couple of burgers and fries to go, and take them back to the homestead. I remembered that I taped the 'Agony and the Ecstasy' a couple of years back, and saved it on my PVR. All this talk about Italy and Rome has kind of put me in the mood to watch it again. Are you up for that?"

Naturally, by now Jesse was up for practically anything. He was really beginning to develop a strong bond of affection toward this amicable, if not somewhat unusual older man, and whatever he wanted to do, was alright with him. "Sure thing, Bob," responded Jesse agreeably. "That sounds great!"

On that note, they began packing away their tools and putting everything back in order. Once everything was locked and secured, they took one last look around, and when satisfied, headed for the Texada Island Inn with the white, Ford F150. As usual, McGuire was performing his Italian-style of driving. Jesse was becoming more accustomed to this, but he was still an uncomfortable passenger. He figured that it would probably be a while yet before he would feel more at ease. At least, there was very little traffic on the road due to the diminutive size of the island and lack of citizenry.

Soon after, they had arrived in Van Anda and ordered their food. With limited patronage at the Inn again today, it was not long before their meals were prepared and packed in Styrofoam take-away containers.

To ensure that their meals remained as hot as possible between Van Anda and Gillies Bay, McGuire did not spare the accelerator pedal of the pick-up truck, and in record time, they were back on Leonard Road and seated at the kitchen table. They thought it prudent to eat first and then shower afterward, since the food was still now at an optimum temperature for consumption. Within the hour, dinner had been completed, their few dishes had been cleaned, and both men had performed their bathroom necessities.

Jesse was now seated on the couch in the living room while McGuire was in the kitchen filling two glasses with the obligatory evening refreshment. Once in the living room, McGuire plopped down in his favorite chair, and with the T.V. remote, began searching his immense library of previously recorded programs. The desired movie was eventually located and engaged. "Okay, Kid. I'm now a bit of an expert on the content of this film, so if you have any questions, feel free to ask. I'll just pause it," was McGuire's offer to Jesse. For this proposal, Jesse gave him an appreciative nod, as the opening scenes began occupying the T.V. screen. For the next 2-hours and 18-minutes, their eyes were fixed on the screen, with McGuire occasionally shouting-out, "I was there!" or "I saw that!". Jesse could tell that Bob was quite thrilled at reliving this episode of his former excursions, and although the film did nothing to stimulate him, he gave McGuire the impression that he had enjoyed it.

After the show had already ended, McGuire still wanted to discuss its merits and dissect the plot. He also wanted Jesse's perspective concerning certain characters and the roles they played in the overall theme. Before long, the night had fully evolved and it was time for bed. Once inside his room, Jesse retrieved and powered-up his laptop computer. Before going to sleep, he wished to perform a few more 'searches', plus he was curious as to whether or not his friend had replied to his previous email. Feeling impatient, Jesse first went to the 'in-box' of his email and indeed found two messages from the mainland. One was from his mother while the second was from his friend. His mother's reply was quite lengthy and contained the usual well-wishes, maternal concerns and advice. The second message from Jesse's friend was brief and succinct. In bold font and upper-case letters, it said…**TRAITOR!**

Jesse was not completely surprised by this response but it still disappointed him. At least now, he knew where they stood with one another, and that was something. Jesse did not feel the need to respond to this message, such as it was.

However, sometime in the foreseeable future he was hoping to address his feelings and concerns with what now appeared to be his former friend. Perhaps they could still come to an amicable understanding. But that was for another day. Tonight, he still had a few inquiries to make, and he got on with them.

When he finally did turn-out the light, sleep did not come as quickly as previous nights. The one-word message in his email still weighed heavily upon him, and this impeded his ability to relax. Eventually the power of Morpheus overtook him in order to settle the impasse.

The next morning at the greenhouse, McGuire seemed to be all business. Without standing on formality, both men embarked on their assigned tasks and by the noon-hour, Jesse had already commenced digging his third hole. Just when he thought he would succeed in completing that task, McGuire exited the greenhouse. He retrieved their lunches from within the pick-up truck, then assembled the folding deck-chairs and makeshift table in preparation for their mid-day repast. "Okay, Kid, get over here. It's time to tie-on the feedbag," was McGuire's informal announcement that lunch was being served. Jesse obeyed the command and both men were soon eating. As anticipated, McGuire once again entered into his recital.

"Considering that I had been sleeping on the ground out in the open in Pescara, I had a pretty good rest. One thing about sleeping outside, you usually get-up pretty early. With the sun shining, the birds chirping and the farm animals making all kinds of noise, there was just no way that the average person was going to be able to stay down for long. No breakfast for Bobby that day, as I headed straight for the highway."

"My good luck tooth was still working, since I got my first ride quite soon. The lift was from a school teacher that was driving to Bari, and he spoke English well. When I told him that I was Canadian, he got quite excited. He was grateful for the fact that the Canadian military had played a key role in liberating Italy during World War II, and he wanted to take me to a special place that was on our route. Not wanting to look a 'gift horse in the mouth', I quickly agreed. As we neared Bari, he pulled off the main highway and we drove further into the countryside. I was beginning to wonder if this guy was perhaps a 'limp-wristed fellow', trying to get it on with some fresh foreign meat, but then we reached our intended destination. The sign posted at the entry said, 'Bari War Cemetery', and now I was really confused."

"The teacher got out of the car and started walking into the cemetery. I followed his lead and as I approached the first line of symmetrically placed headstones above a gorgeous lawn, I saw that they all bore maple leaves. In addition, the names of those buried weren't Italian, but mostly names that I clearly recognized."

"It turned out that this was a Commonwealth Cemetery, and hundreds of Canadian soldiers killed in the battle for Italy, were buried here. I'm not into army stuff, but seeing all those rows of crosses and headstones was sobering. Knowing that many of those buried here were Canadians, and probably close to my own age when they were killed, really put it into perspective for me."

"And here was this Italian guy old enough to be my father thinking that this was something important for me to see. The more I thought about it, the more I realized that if it wasn't for those soldiers, and the thousands of others who fought and died throughout the European Theater during the last war, the freedom I now possessed to casually travel around the world, perhaps wouldn't be possible."

Jesse readily agreed with McGuire's assessment. Although his generation had never known a global conflict, and considering that his own people were still living under certain adversity, he also wondered what the world would now look like had those former megalomaniac oppressors of the mid-twentieth century achieved their goals. He was glad that question would never need to be answered.

McGuire broke Jesse's train of thought. "We stayed at the cemetery for a while and walked in between the rows of grave markers checking out the names. Before leaving, I signed the registry book in the administration office. After that, it wasn't long before we reached Bari, and the school teacher dropped me off near a busy intersection. You know Kid, they always say that it's a small world, but sometimes I think that it's actually true. If you can imagine, the next guy that picked me up, was an Italian driving a dump truck, and would you believe it, he had lived in Canada for a while and actually worked in Vancouver for about 3-months. 'That's freaky, eh?' Anyway, this guy drives me all the way to Brindisi, arriving about 6:30 P.M."

"After dropping me off, it took me a while to find the Youth Hostel. When I finally did, it was closed, and I was majorly pissed! Now, I had to hitch-hike all the way back into town. It was quite late when I made it back, but the town had come alive. There was a big festival going on, and the streets were packed with people."

"It was the day before Lent and everyone was celebrating. Young kids were dressed in costumes, there was music playing and everyone has handing-out food and drinks to anyone who wanted it. I grabbed a few snacks and a drink but didn't spend much time with the crowds. I needed to get some rest, so I started searching for a cheap room. I found one quite quickly, and considering all of the racket outside, I actually had a pretty good sleep."

"The next morning, I headed back out to the Youth Hostel thinking that Phil might already be there. He wasn't. I then headed back into town and went to the steamship office where I bought a one-way, deck-class ticket to Corfu, for the equivalent of ten-dollars."

"While in the ticket office waiting for clearance to board the ship, I picked-up a pamphlet written in several different languages. One of them was English. The subject concerned Spartacus and the Slave Revolt of 72 B.C. I had seen the Spartacus movie starring Kirk Douglas in the Orpheum Theater, downtown, when I was in my mid-teens. I really enjoyed the movie and I particularly remember one scene where Spartacus and his slave army had fought their way to Brindisi, where I was now. Over a period of months, they had raided numerous towns and wealthy villas."

"As a result, they had captured a considerable amount of booty and they were going to use that to buy their freedom. They had made a deal with Cilician pirates to transport them off of the Italian Peninsula and over to Greece, in order to solidify their liberation. The Cilician pirates had terrorized the Mediterranean coastline for years and possessed a considerable fleet of ships. More than enough to carry the entire slave army away from the Roman legions, that were now pursuing them. Having given the pirates all of their wealth, they were shocked when they arrived in Brindisi, to find there were no ships in the harbor. Marcus Lucinius Crassus was the Roman general now closing in on them, and a very wealthy man in his own right. He had made a secondary deal with the Cilician pirates. He offered to pay them more than what they had received from Spartacus, so long as they did not deliver their ships to him. Pirates being what they are, with trustworthiness not being one of their stronger character traits, they betrayed Spartacus and never sailed to Brindisi with their ships. As a result, the slave army was left trapped with their backs against the sea, and three Roman legions bearing down on them. The end game was now a foregone conclusion, and within days they were defeated by Crassus and his legions."

"It was estimated that over 6,000 slaves recaptured by Crassus were crucified along the entirety of the Appian Way, from Capua all the way to Rome. The body of Spartacus was never identified, so, in a very small way, he still found victory over his Roman oppressors. Just think, Kid. I was in Brindisi, and I also walked along the Appian Way. Maybe I stood in the exact same spot that Spartacus once did, in either of those two locations 2,000-years later. Some thought, eh?"

Jesse had to admit that it all sounded pretty exciting and he could only imagine the number of momentous events that took place over the millennia all through that region. In his brain, he was feverishly trying to visualize that final epic battle, and then the gruesome sight of all those hapless souls left to die a slow and excruciatingly painful death on a wooden cross, at the side of the Roman highway.

True, his own ancestors had occupied the earth for thousands of years. However, they could not even come close to competing with the history making and life altering events of the ancient Greeks, Romans, Egyptians, and all of the others that lived and dwelled around the Mediterranean Sea. At that exact moment, Jesse actually felt a twinge of envy toward McGuire. He was able to experience all of these exciting adventures, while Jesse had essentially been nowhere in particular.

McGuire could virtually see Jesse's imagination at work upon hearing his words, and he reveled in the thought that his tales were impactful. After all, these travels affected him profoundly, so why wouldn't they be interesting or meaningful to others? This thought gave him even more motivation than before to impart his tales, so with greater enthusiasm than ever, he continued with his story.

"Okay, so while I'm waiting in the ticket office with some of the other passengers, one of the ship's crew comes out and says that the sailing will be delayed due to mechanical problems. Unfortunately, now, it wouldn't be until this evening before we could board. Looking at the old tub in the harbor, I wasn't surprised. In fact, it would be a minor miracle if the rusty hulk didn't end up sinking half way across the Adriatic Sea. With time to kill, I walked the short distance back into the city center and made myself comfortable in the main square. From here, I could be on the lookout for Phil, and I could also check-out Brindisi's female talent."

"After a while I became bored with people-watching, so I started reading one of the new books I had packed from home. It was called 'The Spy Who Came In From The Cold'."

"Within the first few pages, I was really getting into it and before I realized, it was time to head back to the harbor. Still no sign of Phil, but as I walked into the ticket office, I spotted Howard. He was an English traveler that we met on the ship from Tunis to Sicily. At the moment, we were both traveling alone, so we decided to join forces for the trip over to Greece."

"As we boarded the ship, I was rudely introduced to the meaning of 'deck-class'. Sailing deck-class literally meant that you remained on deck during the entire trip. There was no admission to a cabin or a lounge or really any form of shelter. There was a small toilet stall accessible from the deck, but that was about all. Howard and I hunkered down in a corner near the bow, and remained there for the entirety of the trip, except for once or twice when we need to pee. It was bloody cold out there and we nearly froze our asses off. Under those conditions it was nearly impossible to fall asleep, so we essentially remained awake all night."

"After the sun rose on the morning of February 24th and began thawing us out a bit, we stretched our cramped muscles and started feeling better. Before long, we were back to normal and we could eventually see the island of Corfu appearing before us. What was initially a tiny speck of land on the horizon, gradually became a significant land mass with each passing minute. It was 10:30 A.M. before we finally cleared Customs and Howard and I then parted company. I slowly started walking toward the highway and immediately sensed a different attitude from these islanders toward strangers. Most of the people I encountered were quite unfriendly and some were downright hostile. As a result of this change, rides were practically non-existent. I had walked 10-kilometers without even catching a single ride and I was really beginning to fade. By now, it was quite hot outside and I had stripped down to my undershirt."

"Finally, I caught a ride with a British dude in a Jeep. He was involved in Animal Welfare and Development, and drove around the region plying his trade. After my epic trek on foot since landing on Corfu, this guy was a lifesaver for me. My feet and legs were killing me, and as a result, I've started to limp. I'm hoping that this is only a temporary setback, since that is really going to cramp my style, if I can't walk. Anyway, I'm not too worried about that right now, since my driver takes me 90-kilometers all the way to Ioannina, through beautifully rugged mountain country. He then literally drops me off at the Youth Hostel. To top that, guess what? I'm the only guest there. I've got complete run of the joint. Out of sight!"

"The next morning when I rise, I'm still in pain and I realize that I'm not going to get too far on foot. This is not a good thing for a backpacker, since walking is a big part of the experience. I soon jumped on to the ferry for the short ride to the mainland, and the operator didn't even charge me. That lucky tooth was still potent. After leaving the ferry, I headed into the next town for breakfast."

"All the while, the little squirrel on the treadmill inside my brain, was running like crazy in order to come up with an idea. After breakfast, I saw a 'Beat Cop' standing on the street corner. At that moment, I just got a brainwave and approached him. Luckily, he spoke a little English, so I told him that I was heading to Athens, and I was looking for a free ride with someone going there as well. Wondering what he thinks of me, I'm half expecting that he's going to tell me to get lost. Instead, he lifts one forefinger and using body language, motions for me to wait where I am. He buggers-off and in a few minutes returns with a guy driving a large truck. Well, guess what? The truck driver is going to Athens, and he's willing to take me along."

"Needless to say, I'm really stoked and I give the cop a hearty handshake before climbing into the truck. I never thought that I would ever find myself saying this, but that officer was a real cool dude. Who knows, maybe they all aren't so bad after all."

"The truck driver only spoke a few words of English, so the ride to Athens was pretty subdued. That didn't bother me too much since, I just wanted to rest anyway. Almost 200-kilometers and several hours later, we arrived in Athens. It took me about a half-hour to find the Youth Hostel, and when I did, it was packed. In Ioannina, I had the entire hostel to myself, but here I was sharing a room with sixteen other guys. However, that didn't bother me at all. I was in Athens, the cradle of civilization. This was the birthplace of Democracy. The original font of western culture. The Greeks gave us art, literature, theatre, modern science and medicine, philosophy, architecture and so much more. Scholars, teachers and statesmen such as Homer, Plato, Aristotle, Pericles, Archimedes and Hippocrates, all originated from this small but profoundly influential nation."

"Once inside the hostel, I recognized a few faces from my time spent in North Africa and Sicily. Then I spotted Howard. He beat me here. That night, we decided to celebrate and got royally hammered. As often happened during my travels, by the time we were ready to call it a night, the hostels we occupied, were already locked. The same thing happened in Athens."

"It was about 4:00 A.M. and the doors opened at 7:00 A.M. However, we didn't want to sit there for 3-hours, so we walked to the Plaka. My previously sore legs and feet were now feeling better. Just so you know, Kid. The Plaka is Athens' version of our Gastown in Vancouver. And just like Gastown, it's geared toward tourists, for shopping and restaurants. It was also quite expensive so, I wasn't going to be purchasing anything there, anytime soon. You also have to watch your step when walking, since the streets and sidewalks are paved in marble."

"They're as slippery as hell when they get wet. The Plaka is at the foot of the Acropolis and this was where we were headed. Acropolis means 'high city' and this was where the world-famous Parthenon stood. We only managed to stagger halfway up the hill before Security Guards spotted us and chased us away. Now we were really pissed-off. We couldn't get into the hostel, and now being chased off by the guards, tipped our tolerance scales. Being drunk and stupid, we started creating a bit of mischief as we slowly made our way back toward the hostel. We didn't really create any major damage or serious mayhem, but I'm not really proud of what we did either."

"We returned to the hostel and waited for the doors to open at 7:00 A.M. Once inside, we immediately headed for the showers but the water was freezing cold. It took everything I had to stay under the spout for the time it took to wash, but at least it cleared my head and chased away last night's 'piss-up'. After cleaning-up and having something to eat, Howard and I went to the YWCA to see if we might be able to score."

"It had been quite a while since I'd had a little 'pinch and tickle', and I was most definitely in the mood for it again. Unfortunately, the place was guarded by a Doberman Pinscher of a house-mother, and she wouldn't let us get anywhere near the doors. After a few attempts, we could see that it was a lost cause, so we left. But not before I gave her a few choice words and called her an 'Old Toad'."

"On March 1st, I finally visited the Acropolis. This time I was sober and ready to experience the full effect of the monument. Naturally, what was left to see of the Parthenon was only a ghost of what it once had been in 450 B.C., but it was still extremely impressive. Imagine how those ancient peoples were able to create such a magnificent structure using only rudimentary tools. That building is still standing 2,500 years later, when contractors today using modern technology and sophisticated tools and machinery, can barely keep their structures standing for 50-years."

"As usual, my nights were spent drinking and chasing wild women but one evening I stayed back at the hostel and played poker with some of the guys. My lucky tooth was on fire that night and I ended-up cleaning house by winning a whopping ten bucks! Well, after the game, guess who comes strolling into the hostel? You guessed it, the long-lost, elusive Phil. It turns-out that Phil arrived in Bari not long after I was there but then he became sick from food poisoning or something like that, and ended up having to remain until he felt better. So now the mystery was solved."

"After a couple more days, our time in Athens was coming to a close and Phil and I started preparing for departure. To lighten our packs, we decided to mail a few things home that we could do without. I mailed a bunch of stuff to friends in England. At least now, if I was going to have to walk any distance, my bags weren't going to be as heavy to carry. On March 3rd, we visited the Lebanese Consulate and obtained Visas for our future visit."

"That same day a bunch of us were approached by some Greek guy that claims to be making a newsreel about hippies and backpackers. He asks if we want to be part of the production, and he offers us each 100 Greek Drachmas. That was about ten bucks in our money. Even though we were leaving town soon, easy money was easy money, and we all agreed. The only crappy thing about the whole venture was that we had to be on the bus that would take us to the filming location at 8:00 A.M. the next morning. The last thing I wanted to do was get up at that un-godly hour, but what the heck."

"We managed to find the bus by 8:00 A.M. the next morning, and were driven to the base of Mount Lycabettus, just outside of Athens. When we eventually arrived, a bunch of 'freaks' were already there, so we just blended right in. There wasn't much of a script to this so-called newsreel, and we didn't have any speaking parts. However, they gave us all a really nice lunch and at the end of the day we received 80-Drachmas for our troubles. That was a bit less than what we were originally promised the day before, but after 8-hours of 'acting', we weren't in the mood for arguing. Overall, it was a pretty cool experience, so we accepted the money and split. We decided to have another 'shaker' that night, since we were leaving the following morning. As usual, we partied way past the hostel curfew time, and naturally, the doors were locked when we arrived. Like I said before, we often got pretty foolish when we were drunk, so Nick, one of our pals, started scaling the wall of the hostel and climbed into an open window several floors up. Had he fallen, that probably would have been the end of him."

"Anyway, Nick crept around through the hostel and finally came downstairs to unlock the door for us. We all tip-toed inside and quickly found a place to crash. Later that morning, we slipped out with all of our bags before the caretakers opened for business. As a result, we didn't have to pay for our night's sleep. Woo-Hoo…another victory for the good-guys."

"We headed straight down to the Metro and caught a train that would bring us close to the National Highway. Walking the remaining distance to the highway it was cold and very windy. Thankfully, we didn't have to 'hitch' for very long, and managed to get a series of short rides. Finally, we caught a good ride that took us 260-kilometers to Larissa, and then another, in a meat truck that took us the full 185-kilometers to Thessaloniki. It was late upon arrival, and we were 'bagged'."

"Not finding any accommodation in the area, we stopped at an all-night roadside diner and asked the owner if we could sleep on the floor in the corner of the restaurant. It was pretty cold outside and we definitely didn't want to sit there all night. Thankfully, the owner agreed, and after a quick bite, we immediately crashed."

"Very early the next morning, we left Thessaloniki and hit the highway once again. Rides had practically dried-up for us, so we came up with a bit of a plan. We would 'hitch' in one spot for an hour and if we didn't catch a ride, we would then walk ahead for a few kilometers. We would then stop and repeat the process."

"We did this for a total of 5-hours and still we hadn't received a single ride. We even took turns with one of us hiding in the bushes, while the other 'hitched'. That didn't work either. We were freezing! Finally, we just started walking, and after a while a charter bus pulled over and picked us up. The driver took us to his next stop in Xanthi, 90-kilometers away, and we didn't even need to pay. After a disastrous morning, the afternoon worked out just fine for us. In Xanthi, we entered a small café and approached a priest who was alone there eating his supper. I asked the priest if he knew of any places in the area where we could spend the night for free."

"Sitting at the next table were a couple of businessmen wearing suits. Understanding English, they had been listening to our conversation with the priest. They called us over to their table and told us that they would offer to rent a room for us at the inn across the street. They also offered to buy us dinner at this restaurant. Not certain if they were serious or just pulling our leg, Phil and I agreed and waited for their reaction."

"Instead of laughing or saying that the joke was on us, one of the men leaves the table and walks over to the inn across the street. The other one calls over the café owner and orders dinner for Phil and me. Speaking Greek with the owner we didn't know what he had ordered, but eventually the landlord comes out of the kitchen with a platter of beef, lamb, salad and bread. He then goes back into the kitchen and a couple of minutes later, he returns with two large omelets."

"Phil and I are beside ourselves over this windfall and the businessman is just sitting there smiling at us and nodding his head in agreement, as if to say, 'Go on boys, enjoy yourselves.' Before we think that things can't get any better, the owner comes back to our table with two large beers."

"Half way through our meal the second businessman returns to the restaurant and hands me a room key for the inn across the street. As I'm thanking him, I look over my shoulder, and see that the priest is ready to leave. Before he does, he looks directly into my eyes and smiles. Without a word, he turns and walks out of the café. Now, Kid…like I said before, I'm not one for religion, but now I'm thinking there's something weird going on here. On the bus from Toronto to Buffalo, I was sitting beside a Minister and was on my best behavior. In Palermo, I found a lucky tooth in the Catholic catacombs. And now here in Xanthi while talking to a priest, as if by divine intervention, we not only receive a great place to sleep but we also received a banquet to boot. I don't know for certain, but it was almost as if I had a Guardian Angel or something watching over me. What do you think, Kid?"

By now, Jesse is ready to bust a gasket. He finally blurts out, "You're jerking me around, Bob. I was believing you there for a while, but there's no way that a bunch of strangers in different countries are going to rent you rooms, buy you meals and drive you half way around the world all for free. There's just no way! It's a great story though."

McGuire realized that for an outsider like Jesse, or someone who hadn't 'been there' before, this story would sound incredible and even unbelievable. However, he had no reason to exaggerate or lie and he told Jesse so. Giving him a very sincere look and speaking in a controlled manner he stated, "I hear what you're saying, Kid, and if I were in your shoes, maybe I wouldn't believe it either. But I wanted to tell you here and now, everything I've said so far, has been absolutely true."

After providing his explanation, he stopped and allowed Jesse time to contemplate what he had just imparted. All the while, Jesse continued watching McGuire's body language and expression, as he considered his words.

It wasn't long before he came to the conclusion, that although these events and claims seemed inconceivable, gauging by the sincerity he now saw on

McGuire's face, and having gotten to know him this past week, he decided that in fact he was telling the truth. That was all there was to it and he told himself as much.

Jesse could still not fathom why someone would show such extreme kindness to a perfect stranger, but obviously McGuire was the recipient of such kindness many times during his travels. After all, like he had already stated, "It was a different vibe then."

"Okay, good…so that's settled," announced McGuire. "After dinner, we sat with the businessmen for a couple more hours telling tales and drinking beer. By 10-o'clock Phil and I were pretty tired, so we said good night to our benefactors and headed across the street to the inn. After a great night's sleep, we were back on the highway early the next morning. At first, it looked like it was going to be another tough day 'hitching', but then a VW bus with a couple of Americans inside, pulled-over and picked us up. We told them that we were heading to Istanbul, Turkey and would you believe it, that's exactly where they were going as well. Once again, my lucky tooth was working its magic, and 350-kilometers later, we were driving near the outskirts of Istanbul. By dinner time, Phil and I were booked into the Tourist Hostel for the cost of 4-Lira per night. Just for your information, that translates to about 35 cents. Not bad, eh?"

Chapter 5
From Disaster to Redemption

While Jesse was still digesting this latest installment of McGuire's adventures, the orator took a couple of minutes to reacquaint himself with specific details from within his journal. Jesse was amazed that McGuire was almost seamlessly recounting the entirety of his travels with barely a need to refer to his journal at all. What this documented version seemingly provided him, was exact dates and times, that tied the totality of the remembrance together, and kept it sequential. Otherwise, the story itself rolled off his tongue as though it had occurred only yesterday.

"The next morning, we managed to sleep right until 10-o'clock when the bell rang to clear the building. As we started getting our act together and were ready to leave for our first full day in Constantinople, as Istanbul was previously known, we bumped into a bunch of guys that we had met at different times and locations during our travels. There was Nick, Jerry, Ken, Brian and three other Canadians that were all making a similar tour as ours. After hearty handshakes and pats on the back from everyone, we all headed into the city, and visited a small café where we enjoyed a big feed of sausages, couscous and salad. After eating, we then made our way to the world's largest covered bazaar."

"What a place! It was jam-packed with tourists and merchants and literally, everything under the sun there was for sale. If you couldn't find what you were looking for there, you either weren't looking hard enough or it didn't exist. It was one of the best days of my entire trip. We had a million laughs and I felt on top of the world. After hours of just curiosity shopping, I ended up buying a 4-piece silver puzzle ring for 50 cents and an 8-piece, 14-carat gold one for 8-dollars. After leaving the bazaar, I went to the General Post Office and picked up seven letters. I didn't think I was so popular, but obviously I was."

At this point, McGuire raised his nose up high in a snooty fashion, and polished the fingernails of his right hand on his chest, in a braggadocios gesture. This elicited another burst of laughter from Jesse.

"That night the bunch of us got together and drank a lot of beer. We then all visited 'Whore Alley' to have a look at the skanks." When he failed to elaborate further, Jesse pressed him. "Aside from the little bit of action you had on board the ship, you haven't said too much about meeting any women. What's that all about, didn't you have any luck?"

McGuire merely scoffed at Jesse's question, as though it wasn't worthy of a response. He then finally replied, "Kid, I was getting all the action I could handle. But it was nothing momentous. The chicks were okay, but what's that old saying, 'we were like two ships passing in the night.' It was no big deal. I'm just not going to get into the nitty-gritty, because I don't want to see you drooling over there." Once again, Jesse couldn't help but chuckle at this response. Quickly checking the dates in his journal, McGuire then continued his recitation.

"Our second day in Istanbul was March 8th. I got up a little earlier than the other guys and went out. I managed to hire a local guide with a small bus for a few hours, and asked him to take me to the Syrian Embassy. I was hoping to get a Visa to visit that country after leaving Turkey. Unfortunately, the Embassy was closed. I wasn't impressed. The guide then drove me downtown to the Central Bank where I wanted to cash another traveler's cheque. Can you believe it? They didn't have any American money that day! Now, I'm really pissed. I've hired a guide to help me out, and now I've just wasted my money. Hoping to salvage some of the day, I had the guide drop me off at the covered bazaar again, and then I cut him loose."

"Once inside the market with its fantastic sights, smells and sounds, I started to settle down a bit and got lost in my browsing. Before leaving, I ended up buying a ¾ length woman's suede coat for Judy, with a matching suede coat and vest for me, all for 50 bucks. What a deal! I was going to return the next day for my fitting. I then left the bazaar and went back to the hostel where I spent the rest of the afternoon writing letters. Four hours later, I was developing 'writer's cramp' and decided to give it a rest. I was also getting mighty thirsty so, Nick and I went out to wet our whistles. After getting a bit of a buzz, we then decided to check out one of the many ancient, public bath houses. It had been quite a while since I really had a good scrub, so this actually was a real treat for me. We spent over an hour at the bath house, sweating in the steam, sitting on hot flag stones, scrubbing down with mountains of soap suds and then jumping into a pool of cool water in order to close the pores that had been opened by the heat. I felt great afterward, and hungry as a bear."

"Once eating a large dinner of sausages and rice plus a few more beers, we slowly walked back to the hostel. After that hot bath, I was feeling very mellow and not really in the mood for doing anything, so I wrote a few more letters and then crashed. The next day was more or less as tame as the previous two. However, a little bit more successful."

"This time, I managed to fill-out my application form for a Syrian Visa, and then I mailed eleven letters to my fans, from the Main Post Office. I returned to the covered bazaar for my suede fitting and also visited the Blue Mosque. This was the most popular mosque in Turkey and one of the oldest. It was called the Blue Mosque because of the thousands of blue tiles fitted to the walls and ceiling on the inside of the temple. I thought some of the mosaic designs inside were kind of cool but otherwise, it didn't really turn me on. Again, that night was uneventful. We just sat around the hostel drinking beer, eating peanuts and playing pool. I always fancied myself as a pretty decent pool player but Brian kicked my butt at each game."

"March 10th was going to be my last full day in Istanbul. After all was said and done, I hadn't really accomplished much, and didn't see many sights. After North Africa and Italy, I wasn't really interested in seeing more ruins. They all pretty much looked the same. However, one thing that was becoming more commonplace, was the availability of drugs such as keef and hashish. They were also cheaper than dirt. After my introduction to the 'weed' in Tangier, I was now doing it as often as I could. We were all using it. It's like I said earlier, for us it wasn't some evil, illegal, mind-corroding substance. It was just organic, natural and a social mechanism. It belonged to us, and those of our generation, and it set us apart from our parents and the rest of the Establishment."

"I went to the Syrian Embassy to pick-up my Visa and then returned to the bazaar to pick up my suede clothing and a few other dry goods, such as warm socks, and a couple of shirts. The suede fit well, and looked good too. I was happy with the result, and at a price that couldn't be beat. While I was there, I bumped into three Vancouver girls that I had met the previous day. I invited them to join us at 8:00 P.M., for an evening on the town. They agreed to my invitation, and we all then met later at a 'control café'. After eating and drinking quite a bit, we took dolmus taksis across the river from the city center. In Turkish, the word taxi is spelled taksi and dolmus means to share."

"We continued drinking down by the river and nearly got into a fight with some Turkish guys. Colleen was one of the girls we were with, and slapped one of the Turks across the face, thinking that he had pinched her bum. We found out that the guy she slapped, hadn't touched her. It was one of his buddies. Anyway, it took a bit of doing, but everyone finally managed to calm down, and the Turks split. We then took dolmus taksis back to the city and dropped off the three female Vancouverites."

"By now we all had itches that needed scratching so, me and the guys headed down to 'Whore Alley' again and selected a couple of specimens to accompany us back to our rooms. And that's where I'll leave that my young friend," ended McGuire's dialogue.

On that note, McGuire simply rose from his chair and without saying a word, pointed toward the unfinished hole that Jesse was in the process of excavating. He then walked into the greenhouse while Jesse proceeded in a different direction toward his task. It was now late Tuesday afternoon and the job was still not completed. He was certain that he could finish digging the third hole today, now almost half accomplished. With one additional hole still to dig afterward, and one more clone to plant, Jesse felt that this would be his opportunity to complete the entire task himself. He was now confident of having memorized the potting sequence and was fairly certain that McGuire would allow him the opportunity of performing that chore.

Having now finally finished digging the third hole, Jesse informed McGuire. The process of bedding three more juvenile plants occurred shortly afterward, and as predicted, terminated that day's responsibilities. Not wishing to go directly home after securing the facility, McGuire informed Jesse that he was going to provide him an impromptu tour of the island. Driving southbound after leaving the greenhouse compound, they found themselves on Mouat Bay Road. Traffic was extremely light on the highway and the odd vehicle that did pass them going in the opposite direction, always received a friendly wave from McGuire. On this occasion, Jesse was quite surprised that McGuire had substituted his usual Italian driving-style to a more subdued version. He wasn't exactly adopting the habits of a senior citizen, but at least now Jesse could actually see objects and landmarks from out of the window. This was quite a departure from just the usual blur that Jesse had become accustomed to when traveling with McGuire at warp speed.

There was actually nothing special to observe during this excursion. However, the occasional glimpses of Georgia Strait to the west and the abundance of greenery surrounding them on all sides, placed them both in a relaxed demeanor. After several minutes of driving, Jesse slumped down on his seat and was beginning to doze.

However, before long Mouat Bay Road suddenly terminated, and the Ford F150 was obliged to reverse and return from whence it came. Jesse was not aware that Texada's road system was abbreviated, and that there wasn't a single artery that traversed the entire island. That was the end of the tour. Just as well, thought Jesse. He wasn't much of a nature buff anyway. Driving past Leonard Road, McGuire continued into Gillies Bay, and picked-up a few items for dinner at the Texada Farmer's Market. The sky above was now rapidly becoming overcast and before their supper had been fully consumed, a heavy rain began falling. The now habitual pre-recorded television programs and drinks after dinner, eventually ended, and both men retired for the night. When morning arrived, it was still raining outside so, McGuire allowed his houseguest additional sleep. There wouldn't be any digging outside today and they could always attend the greenhouse later in order to perform the required husbandry of the mature plants. After washing himself and before taking his breakfast, he removed a whole, frozen chicken from within the freezer and placed it in the sink to thaw. That would be their supper tonight.

With the luxury of some quiet time while Jesse slept, McGuire sat down at the kitchen table with a pot of tea and began leafing through several of the travel journals he had accumulated over the years. The travels of his first trip, now being reiterated to Jesse, were nearly at an end. Aside for the need to refresh himself with a few additional dates and some of the more obscure details, most of the events were still fresh in his mind. He frequently thought back to his many adventures and the memories they provided. However, he didn't actually realize how much more meaningful they actually were, until they were openly discussed rather than simply mused. As apprehensive as he first was before Jesse's arrival five days prior, he was now grateful for his presence, as it allowed him to relive his former experiences. Even if they were only through vocalization.

Without having noticed the passage of time, the pot of tea was now empty and Jesse could be heard stirring inside his bedroom. McGuire put the journals aside and stood in order to prepare breakfast for his young charge. It was near the noon-hour before breakfast had been finished and the resulting soiled crockery cleaned and returned to its rightful place of storage. The rain outside had abated somewhat but still falling. McGuire informed Jesse that they would be attending the greenhouse again, but merely for interior work. He thought it might be interesting for Jesse to observe the tending of the mature plants. If Jesse thought the inclement weather would curb McGuire's enthusiasm for fast driving, he was sorely mistaken.

Behaving as if the road was bare and dry, and visibility was unlimited, McGuire's right foot lay heavily on the F150's accelerator pedal, and remained there until they reached their destination.

Before long the pair were back at the greenhouse. Prior to entering the structure, McGuire inspected 'the fifteen young ladies' now planted in the ground outside. The fill was still nice and firm around the tops of the holes and the 'girls' appeared to be enjoying their first natural shower in their new surroundings. Satisfied with their status, he then unlocked the door to the greenhouse and entered. By now, Jesse had become accustomed to the strong herbal aroma arising from within and paid little attention to it.

Walking to the corner where the tools were stored, McGuire selected a chrome pair of Fiskars, non-stick pruning snips. He then motioned for Jesse to follow him as he systematically wandered from plant to plant carefully inspecting each. While doing so, he explained to Jesse what he was attempting to achieve, and what he needed to do in order to reach his goal. He explained that it was necessary to 'train' the plants in order to promote optimum grow and a full, healthy shape. Pruning the plants correctly would ensure that end result. First, one needed to remember that cannabis plants enjoy an abundance of sunlight and adequate airflow. Since the upper portion of the plant received more light than the bottom of the plant, it was there that you wanted to focus most of your attention. Also, if Jesse had noticed, the sides of the greenhouse were maneuverable, and could be raised, if desired, in order to allow a free-flow of air throughout.

You also needed to remember that cannabis plants should only be pruned while within their vegetative stage. Some people may think that a plant should just be allowed to grow without being manicured, but that is not the case with cannabis, or even any fruit bearing plant for that matter. Snipping useless shoots that draw essential nutrients away from developing buds, removing unnecessary or dead leaves from the plant, and trimming unproductive branches from the lower and middle portions of a plant, all assist in the growing process. However, it's important not to over-prune and never prune when the plant is flowering. And of course, don't forget to always water your plants immediately after pruning to stimulate growth and reduce shock. You also want to occasionally incorporate sufficient added nutrients to the soil when necessary. Regularly performing all of these tasks will ensure a healthier and more productive plant.

Jesse had absolutely no idea that the cultivation of cannabis was that involved, and he admired McGuire's knowledge and abilities on the process. For his part, McGuire continued visiting each plant in turn and performed whatever task was necessary. Jesse could see that McGuire's commitment to his crop was unequivocal and watched as he painstakingly performed every aspect of the inspection and pruning process for each of the multiple plants within the greenhouse. He was patient, thorough and meticulous in his work, and it was several hours before he was satisfied that his job was properly completed.

As soon as they arrived home, McGuire washed the thawed chicken and then seasoned it. He dropped it into a roasting pan and quickly popped it into the pre-heated oven. He instructed Jesse to peel some potatoes and carrots while he visited the bathroom to clean-off the marijuana residue accumulated during their visit to the greenhouse. With the peeled potatoes and carrots now simmering on the stove top and the chicken half cooked in the oven, the boys repaired to the living room with a glass of Golden Gage Plum wine. McGuire flipped through his record collection and eventually located the disc for which he searched. It was Texas blues artist Johnny Winter's first album entitled 'The Progressive Blues Experiment'. As the first track began playing at high volume, Jesse picked up the album's sleeve.

"Wow, that boy is really white!" remarked Jesse. The statement caught McGuire off-guard and caused him to laugh. "Yah, he sure is," replied McGuire still chuckling. "He's an albino and so is his brother Edgar, who's also a musician. That boy could sure play guitar. Unfortunately, he's dead now. He was heavy into heroin for many years but no one knows for sure what killed him. I've always liked his music, and I usually throw him on the turntable whenever I wanna rock."

By the time side two of the platter was half played, dinner was ready to be served. McGuire removed the chicken from the oven and placed it off to the side, allowing it to rest. He was pleased with the skin's crispiness and golden-brown color. Draining the vegetables, he placed a generous dollop of butter on the carrots as well as a sprinkling of salt. Adding another dollop of butter to the potatoes plus half a cup of milk, he began whipping them with a hand-held, electric beater. Reaching into the cupboard, he removed a platter for the chicken plus two plates for themselves. Lifting the chicken from the roasting pan, he cut it vertically down the center into two even halves. Placing those on each of the two plates, he then piled on a large helping of mashed potatoes and carrots was well.

"Hey Kid. Do you want me to make gravy with the juices?" questioned McGuire. Jesse merely shook his head to the negative as a response. "Okay, let's dig in then," ordered McGuire. Meanwhile the final track on Side B of the Johnny Winter album had just finished playing. The sound of the turntable arm automatically lifting, returning to its starting position, and then the click as it dropped into the holding cradle, announced the termination of that function.

After supper, the two men returned to the living room to resume their previous positions. McGuire turned-off the power to the stereo and asked Jesse if he wanted to watch T.V., or would he prefer further enlightenment with his travels. To his private approval, Jesse chose the latter.

"Okay, so the boys and me received a little light-hearted entertainment from our rented female friends during our last night in Istanbul, and then we packed it in shortly after midnight. By 8:00 A.M. on March 11th, Nick and I had left the hostel, and we were back on the highway heading southeast toward Ankara. This was the capital city of Turkey and one of its most populous. Phil and the other guys remained in Istanbul. Just like my experience in Algiers, the wind is howling and the rain is coming down in buckets. Luckily, the third car to pass by, stopped to pick us up. The driver took us about 80-kilometers down the highway and then dropped us at a gas station. It was coming down so hard that we didn't want to just stand there at the side of the road waiting for traffic to approach. Instead, we stood under cover at the gas station. After a while, I noticed a bus heading in our direction so, I ran out to the highway and flagged it down. As the driver opened the door, I was greeted by a beautiful young woman. I explained that Nick and I were heading to Ankara but we didn't have any money to pay for a ride. She turned to the driver and after speaking briefly, they waved us on board. We were now on cloud-nine, out of the crappy weather, sitting in a luxurious bus, accompanied by a beautiful girl, and getting a free 400-kilometer ride, all the way to Ankara. We had a great time, and chatted-up the hostess almost the entire trip. We discovered that all long-haul bus excursions carried one of these young, female hostesses to help keep the passengers happy and comfortable. It sure worked for us, and we didn't even have to pay!"

"We arrived in Ankara at approximately 7:30 P.M. and just our luck, it was pouring there as well. We shouted a heartfelt thanks to our hostess and driver, as we exited the bus and started looking for a place to stay. Ankara is another big city, and we had absolutely no idea where we were going."

"We grabbed a quick bite from one of the few street vendors brave enough to stand out in the weather, and then kept walking. Not wanting to be out there for too long, we noticed a building under construction and found our way inside. Just as we made ourselves comfortable and were ready to kip, in walked a couple of Turkish cops and they booted us out. We reluctantly picked-up our gear and then went back out into the weather. Checking over our shoulders to make sure the cops weren't following us, we soon found another building under construction and went inside. We made ourselves comfortable and this time the cops didn't find us. But wouldn't you know it, the night watchman did. Thankfully, he just gave us a quick look-over to make sure that we weren't wrecking anything, and then tipped his hat and left."

"By 6:30 A.M., we were up and out of there. Without first stopping for breakfast, we're back on the highway by 7:00 A.M. Our first ride was a relatively short one with a Turkish Air Force officer, but anything was better than nothing. He dropped us at a small village where we bought a loaf of bread and some sardines. After eating our makeshift sardine sandwiches, we headed back to the highway and started walking southbound, waiting for the next potential ride. We'd only walked a few hundred yards when it dawned on me that I may have left my traveler's cheques in Istanbul."

Hearing McGuire make this statement, Jesse's jaw dropped open. During the course of the recital, he had been attempting to personally place himself within the storyteller's positions and situations, and was imagining himself inside the story. He couldn't even dream of how he would have reacted had that been him in this circumstance. However, he knew it wouldn't be good.

"Yah, by the expression on your face, I can guess what you're thinking. You know when someone says that they felt their 'heart drop into their stomach?' That's exactly what it was like for me. In that immediate second, I had a sick feeling in my stomach and I felt the blood rush to my face. I was trying to concentrate and remember back to when we left Istanbul, but I was really confused. Nick kind of shook me out of my funk and told me to check inside my pack. I dumped everything from my backpack on to the side of the highway and started sifting through my stuff. The cheques weren't there. Then it dawned on me. I could see it clearly now. I had left the cheques inside the safe at the Tourist Hostel in Istanbul. That was 800 bucks, and we were 500-kilometers away from the hostel!"

"While I was still bent over shoving the stuff back into my pack, Nick kicked me hard in the ass and ran across the highway. He was starting to hitch-hike back to Istanbul. I was now slowly getting over the shock of losing my money and then realized, that I still had the receipt for the traveler's cheques in my other bag. I quickly reached inside and pulled-out the receipt. I then waved it toward Nick and told him to come back across the highway. I said not to worry, since when we arrived in Lebanon, all I needed to do, was go to the American Express office in Beirut and get a refund."

Receiving the good news that this updated version provided, made Jesse feel better. He wasn't even there at that time, but he was becoming so wrapped up in the story, it was almost as if he were.

"You know how you get when you've just dodged a major bullet, and start laughing like crazy?" questioned McGuire of Jesse. In response, Jesse hesitantly nodded in agreement.

"Well, that's what happened to us. Nick and me were laughing like hyenas at my stupidity, but also that I had been granted a reprieve from the Travel Gods. I didn't know it then, but I later discovered that I was completely wrong! We were laughing so hard and pushing each other back and forth that we nearly missed the next bus approaching us. When the driver pulled-over and opened the door, the conductor asked if we needed a ride. Naturally, our response was 'yes', but we didn't have any money. Once again there was a mini conference between the driver and his female conductor and in the end, they waved us on board. We couldn't believe our good luck! Unfortunately, there was a strong smell of body odor inside the bus, and it was filled with a bunch of people who seemed like they didn't know what a bar of soap was supposed to be used for."

"This ride took us all the way to the city of Adana which is near the southwest coast of Turkey, very close to the Syrian border. That lucky tooth of mine was still solid gold and doing its job, despite the fact that my traveler's cheques were back in Istanbul. However, in my mind that wasn't going to be a problem getting back that money. A couple of phone calls and I'd be good to go. It was very cold on that last stretch of highway and there were patches of snow on the side of the road all along the way to Adana. We wondered what we were getting ourselves into, but definitely thankful that we didn't have to slug it out on foot, or with short rides. When we finally arrived in Adana, I telephoned right away to the Tourist Hostel in Istanbul. I spoke with the old caretaker there, and confirmed that my cheques were still inside the safe."

"Unfortunately, he also replied that he couldn't return them to me without first seeing the receipt. No matter what I said, I couldn't change his mind. I was thoroughly pissed-off with the old goat! After our phone call, I wrote a letter to him that included the serial numbers of my cheques, and posted it from the Main Post Office. In the letter, I asked him nicely to mail the cheques to me, as soon as possible, to an address in Beirut that I provided him. I had no choice but to leave things like that for the moment. At least the long-distance phone call to Istanbul was cheap. It only cost me 55 cents. After that disappointment, we started looking for a place to spend the night. We bumped into a big, burly guy and asked him if he knew anywhere we could sleep. The guy turned out to be a German ex-pat wrestler. He said that we could stay at his mother's house and took us there directly, since it wasn't too far away. The place wasn't fancy, but the beds had nice thick mattresses and warm blankets. Our host was planning to visit Vancouver in the future so, I gave him my parent's home address if he wanted to write me a letter. The next morning his mother fried us up a nice greasy breakfast, and we then hit the road."

At that juncture, Jesse interrupted the story to ask a question. "If you left your traveler's cheques in Istanbul, how much money did you still have with you? Were you borrowing money from Nick?"

"Hell no, that's not my style," replied McGuire. "I pay my own way and I don't like borrowing money from anyone. However, if someone wants to give me money, or offers me a free meal or a ride, that's fine. I still had about $40.00

dollars in my pocket and I figured that was more than enough to last me until we arrived in Beirut."

"It was still pretty cold outside and I was glad that I bought those extra warm socks at the covered market in Istanbul. They were really coming in handy right now. Before too long, we managed to catch a ride in a commercial truck, and that took us about 30-kilometers out of Adana. The driver dropped us at a major intersection where a couple of highways joined. We could see that there were two other backpackers across the road, so we approached them. Both were from the U.K. and we introduced ourselves. Murray was from England while Jock was from Scotland. We all decided to stick together for a bit, and then started plodding down the highway toward the Syrian border. Hopefully, we wouldn't have to walk the entire way. After the limp I had developed a couple of weeks prior, I wasn't in a big hurry to repeat that."

"Not long after our departure, it started raining. The intensity of the flow increased with each passing minute and it made our lives miserable. It wasn't actually hailing, but it was still an icy rain that stung our skin when it landed. I'll tell you Kid, one of my least favorite things, is when cold water is dripping off my hair and running down the back of my neck, underneath my shirt collar. I'm shivering even now just thinking about it. I began to wonder if it was wise to team up with Murray and Jock, since it was hard enough for two guys to catch a ride, let alone four."

"Over the next couple of hours, we got pretty wet, but we also managed to catch three short rides on buses. It was enough to bring us to the Turkish/Syrian border and we passed through Customs without a problem at 6:00 P.M. They even fed us bread, cheese, olives, onions and tea at the check-point. That was sweet!"

"After leaving the Customs Office, we approached a truck driver in the parking lot who agreed to give us a ride. Unfortunately, we had to ride in the back of the truck, and although it was covered, it was still freezing cold. During the entire 40-kilometer trip, the four of us huddled together under Jock's sleeping bag in order to stay warm. I can't remember the name of the town where we ended up, but we quickly managed to find an apartment complex under construction, and that became our lodging for the night. We left the unfinished building early the next morning, but were having difficulty finding the highway. Practically no one spoke English and those that did manage to provide us some limited information, all pointed in different directions. We finally managed to 'bag' a teenaged kid driving a van, and convinced him to take us to the highway. As soon as we arrived, he was holding out his hand for payment, but we all answered that we were broke. He was seriously pissed-off when we didn't give him any money, and offered us a few choice words in Arabic as he drove off. None of us understood what he said, but we were pretty sure that he wasn't wishing us a pleasant day."

Jesse started laughing and then said, "Why didn't you give him any money for his time and gas? The four of you could probably have scraped a couple of bucks together." McGuire now also laughed at Jesse's naiveté and responded, "That's not how it works, Kid. We were backpackers, vagabonds, transients, voyageurs, men of the road. We don't pay. Have you ever heard of the Indian Sadhus?" When Jesse shook his head to the negative, McGuire continued.

"Sadhus are an order of wandering holy men. They walk along the roads throughout India, begging for food with their tin dinner pails, in return for providing their benefactors blessings and prayers. They grow their hair and beards long and barely wear any clothes. They have renounced all the trappings of traditional society and spend their lives attempting to achieve enlightenment. In India, they are respected and even revered. Everyone there makes it a point of helping a Sadhu, and they never ever receive something monetarily in return. That's just how it is."

"Now, we aren't holy men but we also wander the highways looking for adventure and enlightenment. We grow our hair and beards long, and if it wasn't so damn cold in some of the places we visit, perhaps we too would walk around half naked. I'm not saying that we're exactly like Sadhus, but we have a lot of similarities to them. The bottom line is, that we don't pay for anything, unless it is absolutely necessary. And even then, we never want to pay full price. Understand, Kid?"

Jesse nodded his head in acknowledgement to McGuire's explanation and that ended the matter.

"Anyway, now we're finally back on the highway and catching a few short rides, but not many. Like they always say, 'Two's company but four's a crowd.' McGuire could see that Jesse was about to correct him and say that 'three's a crowd,' was the actual saying, but the look he threw him caused him to stop short. Jesse quickly understood that the incorrect adage was purposely made."

"By the early afternoon, we were only about 60-kilometers further down the highway from where we started. Not great. Eventually, a large truck stopped, but the driver said he would only take two of us. We flipped a coin and Nick and I won the toss. We climbed into the truck leaving Murray and Jock standing at the side of the road. The driver gave us each a can of sardines and some tea. About an hour later we saw a big, black 1956 Caddy pulling up beside us and then passing. Murray and Jock were in the back seat thumbing their noses at us and laughing. Me and Nick also started laughing and convinced the truck driver to catch up with the Caddy, and pull it over. Well, that's exactly what he did, and soon Nick and I were also in the back of the limo with Murray and Jock. The driver was a rich Syrian businessman, en route to Damascus. He was a real friendly guy and we all had a million laughs, until the car started acting funny. We managed to reach the next village and then the driver stopped. It looked like the Caddy was going to need some repairs."

"The driver eventually located the town's mechanic and negotiated the cost of repairs with him. Once that was done, he treated us all to a fabulous lamb and sausage 'scoff' at a local café. It was going to take at least a couple of hours to fix the car, so us four guys started walking through the village. The place was kind of grubby and smelled like cow piss. There were stray dogs skulking around the corners and they took off as soon as we came close to them. We're walking around, but it seems like all the inhabitants have slipped away. As we go past some of the houses, we can see eyeballs peeking through the curtains but we get the impression that no one wants to be seen anywhere near us. It made us feel like a pack of Lepers. As we came walking out of a narrow alleyway that was meant to be a main street, we entered into a large clearing. This must have been their dump, since there were big piles of garbage everywhere, and it stunk to high heaven."

"We figured that this was the end of the road and were about to turn back, when what looked like about 75 to 100 boys of various ages suddenly appeared from behind the piles of garbage and started throwing rocks at us. Now, I'm not just saying a couple of small pebbles here and there, but large stones and chunks of broken concrete. We couldn't believe it! The little suckers were yelling at the tops of their lungs and pelting us like there was no tomorrow. I felt like one of those wooden bowling pins in the Midway 'games of chance' at the PNE. But rather than baseballs, being thrown by hopeful contestants trying to knock me down in order to win a Kewpie Doll for their girlfriends, those little punks with rocks, were really trying to hurt us."

"Over the last couple of months, I've been 'stoned' plenty of times by smoking keef and hashish. But this was taking that term way beyond where I wanted it to go. The four of us immediately crouched down and covered our heads and bodies, as best we could, with our hands and arms. We then made a rapid retreat back into the alley from where we came, and eventually to where the Caddy was being repaired. Thankfully, the small army of juvenile terrorists didn't chase after us, but we all still suffered a few bumps and bruises on our bodies where the projectiles made contact. Luckily, we didn't suffer any serious injuries. When we later informed our driver of the episode, he simply shrugged his shoulders and stated that people in these smaller villages don't like strangers, especially foreign ones. After receiving this casual explanation from the Syrian businessman, we all just looked at each other in disbelief, and shook our heads."

"It took a bit longer than expected to fix the Caddy, so it wasn't until 6:00 P.M. before we were back on the highway. Trying to make up for lost time, the Syrian was now driving like a bat out of hell. Do you remember when I described some of those crappy drivers in Italy? Well, this guy made them look like experts. He was the absolute worst driver that I had ever ridden with, and it was a miracle that we didn't end up in a ditch somewhere. Anyway, I guess heaven wasn't quite ready for us yet, so we managed to make it to Damascus by 10:00 P.M. After that 'ride from hell', we needed something to eat and drink in order to calm our nerves. There were a lot of places to choose from, but we went to one that looked cheap. After the usual sausages and rice, and a few beers, Nick and I shook hands with Murray and Jock and wished them well. They were continuing on to Jordan, while we were heading for Lebanon. It was now 11:00 P.M., and normally, I didn't 'hitch' at night. This was for safety reasons as well as minimal traffic on the roads in the later hours. However, Nick and I weren't tired at the moment, so we thought we would give it a go. Well, go figure, within a few minutes we already caught our first ride. It was short, but it was a start. Before we knew it, we had received several short rides, and they were all getting us closer to our goal. Our last ride brought us to a police check-point. The driver got out of the car and spoke with the cops. One of them started talking on his police radio. We thought that they were going to pull a fast one on us, but less than 10-minutes later, a large truck pulled up and the driver told us to get in. He's heading in the general direction of Beirut. Well, before he can tell us twice, we're jumping into his truck."

"This ride took us within 6-kilometers of Beirut and a free taxi ride after that, took us the rest of the way in. We were now in the 'Paris of the Middle East', and dead tired. It was the banking and business center of the Maghreb, and a very cosmopolitan city in its own right. We didn't know what time it was, but we did know that we sure needed to get some sleep. It seemed like every city we visited was under construction, and Beirut was no exception. Walking only a few minutes, we soon came across a semi-completed shopping center. Easily finding our way inside, and after chasing away a pack of curious rats, we bedded down for what was left of the night. Our nocturnal bliss was short-lived when we were rudely rousted by the watchman at 6:00 A.M. Half awake and with brains screaming for more sleep, we stumbled down the main street like a couple of drunks. We eventually made our way into the city center. Inside the main square, people were already milling about, heading to their jobs, or wherever else they might be going. Nick and I sat on the steps of a then silent cabaret and watched the steady parade of cute fluff sashaying by."

"We looked at each other approvingly, and I gave him a quick wink. I immediately surmised that I was going to be enjoying my visit to Beirut."

"We sat on those steps people-watching until 8:00 A.M. and then walked to the Canadian Embassy, where we obtained documents confirming that we were Christians. We thought that was kind of silly, but apparently, we needed them here as well as in Jordan, if we decide to go there later. After leaving the Embassy, we went to the Main Post Office where I then mailed home the suede clothes that were made for me in Istanbul. It was going to take about a month to arrive since it was traveling by ship, but the whole shebang only cost me $2.10 to mail. Such a deal! Next, we checked into the Youth Hostel and took a cold shower. Drinking a beer for breakfast, we now had our 'second wind', so we headed back into the city. We trucked around the city all day on foot, and were essentially running on adrenalin. We had only 2-hours sleep the previous night but that didn't matter. We figured we would sleep enough, when we were dead. In the mean-time, there was just too much to see and do. We noticed that there were street vendors selling very reasonably priced roasted chicken all over the city. Nick and I both grabbed one for dinner. It was delicious and went down even better with a few gulps from a bottle of Vodka that Nick had purchased earlier."

"After eating our chickens and drinking about half the bottle of Vodka, all was right with the world. By 7:00 P.M., we were beginning to flag and started walking back toward the hostel. On the way, just outside a fancy restaurant, we spotted a big, bodybuilder type guy wearing a suit, and a beautiful, well-dressed woman just about to enter. Nick made a bit of a wolf whistle that was overheard by the hulk. He came storming over to us, got in our faces, and just glared. I don't consider myself a small guy but this brute made me look like a scarecrow. He was huge! He didn't say a word, but just kept staring at Nick. Although we were both a bit tipsy, we were smart enough to know that it was best we kept our mouths shut. I think the guy was just waiting for us to beak-off, and then he was going to wipe the floor with us. Eventually, seeing that we weren't going to challenge him, he made an about face and returned to his girlfriend. We learned a valuable lesson that night and came to the realization that this part of the world was much different than what we were used to. We weren't going to change who we were, but in certain circumstances, we would be minding our P's & Q's a bit more. The next morning at 10:00 A.M., we were rudely awoken by the eardrum shattering sound of three loud bells announcing that everyone needed to vacate the hostel. Breakfast was a sardine sandwich and a glass of chocolate milk. We then headed over to the Canadian Embassy."

"I wanted to check on the status of my traveler's cheques and Nick needed to renew his passport. Nick got his paperwork started but I drew a blank. There was no one there that spoke Turkish, so they wouldn't be able to contact the necessary people required to assist me. What a drag. I felt that I was still okay for cash, but that wasn't going to last forever."

"The next few days were repeats. Nick and I would leave the hostel first thing in the morning and go to the Canadian Embassy. We'd receive the same responses to our questions. Nick's passport wasn't ready yet and they still didn't have anyone who spoke Turkish that could help me. One clerk informed me, that if the cheques were in a bank account, then it would be easy to have them sent to Beirut. However, because they were inside the safe at the Tourist Hostel, that created a problem. Disappointed again, we went into the city center and started checking out the usual haunts. Our diet for breakfast, lunch and supper was pretty consistent. Roast chicken, sardine sandwiches, chocolate milk and beer. Daylight hours were usually spent in the main square while evenings were drinking, smoking dope and going to the movies."

McGuire now took a moment to review his battered travel journal. Situations and circumstances were beginning to blend together so, he wanted to refresh is memory in order to keep events accurate. In a moment, he was satisfied.

"So, now it's March 17th and if you were Irish, you would know that's St. Patrick's Day. I can't forget that date, since my middle name is Patrick. By now, we're getting to know our neck of the woods pretty good and the regulars are recognizing us as well. There's a Spanish guy named Jack that is regularly selling his paintings in the main square. I talk to him once in a while and discover that he makes pretty good coin. He's never going to become a millionaire, but he does alright. That afternoon, me and Nick meet a couple of American sailors and they offer to buy us a beer. The 7th Fleet is in the harbor. Well Kid, I guess you know me by now. One beer leads to another and then a glass of wine. A glass of wine becomes a bottle of wine and then several bottles of wine. By midnight, we're still with the sailors, and we're bombed."

"As we shut down the bar, the mariners tell us that they want to smoke some hash. Well, it just so happened that we had some with us and peeled off into a dark alley in order to light-up. Unfortunately, it was so dark in the alley that we couldn't see what we were doing."

"We decided to take the sea-dogs back to the hostel and we all snuck-in through a ground floor window. While we're smoking and enjoying ourselves, one of the other travelers in our room rats us out to the manager. However, before he can come down to give us the boot, we hustle the Yanks back out of the window and send them on their way. The manager never did come down. The next morning we're back on the street by 10:00 A.M., and go straight back to the Canadian Embassy. Nick was in luck and received his new passport. Unfortunately, my situation hadn't changed. No traveler's cheques and no Turkish speaker to assist me. After leaving the embassy, we decided to change things up a bit and made our way out to the American University of Beirut, in the Hamra District, rather than going downtown. The university was right on the beach and we felt like going for a swim. Before hitting the beach, we stopped in at the school's cafeteria and ate lunch. I had a deluxe burger with fries and a strawberry milkshake. It was a little taste of home and a nice change from the usual 'growlies' that we were getting in this part of the world."

"The breakers on the beach were monstrous and they threw us about as if we were made of cork. Tired of getting the crap kicked out of us, at one point,

Nick and I swam past the surf and into the open water. What a mistake! We both got caught in a 'riptide' and it was taking us further out to sea. We were stroking like crazy but couldn't seem to make it back to shore. We were becoming more tired by the minute and eventually, I thought we were goners. What a way to end my adventure. But like I said earlier, heaven wasn't quite ready for us yet, and after a massive struggle, we finally made it back to the shore. We were both exhausted and I had swallowed so much salt water that I nearly puked. We decided to stay out of the water for the rest of the afternoon and spent our remaining time there just sun-tanning."

"That night we slept the slumber of the dead. Earlier in the day we came pretty close to making it permanent. The next morning started off with thunder, lightning, torrential rain and even some hail. We wandered around searching for places to stay dry. One game that we always enjoyed playing, was conning the local merchants for a free sample. If someone was selling juice from a large fountain-like decanter, we would stand there gazing at the container in amazement and telling the vendor how impressive it was. He would tell us the cost of the juice but we would always answer that we didn't have any money. We would just keep staring at the dispenser with a wondrous expression until the vendor would eventually feel sorry for us and give us a free drink. This trick worked quite often and not just for juice. It also worked for fruits, nuts, vegetables and more. Remember the Sadhus of India? Well, we were the Sadhus of Beirut."

Jesse caught this clever comparison to a previous reference and laughed. You had to hand it to McGuire. He possessed a wicked sense of humor, and he was a heck of a storyteller.

"One thing that was really getting on my nerves was that wherever we went, or if we sat down even for a minute, it wouldn't take long before we attracted an audience. It was usually older Arabic men wearing djellabas and sandals. The djellaba was their traditional garb that always reminded me of men wearing dresses. Anyway, these old farts would just hang around and stare at us. I don't know what they were expecting to see, but they just wouldn't leave. We tried to ignore them, but after a while we couldn't hack them anymore, and we would shout, 'YALLA'. This meant 'get lost' or 'take a hike' in Arabic, and it usually sent them on their way."

"I also noticed that some of the old codgers wouldn't trim the nail of their pinky finger on the right hand. When I asked one the younger locals about this, the answer was, that they were 'true believers'. Not only did they grow the nail of their baby finger, they also wore a form of underpants that sagged in the crotch almost to their knees. The logic for these two unusual things was that they believed the next Muslim prophet was going to be born of a man. Having a saggy crotch in their 'gonchies' meant that the baby, when born, would be caught safely in the underwear of the 'father'. The umbilical cord could then be cut with its parent's long, pinky fingernail. I've heard some strange stories before, but this one was probably the weirdest. I don't know what those guys were smoking to dream up something like that, but I sure wasn't going to try any of it."

"Seeing our long hair, some of these guys also thought that we were homosexuals, and they would try to proposition us. Whenever they tried to grab my hand or touch my face, I would go ballistic and more than once, I squared off with those suckers ready to beat the crap out of them. Whenever I became this angry, I tried to quickly settle down. I always kept a small piece of hash in my wallet, and all I needed was to get into a fight, and have the cops attend. If they searched me and found the dope, I'd be busted."

"It rained for the next couple of days, keeping us close to the hostel. We just hung-around, shooting the breeze, reading or catching a few extra winks. I also managed to catch up on some letter writing. By now, I was starting to become worried, since I was getting short of funds. If I didn't get those traveler's cheques soon, I'd be in dire straits. The last thing I wanted to do was call home and ask my parents to send me money."

"By March 21st, I'm getting pretty desperate. One of the guys at the hostel tells me that the local hospitals pay donors for their blood. That sounds promising so, right after breakfast, I boogie down to the Khoury Hospital on Makhoul Street. Unfortunately, they have enough blood. They told me to come back in a few days. The Najjar Hospital on Mohammad Abdul Baki Street is nearby, but I got the same response there. Come back in a few days. I met up with Jack downtown, later in the afternoon, and we started drinking. Before we realize, it's late and the hostel is locked for the night. As we try to climb into one of the ground-floor windows, the manager catches us. I guess he's looking forward to see the back me, so he tells me to leave, and says I'm barred from returning. With all the stress I'm under, I blow my cool and tear a strip off him. Well, that sealed the deal. There was no way that I would ever be allowed back there after that. I took Jack to the construction site where Nick and I slept our first night in Beirut, and we bedded down. At 6:00 A.M. the next morning, the watchman gives us the boot and we're back out into the rain. Jack informs me that he and Nick are leaving Beirut today and heading for Damascus. Now, I'm really feeling miserable. It's raining, I'm homeless, I'm nearly flat broke and now, two of my best buddies are leaving town."

"I headed into the city center and waited in front of the American Café until it opened. I checked my pockets and pulled-out 4 Lebanese Pounds. That was only about $1.30, and all the money I had to my name. When the café opened, I drank a cup of tea and ate a couple of pastries. By the time I finished, the rain had stopped. At least that was something. I hung around the main square all day just shuffling around. Later on, I bumped into Brian who was another traveler. He owned a motor scooter and also had a room in town. I asked if I could spend the night at his place. Brian spoke to his landlord and for 1 Pound, he let me sleep on the couch in the hallway. Now, I only had a few pennies left to my name. The next morning, Brian and I were out of his place and off to the Khoury Hospital on his scooter. At the hospital, the nurse pricked my finger and tested my blood. She told me that I had A-Type blood and they didn't need any at the moment. Seeing the dejected look on my face, she suggested that I go to Hospital Wardieh near the harbor. Thankfully, they needed A-Type blood there, and they drained 350 ML from my veins. Brian donated the same amount, and for our vital fluid, we both received 30-Lebanese Pounds. I was finally back in the black!"

"When we returned to the scooter it wouldn't start, so we took turns pushing it. Kid, I don't know if you've ever given blood before but you're supposed to rest afterward."

"The scooter wasn't very heavy, but it wasn't long before Brian and I were getting really light-headed and weak. We had to rest at the side of the road several times before we finally reached Brian's place. With our new found wealth, we decided to splurge and treated ourselves to a fabulous bowl of tomato soup and a deluxe burger. After supper, we walked to the Groovy Owl record shop on Sadat Street. It was near the American University and the proprietor was a cool dude. I can't exactly remember his name but I think it was Ibrahim or something like that. He had a good selection of music and if you asked him to play something over the P.A. System, he always obliged. He also spoke Turkish and agreed to help me retrieve my traveler's checks. After he closed shop at 8:00 P.M., we all headed to the PTT. The PTT was that era's version of what today is an Internet Café. You could make a phone call to anywhere in the world for a designated fee. Ibrahim tries to place the call to Istanbul but he's informed that bad weather between Beirut and Istanbul is interfering with the lines. We're told to come back tomorrow."

"I thanked Ibrahim for his attempt, and said goodnight. Brian and I then headed to the docks because I heard that the S.S. Independence was in the harbor. It was like seeing an old friend again when I spotted her tied-up to the wharf. She was as elegant as ever. Security wouldn't let us get anywhere near her, but I recognized a couple of her crew members standing on the pier, and we chatted for a while. Brian and I left the docks and headed back to his room. After a bit of tinkering, he managed to start the scooter and then drove me out to the construction site where I bedded down. The watchman woke me at 5:30 A.M. the next morning, but he was gentle with me. I had taken his picture the day before and he liked that. I made my way back to the American Café and waited out front until it opened. After a light breakfast, I wanted to go back to the Canadian Embassy to check on my traveler's cheques situation. The café manager allowed me to store my bags in their broom closet. As I'm walking through the main square, I spotted a couple of long-haired guys sitting at the base of the Martyr's Monument. They've got a bunch of pencil sketches spread out on the pavement, with an inverted cap containing a few Pounds, in various denominations. I walked over to them and bent their ears."

"It turned out that they were a couple of German brothers traveling the region and selling their sketches to augment their cash flow. I examined their sketches, and although they weren't bad, I was certain that I could do much better. After speaking to them for a few minutes, I started back toward the Embassy. But in the mean-time, a seed had been planted. The closer I got to the embassy, the greater that seed began to germinate."

"After being told once again at the embassy that my cheques had not arrived, that germinating seed was now maturing into a full-grown plant. I had made my decision. Obviously, it was going to take a while for my traveler's cheques to arrive. Maybe they would never arrive. Meanwhile, I needed to support myself. I was no Vincent Van Gough, but I knew that I had a certain amount of artistic talent. After seeing what the German brothers were offering, and making decent money doing so, that was going to be my deliverance. I was going to create some of my own works of art, and sell them to the public.

Naturally, I was still going to continue working on getting back my 800 bucks, but at least for the moment, I wasn't going to starve."

"As I made my way back into the city, I stopped at a stationary store and picked-up a few supplies. I purchased a block of plain white sketch paper, a package of felt pens in various colors, an 8-color, tin palate of water paints and two fine tipped paint brushes. I took all of these items to the Groovy Owl and sat in the corner of the shop commencing to put together a few drawings. Ibrahim had the latest Beatles album playing over the P.A. System and I was really starting to get into it. I asked him what it was called, and he answered, 'Rubber Soul'. After listening to a few more tracks, I decided that one day, I should buy a copy. After a while, Brian showed up at the record store and told me that he had sold his scooter for 100 bucks. There goes my ride to the construction site. While his girlfriend watches the shop, Ibrahim and I go back to the PTT. Unfortunately, the telephone lines were still down to Istanbul, so that put a quick end to that."

"I left the record shop alone and walked back into the city center. I met a couple of Americans and a Frenchman there, and we all go for the usual chicken with rice supper. After eating, we decided to go to a movie house and for 38 cents, watched 'The Sound of Music'. What a great flick! I really enjoyed it and told myself that I wanted to see it again before I eventually left town. After grabbing my bags from the restaurant, I went back to the construction site to catch some welcomed sleep."

"Speaking of sleep," announced McGuire. "What do you say we hit the sack, Kid? It's getting pretty late."

Jesse agreed that it was time for bed, but wanted to ask one last question. "Bob, weren't you worried or scared to be sleeping alone at that construction site every night? Anything could have happened to you."

McGuire scoffed at that notion and responded, "What's there to be afraid of, Kid. You do the best you can to stay safe and protect yourself using your wits and intelligence. But if something happens, then it was meant to be. Anyway, I wasn't worried. I had my lucky tooth to protect me."

With that final remark, both men started laughing and continued doing so as they made their way toward their respective bedrooms. Thursday morning broke brightly and the previous day's precipitation was only evidenced by a few placid puddles of rain water, filling the occasional divot, on the otherwise hard-packed earth of the McGuire driveway. Jesse had now been on Texada

Island for six days, with his recently inherited mentor. As he washed the sleep from his eyes in the bathroom sink, he wondered how much longer he would be staying. There was only one more hole to dig at the greenhouse compound, and surely that would be finished today. Regardless of how long it took for that noon-hour's histrionic conversation, his commitment to labor for Mr. McGuire, will have been completed. What then?

As Jesse predicted, his final hole was completely dug soon after their arrival at the greenhouse. The previous day's rain had loosened the ground somewhat, allowing the crater to be excavated quicker than the previous fifteen. Although, the dousing had made the job a bit messier. Afterward, Jesse just sat on the damp ground for a while. He finally stood and walked to the greenhouse in order to report his news to McGuire. There were still things to do inside, so McGuire informed Jesse that he would be a while before the final transplant could be affected. Testing Jesse's recollect, he then instructed him to gather together the single plant sitting alone in the corner of the greenhouse, plus the required nutrients, and then load all onto the cart. McGuire then directed Jesse to clean-up behind him, as he visited each mature plant and performed his pruning chores. With that task eventually completed, and with Jesse then pushing the loaded cart outside, he asked McGuire if he could plant the final clone. Secretly, McGuire had hoped that Jesse would ask this boon of him. It meant that he wasn't just there killing time or working because he was told to do so. It indicated that he was developing a genuine interest in what they were attempting to accomplish, and the acquiring of an additional skill-set and knowledge. Naturally, McGuire agreed to Jesse's request and watchfully stood-by as the novice horticulturalist carefully and methodically embedded the sixteenth cannabis plant into its new home.

The pace of Jesse's efforts had been slower than that of his teacher. However, the approving regard on McGuire's face and firm pat he placed on his shoulder once the job was finished, told him that he had done well. Aside from getting good grades in school, Jesse felt that he hadn't accomplished much in his life. His parents had always seen to his needs. Digging those sixteen holes and now planting a perennial, that would eventually benefit others, gave him a sense of accomplishment and pride. It was now approaching the noon-hour and time of their daily ritual. Seated in their rightful places and beginning to devour their cheddar cheese and sweet pickle sandwiches, the oration once again commenced.

"Considering that I was lying on a concrete floor, I was sleeping quite soundly at the construction site. That was until 5:15 A.M. when the watchman woke me. He's waking me earlier each morning and I'm beginning to wonder if this is his way of discouraging me from returning every night. If that's his plan, it's working. I need to get more sleep, and it just isn't happening there. My day began with the usual routine. I went into the city center and waited for the American Restaurant to open. Once it did, I ate a few pastries, drank a cup of tea, and left my luggage there. However, I took my art supplies with me in a plastic bag. I boogied up to the Canadian Embassy. Still no traveler's cheques. I can tell that they're tired of seeing me there every day, but I'm not letting them off the hook. Eight-hundred bucks is a lot of loot, and I'm not just going to let that slide."

"I'm fuming as I leave the embassy, so this motivates me to go down to the PTT. I sent a scathing 21-word telegram to the old caretaker at the Tourist Hostel in Istanbul. I demanded that he immediately post my traveler's cheques to me in Beirut. By now, I'm really fed-up with this cat and mouse game, and I can see now that I'm not going to receive help from anyone else. After leaving the PTT, I went directly to the Groovy Owl. I knew that speaking with Ibrahim and listening to some music would calm me down. That's exactly what happened. After a few minutes, I parked myself in a corner of the shop and began creating some pictures."

"Before I knew it, I was attracting a group of curious onlookers watching me work. Three of the group were young prostitutes, and were they ever cute! What would I have done to have those traveler's cheques with me, right there and then! I would have treated those girls to a delicious 'sausage lunch'. He rose his eye brows salaciously as he uttered the word 'sausage', in order to emphasize his actual meaning. Jesse caught the anatomical reference of the term, and what it actually signified. He just laughed and shook his head."

"They watched me for about an hour and then split. They must have provided me some inspiration, since by supper time, I had completed four drawings. Overall, for a first attempt, the drawings were pretty good, if I do say so myself. As I'm leaving the Groovy Owl, I bumped into Steve and Brian. Steve is an American traveler from Anacortes, Washington that we recently met."

"After eating our usual chicken and rice 'grunts' in a little dive downtown, we all headed over to Steve's place that was nearby. His single room was on the sixth floor of a run-down apartment building, and really shabby. But it was still better than the construction site where I had been hanging my hat. After smoking, drinking and shooting the breeze for a while, Steve agreed to let me crash there for the night. I ran back to the American Restaurant to grab my stuff, and was back at Steve's within a few minutes. I couldn't believe it, I slept until 10:00 A.M. the next morning. What a luxury, and quite a change from the early rousting that I was receiving daily at the construction site, from the overzealous night watchman."

"Steve introduced us to Tony's Bar. It was another local establishment sympathetic to people like us. He also introduced me to Preludin. This was a weight loss supplement that came in capsule form. It not only curbed your appetite, it also kept you going when your energy was flagging. And if you took enough of them, you could also achieve a decent 'high'. It wasn't as good as keef or hash but it was alright in a pinch. After all, the Beatles used it. It must be okay. If you asked me today if I were still interested in 'dropping' those pills, I would tell you 'No'. But back then I was 20-years old and just looking for a good time. How I got to the good time wasn't as important as just getting there."

"March 27th was the birthday of my older brother, Mike, back in Vancouver. It was also the first day I decided to expose my art work to Beirut's buying public. I had now created five drawings. A couple were landscapes while the others were merely abstracts, colorfully illustrated. To help sell the pictures, I painted the top flap of my backpack white and printed 'CANADA' on it, in large block letters. Everyone loves Canadians. I also drew a map of the region and ran a red line through all of the spots I had visited, tracing the termination point in Beirut. Finally, with the assistance of a few new-found friends, I produced a sign stating that I was a Canadian, traveling through the territory, seeking donations. This sign was written in English, French and Arabic. Packing up the entirety of these items, I then headed into the main square and established myself in an area heavily trafficked by pedestrians."

"I laid out my art work, my signage, as well as my overturned Tam O'Shanter, and then waited to see how it would be received. Beyond my wildest expectations, I earned 33 Lebanese Pounds within three and a half hours. That converted to $11.00 American dollars. I was over the moon with the results, and deemed this day to be a complete success. I had received donations from children, cute young girls wearing Hijabs, a few middle-aged shoppers, and even a few poorly dressed older men in their Djellabas. And to beat it all, only one contributor took a drawing. I still had four left! That night, I celebrated with hamburgers and fries, a half dozen large beers and a hot shower at a public bath house. You can't even imagine how I felt after that shower. Living my current life-style meant that my personal hygiene was not always where I wanted it to be. I often went days without a decent scrub. I couldn't always brush my teeth, and even wore the same clothes for several days, in a row. Using a toilet may occur in a restaurant, a public convenience or even occasionally at the side of a road or in a dimly lit alley. That's just the way it was. And believe it or not, you actually got used to that form of living."

"We got pretty blitzed that night. As we were leaving the bar, this weird local guy that I had seen around before, gets right in my face. He's got this wild look in his eyes like a madman. He suddenly stretches out his right arm in a Fascist salute, and yells 'Heil Hitler!' I'm debating whether or not to deck him, but then decide to just push him out of my way. Well, he suddenly goes ballistic and pulls an unprotected syringe out of his jacket pocket. He then starts jabbing it toward me in a threatening manner, shouting something in Arabic. By now, I'm really pissed-off and I want to beat the shit out of him. Instead, Steve grabbed my arm and rushed me away from there. The weirdo didn't follow us. It took me a few minutes to calm down but when I did, I realized that the goof was just some nut, and I just happened to be in the wrong place at the right time."

"I've now been sleeping at Steve's place, and could finally say goodbye to the construction site. I tried a different spot to 'expo' the following day, but it proved to be unproductive. I then returned to my original location, and within a couple of hours, I was the proud recipient of an additional 30 Lebanese Pounds. At this rate, I would not only be earning enough money for my daily expenses, I would actually be able to save a couple of bucks. It was then that I came to the conclusion, that this was exactly what I might have to do. I was still checking-in daily at the Canadian Embassy but receiving the same negative responses. If push came to shove, I would have to make my own way back to Istanbul, and retrieve my traveler's cheques in person. From then on, I made it a point to set aside a couple of Pounds from my total daily earnings, in case of that eventuality."

"I'm having mixed feelings about using the Preludin caps so frequently. Sometimes, I feel alright but other times, I'm edgy and depressed. One night I couldn't sleep at all. Since backpackers are constantly entering and leaving Beirut, I got to meet a lot of people. Some I liked more than others. When it came to drugs, I always leaned toward the softer ones. But recently I met a group of Frenchmen that were intravenous drug users. Watching them crank that poison directly into their veins, is where I draw the line. I'm a pretty liberated fellow, but just looking at those guys, I could tell that they were all 'dead men walking'. I'm not judging anyone and people can do whatever they want. I just see that activity as being bad news."

"By the end of March, I still didn't have my traveler's cheques, and I'm still receiving the same feeble excuses at the Canadian Embassy. I even tried calling Istanbul myself, but the telephone lines are still out of commission. Can you believe it? I'm thinking, how does anyone get anything done in this part of the world, with that lack of service?"

"A few times now, I've left my bags at Tony's Bar for safe keeping. One night it was too late to grab my stuff, so I waited until the next morning. When I finally picked them up, I could tell that someone had been inside the bags. When I checked, I discovered that my new warm socks and a couple of shirts that I purchased at the covered bazaar in Istanbul, were missing. Some bastard had stolen them! I asked the manager if he had seen anyone near my bags last night, but he answered that he didn't. I was still 'seeing red', when I met Steve and Brian. We then jumped into a taxi that took us to a full-service laundromat just outside of the city. For 75-cents, I had all of my dirty clothes cleaned and folded. Unfortunately, the color from my socks 'ran', and my entire wardrobe became a soothing shade of light blue. Luckily, blue is one of my favorite colors."

With that last comment, in what would otherwise have been a disastrous moment for most people, Jesse burst out laughing and couldn't stop for a considerable time. McGuire patiently waited until he eventually stopped, and then continued with his dialogue. "Although, my traveler's cheques still haven't appeared by March 31st, I'm feeling pretty good with my overall circumstance. My drawings are selling well, and at the end of every day, I've got money in my pocket. Even after my daily expenditures of food, drugs, liquor, movies and art supplies, I'm still able to tuck a couple of pounds into my 'savings pocket'. Steve's landlord wouldn't let me stay in his room any longer without paying rent, so now I'm back to sleeping wherever I can. This includes rooftops, building alcoves, construction sites and even occasionally on the beach."

"One night at my construction site flop, a big fat rat scurried across my sleeping bag and buzzed right past my head. It scared the crap out of me and I couldn't fall back asleep, thinking he was going to come back and chew on the end of my nose, or something. Near midnight the wind suddenly picked up and started blowing sand and debris all over me. I tried hiding inside my sleeping bag but it became so stuffy, I could hardly breath. When I finally stuck my head back out, the wind was so hot and choked with sand, that breathing outside of the sleeping bag was also difficult. What a mess! I finally left the site and walked to the beach to hear the surf crashing upon the shore. That thundering sound always seemed to relax me. Eventually, I made my way back into the city and waited for the day to begin in earnest."

"While I'm creating new drawings, my buddies, Brian and Steve are manning the expo site. So, even when I'm not there, I was still earning money. In fact, business became so lucrative that I was able to take a day-off whenever I wanted. How's that for enterprising? Because I always displayed my wares in the same spot near the Martyr's Monument, we soon became minor celebrities. Groups of girls from all over started making donations, and asking if they could snap pictures with us. Naturally, we always agreed. Though, it got me to thinking."

"Maybe they wanted photos of us because they considered us 'rock stars' or maybe they thought we were 'circus freaks'. Either way, I didn't care. Their money was as green as anyone else's. I'm now spending quite a bit of time at the Groovy Owl. I like the vibe there, to work on my drawings. The music is great and there's always an unusual group of people hanging around. Some came to buy records. Some just wanted to sit around and listen to the music. Some came there in hopes of meeting their 'soul mate'. While others just came to watch everyone else. Either way, Ibrahim loved it. The shop was always full and sooner or later, someone's going to buy some music. One day, I even ended up buying the Beatles, Rubber Soul album. Don't ask me where I thought I was going to be able to play it."

"One of the benefits of backpacking and living in hostels, was that occasionally you were bound to bump into another traveler you came to know from a different location. This was also true in Beirut. During my stay, I recognized several people, men and women, where our paths had crossed before. We would always grab a cup of coffee or a Coke, and catch-up with what each other had been doing. Then they'd be off on their way again. It was kinda neat. It gave me a sense of connection."

McGuire quickly scanned a couple of pages in his journal in order to gain his bearings and began once again. Jesse was still all ears.

"Okay, so now it's April 6th and kind of a big day for me. First, it's my best day at the expo site. Between 10:30 A.M. and 1:00 P.M., I made 55 Lebanese Pounds. I closed shop and headed to the Groovy Owl. As I'm sitting there creating more drawings, Ibrahim approaches me accompanied by a gorgeous young woman. He wants to introduce her to me. I quickly looked over my shoulder to make certain he was actually talking to me. Not seeing anyone standing behind me, I guessed that he was. Her name was Antoinette, and she was absolutely ravishing. She was of Armenian descent and 24-years old. She stood approximately 5' 8" tall, with long, dark brown hair and was built like Sophia Loren."

"Judging by her outfit and jewelry accessories, she came from money. According to Ibrahim, she spoke five languages fluently. Unfortunately, English wasn't one of them. However, between her broken English and my broken French, and a lot of hand gesturing, we were able to communicate. Seeing that we were beginning to get along, Ibrahim returned to his service counter. I soon found out that she was an only child and lived with her parents in a large apartment in Beirut's affluent quarter. Her father was a successful dentist and she was his little princess. Once completing her formal education, she didn't know what to do with herself. She asked her father to buy her a medical clinic that she could manage. Well, guess what? He bought her a medical clinic."

"I got the distinct impression that she was flirting with me, but I couldn't understand why. Now, don't get me wrong. I know that I'm an attractive, viral hunk of Canadian manhood, but this chick wasn't in the same category of birds that I usually hunted. Was she just playing me or was she being sincere? I'd have to wait and find out. We spoke for about two hours, and then she intimated that her girlfriend would probably love to meet me as well. She described her as being Arabic, pretty and with mocha-colored skin. She was starting to get my attention, if you know what I mean. We left the record shop and starting walking. I wasn't allowed to walk beside her, but rather followed about twenty feet behind. Beirut society and most other Arabic countries forbid women to be seen in public with men, other than their brothers, cousins or husbands. There was a good chance that the police would stop and question us, if this rule wasn't followed."

"After walking for several minutes, we arrived at an apartment building. Unfortunately, the friend wasn't home. Antoinette motioned that I should meet her around the corner, so that we could speak. She stood very close to me and seductively suggested that I call her Toni. She also wanted to meet me again at 10:00 A.M. the following day, at the record shop. I tried to pinch myself to make certain that this was actually happening. Naturally, I agreed, and we went our separate ways. The first thing I did after that, was to head directly for the shopping district and bought a new, non-iron shirt. The next day after leaving the construction site, I went to the public bath house for a hot shower and to clean-up my beard. It was the first time I had touched my whiskers since leaving Istanbul. Yesterday, I was caught unaware, but today I wanted to make a good impression. Now that I was all sparkling fresh and wearing some new duds, I would be irresistible. I headed to the Groovy Owl well before 10:00 A.M., and waited."

"Just like clockwork, Toni walked into the shop at 10:00 A.M. However, she wasn't alone. Accompanying her, was another bombshell that was introduced to me as Jaqueline. The girls suggested that we leave the shop and go to a café. Like before, they walked ahead, while I followed from an acceptable distance. We then all boarded a bus, and went to the Achrafieh District of East Beirut. This area was a bit above my status level. Sitting at the back of the bus, with the girls sitting in the front, made me feel like a black person in segregated Mississippi. Talk about your 3rd class citizen! I wasn't thrilled with this arrangement, but I didn't have much choice. We eventually exited the bus and entered the Cantina Sociale Restaurant on Independence Road. It was sort of like a Denny's Restaurant back home, but only pricier. I was sweating bullets wondering how I was going to afford ordering anything. Toni resolved that dilemma by immediately announcing that she would be picking-up the entire tab. That was a definite weight off my shoulders."

"I sat at one side of the table while the girls sat on the other. After the waiter took our orders and left, I nearly fell out of my chair when Toni pulled a container of Preludin out of her purse and placed it on the table. Flashing me a knowing smile, Toni opened the container and we all started popping the pills. We literally sat there for 11-hours talking, eating, drinking tea and revealing our life stories. The girls complained of how little freedom they possessed and that their fathers controlled every aspect of their existence. Eventually, both would be relegated to marriages arranged by their parents and that was that. I couldn't believe what I was hearing, and tried to convince them that at 23 and 24-years old, they were old enough to choose their own paths in life."

"The girls both resignedly chuckled, and wished that were the case. Although neither embraced the thought of domestic imprisonment, likewise, neither would dishonor their families by refusing."

"As we left the Cantina Sociale late that night, we agreed to meet there again the next day. I rushed back to Steve's place and described my day to him. I also convinced him to come back to the restaurant with me tomorrow, so that I could focus my attention solely on Toni. The next morning, Steve and I were waiting inside the Cantina Sociale, when Toni and a different girl entered. This new bird was also a knockout, and introduced to us, as Debbie. As we ordered our food, Toni once again informed us that she and Debbie were paying for everything, since we were considered their guests. What a couple of class broads. I had my Beatles album with me since Jaqueline, from yesterday, said she would buy it from me. I couldn't play here it anyway."

"Toni suggested that she and I should leave and take the album to Jaqueline's apartment. Debbie and Steve would wait for us inside the restaurant until we returned."

Once again, Jesse could not believe McGuire's good fortune as he wiled-away his days in these foreign lands. Strangers donating money toward his transient cause, beautiful women paying for his meals and drinks. Where did it end?

"We took another bus to Jaqueline's apartment, and luckily, she was at home. She wasn't feeling well after popping the Preludins yesterday. We stayed and chatted with her for a while, and she did indeed buy the album from me for its full price. Toni and I were sitting closely together on the couch and starting to become pretty chummy. I held her hand and surprisingly, she didn't pull it away. I took that as being a great sign. After leaving Jaqueline's place, we returned to the restaurant with Steve and Debbie still there. They were both quiet when we arrived, so I surmised that they weren't hitting it off very well. Toni and I sat together at the table and were still holding hands. Debbie observed this and went 'off' on Toni. They had a heated argument for several minutes, and although I couldn't understand what they were yelling about, it was pretty clear that it concerned our public, intimate display."

"Steve and I just sat there keeping our mouths shut. Once things settled down again, Debbie said goodbye and left. Now that the argument had put a damper on the outing, Steve, Toni and I, also left a few minutes later. I informed Steve that I would be taking Toni home, so he went off on his own."

"We started walking with the usual appropriate distancing but then I would start teasing her, by running up from behind and touching her, or kissing her on the cheek. We started making a game of it. Whenever we saw a policeman or group of people approaching us, we would separate. Otherwise, we walked side by side, touching and sneaking quick kisses."

"At one point, we were nearing an office building, when Toni suddenly grabbed my hand and pulled me inside. We took the elevator to the 5th floor and found it empty, and very dimly lit. Obviously, Toni was familiar with this location. Anyway, things started heating up between us mighty quickly as our hands began frantically exploring each other's bodies. I almost froze for a moment remembering that I was with this exotically gorgeous, wealthy, high-class woman. But being the hot-blooded boy that I was, it didn't take me long to recover. As I reached underneath her dress, I was shocked to discover that she wasn't wearing any panties. I just about stopped breathing!"

"The next thing I realized, Toni was slowly slipping downward through my arms. As I felt my belt buckle unfasten and the zipper of my jeans start to open, I wanted to scream. Was this really happening to me? In the blink of an eye my drawers were around my ankles and an unbelievable sensation was coursing throughout my entire body. As Toni kneeled humming a merry tune, all I could do was stare at the ceiling and repeatedly thank my lucky stars. All conception of time and space were completely lost to me until Toni finally regained a standing posture. It took a moment for me to realize that my balls were still blowing in the wind, and my once proudly erect manhood, was now as limp as an overcooked noodle, dripping the last few drops of essential seed onto the floor below. I quickly pull-up my pants and we left the building as though nothing ever happened. I remained in a semi dream-state, until we arrived at Toni's apartment. We agreed to meet the following day and said goodnight."

"On my way back to the city center, I ran into Nick and we headed out to the Hamra District. We bummed around there for a while and then hitch-hiked out to the airport. Before we realized, it was 2:00 A.M. and we were stuck. Our gear was stored at Tony's Bar and we were not going to get back in time before it closed for the night. As a result, we ended up just walking around the town looking for a warm place to bunk. Unfortunately, it was a holiday and everything was closed. We ended up going to the Lebanese Immigration Office and arrived at 5:00 A.M. I needed to renew my Visa, so we just waited until it opened. We finally gained entry at 9:00 A.M., and I managed to obtain my new Visa."

"Later, I met Toni at our pre-determined neutral location, and she took me to an exclusive club downtown. Once inside, I could barely see my hand in front of my face. There was only the glow of single flames positioned randomly throughout the space. I correctly assumed they were candles on table tops. As if utilizing radar, Toni grabbed my hand and navigated herself within the gloomy interior, until she located an empty table in the corner. Once again, I couldn't help but feel that she had been here before. Probably with someone else."

"Miraculously, a darkened figure eventually attended our table and we ordered cocktails. As we sipped our drinks, Toni explained that the interior of the club was kept dark because it was frequented by lovers, adulterous couples, homosexuals engaged in forbidden love trysts, and everything else in between. Before I could even acknowledge her, she had disappeared. Within moments, I felt activity between my legs, and with the practiced movements of a true professional, Toni was once again servicing my plumbing. What a trooper!"

"The subsequent days ran into one another with repetitious activity. Visits to the Canadian Embassy and the Groovy Owl in the early mornings, displaying my drawings in the late mornings and early afternoons, meeting Toni for our daily liaisons, in the later afternoons, and then drinking or getting high and going to the movies at night. It was a great life! Activities were interchanged once in a while like going to the beach of the American University but otherwise, I had established a definite life pattern."

"Near mid-April, I was tired of sleeping rough at the construction site or in building stairwells and alcoves where we were constantly being rousted and pushed on by policemen, shopkeepers or watchmen. I asked Toni if her parents would let me sleep in the outside hallway of their apartment building. Not expecting a positive response, I was truly surprised when Toni informed me that her father had agreed to my request. I now had a safe and dry place to crash. I felt a bit guilty knowing that her parents were completely oblivious as to who I really was. If Toni's father were to discover what I was doing with his daughter practically every day, I would be dead."

Having said this, McGuire looked Jesse directly in the eyes and re-emphasized, "I'm not kidding. I would have been dead!" Making this spine-chilling statement a second time, McGuire then drew his finger across his neck in a slicing motion. Jesse captured the gravity of his words and realized that they were made in earnest.

Jesse wasn't overly knowledgeable with all aspects of Arabic culture, but he was certainly aware that promiscuity, infidelity or dishonoring the family name usually suffered grave consequences.

"Sleeping in the hallway of Toni's apartment was working out well and she was constantly sneaking out food and even money to me. What a great kid! I know that the cleaning lady didn't like this arrangement because she was constantly making unnecessary noise and waking me whenever she came to clean. One morning she even soaked my sleeping bag, while washing the hallway. I guess my charm wasn't working on her, or maybe she was just jealous that she couldn't have me."

Once again, McGuire's form of dry-humor caused a laughing jag within Jesse, and it took him a few moments to recover. McGuire just seemed to have a punch-line for anything and everything.

"As always, where there's a 'yin' there's also a 'yang'. And where there are good times, there are also bad ones. That's what I encountered in early May."

"Things were going great! Toni and me were still an item and my drawings were selling well. I had already saved a considerable amount of money from my expos, and decided to return to Istanbul myself, rather than depend on Canadian Embassy assistance, or should I say, the lack thereof. For some reason that I can't remember right now, one day, I decided to change my expo site from the base of the Martyr's Monument, to the sidewalk in front of a busy shop further into the Square. I'd only been there a few minutes when the shopkeeper came out and told me to 'YALLA!' I waved him off and told him that I was perfectly fine where I was. Over the next few minutes, he came out a few more times, and attempted to chase me away. I just ignored him. Finally, he came out with a sprinkling can and started dousing my drawings with water. Before I could jump up and stop him, he had already destroyed about a half-dozen of the drawings. He stood there staring at me with a stupid smirk on his face and all I could see was 'red'. Before he could react, I hauled off and slugged him square on the chin. He went down like a sack of shit, and just lay there completely unconscious. For a minute, I was a little worried, until I saw his chest rise and fall. At least I didn't kill him. Before long the police arrived and I thought my 'goose was cooked'. Here I was, some long-haired foreigner that had just assaulted a local businessman, and probably a pillar of the community. I didn't give myself great odds of getting out of this one unscathed."

"The police started questioning witnesses, and to my amazement, they were all supporting me. They said that I was just sitting on the sidewalk minding my own business trying to sell my drawings. Completely unprovoked, the shop owner came out of his store with a can of water and soaked my drawings. They all felt that I was perfectly justified in defending myself and protecting my property."

"I couldn't believe it. Once again, I was benefiting from the kindness of strangers. The policemen compared notes and examined their evidence. After a few minutes, they broke from their huddle and informed me that I should pack-up my property and leave the area. I was free to go. No charges. They didn't even search me. Had they done that, they would have found the half gram of hash tucked inside my wallet. Had that been the case, things would have definitely turned-out much differently for me. That afternoon, I dodged two bullets, and for me, it was a bit of a wake-up call. Perhaps my time in Beirut was coming to an end, and I should consider moving on. It wasn't merely by choice that I made this decision, but rather through necessity."

Once again, Jesse couldn't believe McGuire's luck. Was it a case of "the luck of the Irish?" Did his lucky Palermo tooth have anything to do with it? Or did he just have a star between his eyes. Whatever it was, it made him unique. If he were a cat, he would have eighteen lives not just nine.

"I decided that Tuesday, May 10th would be my final day in Beirut. On the 11th, I would be taking a bus to Istanbul. I had now saved over $300.00 dollars from donations and selling my drawings. I was eager to retrieve my $800 bucks from the safe at the Tourist Hostel. In this instance, I wasn't going to waste anytime 'hitching', but rather make the 17-hour bus ride directly there."

"A couple of days later, I treated Toni to a luxurious western hemisphere supper of deluxe burgers, fries and milkshakes at the American University.

After eating, we found a secluded spot on the beach and went skinny-dipping. We then found a place on the sand hidden by large boulders and made love. As we were leaving the strand and walking toward the bus stop, I informed her of my plans. She seemed to take it well. Again, I had the premonition that she might have had this, or several similar conversations, before. I don't think that Toni and I were in love with one another, but we were very attached. Both physically and emotionally. I promised that I would return to her the following year, but she responded that she might be married by then."

"Her father had recently been searching for a suitable beau. I told her that I would make her pregnant and then she would have to marry me. She just laughed since we both knew that was never going to happen. When I dropped her off at the front door of her apartment an hour later, she gently touched my cheek and kissed me softly on the lips. I quickly looked around to see if anyone had seen us, but there was no one there. Toni just giggled and went inside. I never saw her again."

Jesse sat quietly for a moment and envisioned the scene. He had never known such intimacy in his life before and wondered if he ever would. Hopefully, he would. For all of his machismo, bravado and haughty attitude, Jesse detected a hollow tone to McGuire's latter sentences. Was he actually that aloof, or was he hiding some unspoken truth? Did that short-lived but intense relationship simply pass without consequence, or did he perhaps consider Toni, 'the one that got away'. Now, fifty plus years later, that question will likely remain forever unanswered.

"My last couple of days in Beirut were spent purchasing a bus ticket, getting my affairs in order and saying goodbye to the many friends I had acquired since arriving there. These included fellow travelers, as well as a few Beirut citizens."

"At 2:00 P.M. on May 11th, I boarded a coach belonging to the Ataturk Bus Lines, and watched from behind as the Beirut harbor, and then its skyline, disappeared over the horizon. The trip to Istanbul was uneventful and I spent most of the 17-hours sleeping. We arrived in Istanbul at 7:00 A.M. on the 12th, and I immediately started making my way toward the Tourist Hostel. I arrived quite early and the old caretaker was not yet there. When he eventually arrived, I greeted him at the front door as he attempted to enter. He immediately recognized me. With a startled look on his face, he quickly glanced at my hands. I suspected that he thought I was there to hurt him. Despite my obvious displeasure with the old goat, I greeted him politely. This seemed to place him at ease. I followed him into the hostel and then to the registration desk. Although he knew exactly why I was there, I reminded him once again. I produced my traveler's cheque receipt and he promptly entered the office. Within moments, he returned with my cheques and suggested I count them. I soon determined that they were all there, and I left without saying another word. All those weeks of stress and anguish, and it was over in the blink of an eye."

"To say I was now relieved would be a massive understatement. With my traveler's cheques, plus the money earned from my expos in Beirut, I had over $1000.00 dollars in my pocket. It was time to celebrate. I booked a room for a few days and decided to splurge. I visited the covered market a few times and bought some more suede clothing. I also ate and drank my fill. On the 15th, I made my way to the National Highway and pointed myself in a northwesterly direction."

"Being alone now, I was having reasonably good luck catching rides. Most of them were lengthy, to boot. Within a couple of days, I had skipped through Northern Greece and arrived in Yugoslavia. I exchanged a $50-dollar traveler's cheque in Skopje. If you can believe it, that $50 bucks became 62,000 Dinars. What a windfall! The road system in this area was very poor, so I took the train from Skopje to Belgrade. Now, I was on a bit of a 'mission', so I didn't remain too long in any one place."

McGuire took another quick scan of his diary in order to keep his details accurate. "On May 19th, I arrived in Tuscany, Italy and decided to remain a few days. Although, I had pretty much had my fill of site-seeing these last few months, Italy still offered much to see and do."

"Florence was fantastic, but occasionally, I would encounter situations where people would spit on the pavement as I walked past them. I couldn't figure out if they didn't like hippies, or just foreigners in general. Or perhaps because of my fair skin and red hair, they thought I was German?"

"Many Italians were still very bitter over the last World War, and although 20-years had passed, Germans were still immensely unpopular there. After leaving Northern Italy, I ventured into Switzerland. It was a very small country but quite beautiful. The only memorable recollection I have from there was catching a ride with a Swiss doctor and driving through the Alps. What fantastic scenery! Anyway, we're driving along these winding roads, that are snaking their way through the mountains. We're climbing and descending and going through tunnels cut right into the rocks. Great stuff. All of a sudden, we came across an old lady at the side of the road. Her car had broken down. The doctor took a quick look under the hood but couldn't detect the cause of the problem. He then placed the woman in the back seat of his car, and he tells me that we're taking a short detour, in order to bring her home. The short detour turns out to be a major one, and we're driving seemingly forever. And practically straight up to boot. When we finally get to her house, its literally near the top of the mountain. I had never seen such a view in my entire life."

"It's like we hiked to the top of Mount Everest, except the car did all the climbing for us. I didn't expect to see it, but I'll never ever forget that view."

"The doctor eventually dropped me off near the outskirts of Zurich, and catching a good ride shortly thereafter, within a few short hours, I was fortunate to have arrived in Southern Germany. Growing up, I had only heard negative comments concerning Germany and the German people. The word Nazi was constantly bantered about, and how they were responsible for causing the Second World War. I wasn't certain what to expect from visiting the nation of our former enemies. What immediately struck me was the absolute beauty of the place. The Black Forest of Bavaria, set to the backdrop of the Alps Mountains, was breathtaking. Serene valleys and meadows nestled between lakes and forests all along my route, were truly magical. It was the 23rd of May and the entire region was in full bloom. If Mother Nature had been an artist, the scene presented before me would have been her masterpiece."

"It was late afternoon when I arrived in the Bavarian capital city of Munich. This was the home of the world-famous Oktoberfest celebrated each year during the Fall, and showcasing many of the world's finest breweries, purveying their liquid gold."

"It was also the birthplace of the BMW Automobile Company, where many exclusive and classic vehicles were made. The multi-winning European Soccer Champions, Bayern-Munich FC called this place home plus many famous and infamous personages such as; classical composer Richard Strauss, medieval artist and sculptor Albrecht Durer, neuropathologist Alois Alzheimer, Nazi S.S. Chief Heinrich Himmler, Adolph Hitler's mistress Eva Braun and even blue jeans maker Levi Strauss, were all born here."

"I soon discovered that the German people weren't all monsters like I had been led to believe. In fact, most of the locals I met were gracious and welcoming. There were a lot of things to see and do in the city and I couldn't get enough of the schnitzel, fried potatoes and pickled red cabbage, that was served everywhere. And of course, washing it all down with several liters of delicious beer. I was enjoying myself immensely, and decided to remain for a couple more days. My decision might also have been influenced by the scores of attractive, blond haired 'frauleins' all over the city. Before leaving Munich, there was one more place that I needed to see. I've said it before and I'll say it again, I'm not big on things military or events dealing with conflict or war."

"But not since the persecution and slaughter of early Christians during Roman times had the world witnessed what the Nazis did to the Jewish people, and to others they deemed to be 'untermenschen', or sub-human. Twenty-eight kilometers outside of Munich was the Dachau Concentration Camp. It was established in March 1933 soon after Adolph Hitler became the Chancellor of Germany. It was originally built to house dissidents to the Nazi Regime and so-called enemies of the State. However, not long afterward, it also began incarcerating men, women and children of the Jewish faith, and they were then systematically pressed into forced labor, tortured, starved, and then eventually murdered. Although this was hardly an exciting place to visit, it was something I needed to see."

"Just as I had suspected, it was an extremely depressing complex with a sarcastically ironic sign above the entry gate. It said, 'Arbeit Macht Frei', which in English translated to 'Work will set you free'. My tour group was relatively small and no one spoke above a whisper, if at all. Our guide took us throughout the facility, including the gas chambers and crematoria, where the victims were first murdered and then incinerated. Some of the women in our group were crying. I felt cold inside, but at the same time a burning anger seethed inside of me, toward the animals who could have perpetrated such grievous offences on fellow human beings. I just couldn't wrap my head around it."

"I left Dachau feeling completely numb, and after collecting my bags in Munich, traded the city for the highway. After that, I didn't remain in Germany any longer than necessary, and gradually made my way through the rest of the country and into Holland. I'm not certain of the exact date, but about May 28th, I arrived in Amsterdam. I was now back in my element and spent my days looking for a good time. It was easy connecting with likeminded travelers, and I hung around for a few days smoking my face off and courting the local fluff.

The night before leaving, I visited the famous 'De Rosse Buurt', otherwise known as the 'Red Light District'. After partaking of my 'Dutch treat', I was ready to move on and gradually made my way to Ostend, Belgium on the northwest coast of the North Sea. From Ostend, I took a ferry over to Dover, England and it was thrilling to view the famous 'white cliffs' as we approached landfall. It was now June 1st, and with my feet firmly planted on 'Old Blighty', I was about to embark on my memorable journey to Londinium. This is what the venerable old city was called 'back in the day', by the ancient Romans. By me and my generation, it was now known as London Town."

"As the sea birds screeched and squawked above the shoreline in search of some delectable morsel, I plodded toward the A2 Motorway that would be my starting point. The boldly printed CANADA on the top flap of my backpack worked its magic once again. Within a few minutes, I caught my first ride in a classic Jaguar piloted by a smartly dressed middle-aged businessman. I actually spotted a bowler hat and black umbrella in the back seat as I loaded my bags into the boot. I thought those particular items were just stereotypical perceptions of common English attire, but they were obviously real. This distinctly educated and well-spoken gent took me as far as Dartford, where I then baled-out."

McGuire paused at that moment in an attempt to clarify certain terms for his youthful audience of one. "Okay, Kid. Just so you know, they speak English in England but it's a bit different than our lingo. For example, the trunk of a car to us is the boot to them. The hood of a car to us is the bonnet to them. The windshield to us is the windscreen to them. And so on and so on. And don't forget, they drive on the opposite side of the road compared to us. Where we first always look to the right before we cross the street, they first look to the left. Are you confused yet?"

Jesse gave McGuire a sneer and informed him that he wasn't completely brain-dead. He was fully aware of the English driving habits and was quite familiar with the many variances between U.K. and Canadian English. "Okay, just so you know," replied McGuire.

"From Dartford straight into London center, my next ride was in a Warburton's bread truck. The driver was making deliveries to various locations within the city. Man, did it ever smell great inside that van. There's almost nothing more tantalizing than the smell of fresh bread. Except maybe the smell of a beautiful woman."

Again, another reference to women, thought Jesse. This guy had a one-track mind. McGuire was fully aware that these continued remarks and comparisons to the female gender were grating on Jesse's nerves but he enjoyed bugging him. He liked to twist the knife whenever he could.

"The van driver dropped me near the Elephant and Castle train/bus station, and went on his way. My friend, Colin lived in the Earl's Court neighborhood of the Kensington District. This was on my opposite side of the Thames River so, I started asking passers-by for directions. After speaking with several different people, the consensus was for me to take the Underground, or subway in North American terms, from Elephant and Castle to Earl's Court. They explained that this would be my best option."

"After exchanging another traveler's cheque for English Pounds at a nearby bank, I entered the Underground Station at Elephant and Castle. I then scanned the giant map on the wall depicting all the various Underground lines and their related stations. I also discovered that most Londoners simply referred to this conveyance as 'The Tube'. The diagram initially appeared quite daunting. However, upon closer examination, and then by visually separating the different colored routes, it actually became quite easy to decipher. Purchasing my ticket and passing through the turnstiles, I promptly located the correct train and descended to the platform below, via a massive escalator. Within minutes, I was on the train and soon afterward deposited at the ascending stairwell to Trebovir Road in Earl's Court. Colin's flat was in Spear Mews, which just happened to be a short walk from where I was, according to the pleasant older matron I queried. Her directions proved valid, since I soon found myself standing in front of Colin's building. According to a neighbor Colin had left earlier that afternoon but was usually home by dusk."

"I plunked myself down on the building's concrete steps and waited. At least it wasn't raining. Within a couple of hours, I observed Colin walking toward me on the pavement carrying what appeared to be two cloth sacks of his weekly shopping. As he came closer, I rose. It took him a moment to recognize me. He hadn't seen me with long hair and a beard before, so I couldn't fault his hesitation. After a heartfelt embrace and a firm handshake, we trotted up to his room. Over a couple bottles of room temperature Watney's Red Barrel beer, I began describing my travels from the past several months. I was barely into my monologue when Colin suggested that we grab a bite to eat and then go to the Troubadour coffee house for some musical entertainment.

There was an American country-rock band called 'Eggs Over Easy' playing there, and according to Colin, they were great. They played mostly cover tunes from other popular musicians, but they also performed a few original tunes of their own."

"At the King's Head Pub, a couple of blocks from Colin's flat, we both ate a greasy plate of fish and chips with mushy peas on the side. The fish had an odd smell to it but I was famished, so I wolfed it down regardless of the unpleasant odor. Afterward, we spritely strode the half-dozen blocks toward the Troubadour on Old Brampton Road, in an attempt to secure a good table. The fish from the King's Head Pub must have been off, since it wasn't sitting well in my stomach. As far as the mushy peas were concerned, they could shove those 'where the sun don't shine'. They looked and tasted like a baby's diarrhea. What a waste of money."

"Colin spoke true when he raved about 'Eggs Over Easy'. I'm not much of a country music fan but they really pepped-up what they played and mixed it with some solid, hard-rock licks. For the cost of a few lukewarm pints of bitter, it was a pleasant evening."

"Over the next week, Colin guided me around the city and touched most of the popular sites. The itinerary included the usual well-known places such as Westminster Abbey, Buckingham Palace and the Parliament Buildings. But it also included some of the more obscure locales, such as the Sherlock Holmes residence at 221b Baker Street, Brick Lane where Jack the Ripper stalked and savaged his victims, plus the Old Bailey where the Elephant Man, Joseph Merrick lived and where Braveheart, William Wallace was tortured and executed. I was now really enjoying myself. Here, I could speak the language, I tolerated the food, and I didn't have to sleep in a construction site or building stairwell. Colin was also a good source for female companionship, for which I was grateful."

McGuire could see that Jesse was eating up his words like an infant gobbling down its Pablum, so he kept the momentum going.

"London in the mid-1960s was probably the most exciting place in the world for a young person to be. The British Invasion was still at its height and musical acts such as the Beatles, the Rolling Stones, the Kinks, the Animals, the Zombies and the Yardbirds, were dominating the airwaves. In the fashion industry, all eyes were on Carnaby Street and Saville Row, for the next 'big thing' in clothing. In fact, the mini-skirt was created in London by Mary Quant.

It was then made a house-hold name by frightfully slim, English model Twiggy, with her flat chest and short cropped, boyish hairstyle. Overall, if you were cool or hoped to be so, you had to be in London during that time."

"Just as our days were spent tripping around the city, so our nights were spent bopping from one club or pub to another. Half the time, I couldn't remember if I was drunk, stoned or straight. As I neared my 21st birthday in mid-June, I started thinking about slowly winding-down things and beginning to consider a return-to-home date. I had no problem living the party life, since if I could, I would have made that my profession. Unfortunately, my financial solvency was dictating my actions. I was still in reasonably good standing, but that wasn't going to last forever. Before going home, I still wanted to visit Ireland, so that was to be considered as well."

"June 15, 1966 was my 21st birthday. Colin and I had started celebrating early and by mid-day we were already feeling no pain. I can't remember where we were going to or coming from, but at one point we found ourselves traveling on the Tube. Sitting directly across from us was a cute little blond in her late teens or early 20s. She sported a close-fitting top and one of those new mini-skirts. The skirt was already short enough on its own, but sitting down on the padded train bench, made it ride-up even further. She had so much of her legs exposed, that I could almost see what she had for breakfast."

Jesse conjured up an image of what McGuire had just described and burst out laughing. This guy was killing him, but he was enjoying every minute of it.

"Now, if you'll stop laughing so that I can continue. We started chatting with this little turtle-dove, and were picking-up good vibes. She said her name was Suzanne. I asked her if she had a friend, and naturally, she answered, 'Yes…Why?' I told her that it was my 21st birthday and I wanted to celebrate it in style. If she and a friend were interested, they could come over to Colin's flat tonight. I guaranteed that they'd both have a great time. She smiled at us salaciously, and asked for the address. Back at the shack, we had prepared a few snacks that would accompany the beer, scotch and hash that we had previously acquired. We then sat back and waited."

"Just before 9:00 P.M. there was a knock on the door. Colin opened the door to find Suzanne in the hallway, with a cute brunette standing beside her. We reintroduced ourselves and discovered that the friend's name was Kathy. We got straight into the refreshments and within a couple of hours we were floating near the ceiling of Colin's shabby little room. I began encouraging the girls to loosen up and give us a bit of a show. They both commenced giggling, and before I knew it, they had removed their tops and bras. Motioning with my hands that they should move closer together, they quickly got into the spirit of the thing, and started rubbing each other. Next, they started necking. By now, me and Colin were super turned-on and we decided that it was time to make our move. It was pretty tight in the one-room flat, so we cut cards to see who was going to stay inside the flat, and who had to leave. I disappointedly lost the cut, so I left with Suzanne while Colin and Kathy remained."

"Shortly after, I was walking in Earl's Court with Suzanne wondering where we could go. She suddenly pulled me into a nearby alley and started going nuts on me. I figured that the alley was just as good as anywhere else, so, from a standing position we consummated our 'knee trembler'. Afterward, we walked around the neighborhood for a while exchanging small-talk and then returned to Colin's flat. As we entered, we found Colin and Kathy both fully dressed and casually speaking. We drank and smoked a while longer, and then the girls left about 1:00 A.M. It's funny when they split, there were no 'Call me's' or 'I love you's'. It was essentially just goodnight and thanks for the fun. In other words, 'Wham bam, thank you, Ma'am.' Hey, that was alright with me."

"I remained in London for the next 5-days, before deciding to take my leave. Colin had shown me true hospitality and a great time. I let him know that I was ready to reciprocate that kindness whenever he wanted. I left the city on June 21st and made my way toward the west coast of the island. Hitch-hiking in southern England along the Motorways was good. In addition, with England being a very small country, it didn't take me long to reach the port town of Holyhead in Wales. As I was making my way westbound along the Motorway, one thing started to bother me. Whenever I tried to relieve myself, there was a burning sensation whenever I peed. By the time I reached Holyhead, I also noticed a thick, pale green discharge, dripping from the end of my penis. Then it finally dawned on me. Suzanne, in London had given me the 'clap'. What a dirty little skank! That's all I needed."

"Despite my unexpected medical revelation, by late afternoon of the 21st, I was on the fast ferry from Holyhead, Wales to Dublin, Ireland. The trip took just over 3-hours, and it was supper time when I found myself walking up Grafton Street, in Dublin's city center. It's funny how the brain can mess you around. Up until then, I was feeling pretty good. Now that I realized I had caught the pox, my physical and emotional stamina suddenly began waning. Over the past seven months, I had traveled thousands of miles. I suffered hunger, thirst, extreme heat, intolerable cold and sleeping out in the elements. I was victimized by theft, threatened by a madman with a syringe and nearly lost all of my money. One by one, I conquered each of these hardships. But in the end, it was a pretty little English 'sparrow', with a syphilitic snapper, that laid me low. Somewhere between Beirut and London, I had also lost my lucky Palermo tooth. I should have expected some bad luck as a result."

Jesse took that break in conversation to ask a question. "Bob, I've only known you now for a week. However, I've quickly discovered that you are a very intelligent and street savvy person."

"Obviously, I know that you had a lot more contact with women during your travels, than you've alluded to, these past few days. Didn't you ever consider using 'protection'? Not just to prevent disease but also to avoid impregnating the girls."

McGuire didn't even have to ponder the question but rather, responded immediately. "I couldn't. I'm prophylactic intolerant."

Because he answered so quickly and definitively, it took Jesse a moment to digest the words. He then cautiously replied, "I'm sorry, I don't understand. I've heard of people who are 'lactose intolerant' but I've never heard of anyone being 'prophylactic intolerant'. Can you please explain?"

"That's easy," exclaimed McGuire. "I can't stand using a 'rubber' when I'm dancing the horizontal mambo. It's like taking a bath wearing your socks. It just doesn't feel right."

It was as simple as that. For McGuire, it was just a personal choice. He was fully aware of why a condom should be worn during sexual intercourse. He just didn't want to. He had thrown the proverbial dice many times before during the sex act, and had always been lucky. However, on this last occasion, he came up short. Hearing the response, Jesse struck his forehead with the heel of his right hand and merely rolled his eyes. He should have known that the issue was just Bob McGuire being Bob McGuire.

"I stopped for something to eat at Bewley's Coffeehouse and asked the waitress if there was a cheap place nearby where I could get a room for the night. She answered that it just so happened her aunt Frances operated a little B&B on Balfe Street, not far from the restaurant. If I could wait, she would give her a quick call and see if she had anything available. She returned to my table within a couple of minutes and informed me that the attic bedroom at Aunt Frances' was available. And because it was already quite late in the day, she was willing to give me a special price. She told me to say, 'Fiona sent me'. I agreed to the offer and received directions from Fiona to the B&B. I arrived there a few minutes later. It was an older, narrow house constructed in typical, Victorian red brick and sandwiched in between a hair salon and an antique shop."

"The stairs leading toward the attic were nearly vertical, and creaked underneath my every stride. With my heavy bags, and the tight space requiring navigation, it took considerable effort to reach the top landing."

"My room was very small, with a gabled ceiling that forced me to duck my head throughout most of the space. However, it was spotlessly clean and the single bed was comfortable. There was an old-fashioned porcelain water pitcher and basin on the side table, complete with a fresh face cloth and towel, and a bar of Irish Spring soap. Nice touch. Being curious, I checked underneath the bed and sure enough, I found an empty chamber pot. I was set for the night and promptly hit the sack."

"When I awoke the following morning, I was feeling worse than the previous day. My penis was still dripping, and despite a good night's sleep, I was fatigued. I also noticed that there was a small pimple developing on the end of my shaft. I didn't feel right about peeing into the chamber pot, especially with that pus leaking from my member, so I just held it inside. As I made my way downstairs, Aunt Frances was just putting the finishing touches on my full breakfast. It smelled great and looked even better. Half way through my meal, I decided to stay another day. If I had been feeling better, I would have spent more time in my ancestral homeland, but under the present circumstances, I was ready to return home. However, there were two places I needed to see before leaving. One was the Guinness Brewery and second was the town of Killarney. I had attended Killarney High School in Vancouver for six years and told myself, if I ever got over to Ireland, the country, I would visit its namesake."

"As soon as I left the B&B, the first thing I did was walk to the nearest place where I could take a pee. That place was a small coffee shop just around the corner. When the hot, yellow stream finally left my body, I nearly screamed out loud. What yesterday just felt like a burning sensation, now felt like razor blades ripping through my dick. I couldn't believe it! If I could have gotten my hands on Suzanne at that moment, I would have strangled her."

Jesse quickly jumped-in. "Did you try to see a doctor or go to a Walk-In Clinic?"

"No, I didn't," was McGuire's response. "I don't know if I was embarrassed or scared or just too cheap to spend the money on the treatment and medicine. I just thought I could ride it out until I got home."

"Anyway, that day I visited the Guinness Brewery and took the public tour. It was really interesting. The few pints of that black gold, I pounded back afterward, was the icing on the cake. I spent the rest of the afternoon in a neighborhood pub near the B&B to drown my sorrows. There were a few young Dubliners constantly eying me from a table nearby."

"Eventually, one of them asked if I was American. When I answered that I was in fact Canadian, you would have thought they had just witnessed the 'Second Coming'. Suddenly, I was the man of the hour and immediately became their honored guest. They were a great bunch of guys and it turned out to be one of the best nights I had during my entire trip. I ended up having to wake Aunt Frances at the B&B by banging on the door. She didn't blink an eye seeing me in my drunken state, which only meant, that this was probably not the first time she dealt with an intoxicated guest. She even stood at the base of the stairs while I stumbled upward, to ensure I reached my room safely. Thankfully, I took a long, painful pee on the side of a building before reaching the B&B, so, I would probably good for the rest of night. I blacked-out as soon as my head hit the pillow."

"After another full breakfast the following day and with bags in hand, I said goodbye to Aunt Frances and hit the pavement, intent on reaching Killarney before evening. I ended up taking a bus out of Dublin proper, and near the N7 Motorway, jumped off. It was a little over 300-kilometers from Dublin to Killarney in the southwest of the country. Before long, I was plying my trade on the N7. I found the Irish to be a very hospitable people and I encountered little difficulty catching rides. My first ride took me as far as Portlaoise, and the second to Limerick. It was now shortly after the noon-hour, so I stopped at a nearby truck-stop outside of Limerick, for a sandwich and a badly needed whiz. Once again, it took everything I had not to shout the roof off the place when I finally drained my bladder. Thankfully, the communal bathroom was empty at the time I performed my business."

"My third ride ironically was in a Guinness Brewery delivery truck, and one of the driver's stops was in Killarney. It wasn't yet dinner time when the truck driver dropped me in front of the Tatler Jack Pub on Plunkett Street. This was his first delivery point. As well as acting as a pub and restaurant, the Tatler Jack also rented rooms for the night. The price of a room per night was more than I wanted to spend, but since my time abroad was coming to an end, I decided to splurge. Anyway, this was one-stop shopping. I had the pub, a restaurant and lodging all in one spot. I didn't have far to go to find my bed at the end of the night."

"I wasn't planning on spending too much time in Killarney. I just wanted to be able to say that I'd been there. But now, I wish that I had stayed longer. It was a beautiful place. The next day, rather than just walking around, I paid 10 bucks for a day tour on a motor coach of some city highlights. This also included a quick jaunt through nearby Killarney National Park, and then out to Lough Leane, before returning to the city. For me, spending 10 bucks on a day tour was quite expensive, but afterward, I was glad I had paid the money. One more night at Tatler Jack's Pub, and I decided to split. I won't say that I was feeling homesick, but I'd been gone a long time. And now feeling sicker by the day, I just needed to call it quits."

"Returning to Canada was not going to be by ship. I needed faster transport. Flying out would be the best way to go. The nearest International Airport to Killarney was Shannon, approximately 130-kilometers to the north. After checking out of the pub, I made my way back to the highway and pointed myself in the direction of Shannon. Once again, the CANADA logo on my knapsack was like a magnet to vehicles, and in just a few hours, I arrived in Shannon. Finding the airport was fairly easy, and I began enquiring for flights to Canada. The earliest flight I could get was the following morning, with a charter airline. It would take me as far as Montreal, and then I would need to find an alternative, the rest of the way to Vancouver. I purchased a one-way ticket for the flight and would worry later about finding the rest of my way home from Montreal. I wasn't going to spend another dime if I could help it, so rather than finding a room for the night, I just slept on the floor at the airport."

"At 9:00 A.M. on the morning of June 26th, the wheels of the Douglas DC8 began retracting into the fuselage, as we rose above the tarmac of Shannon Airport. As we increased altitude and banked southward, I could see why Ireland was known as the Emerald Isle. I had never before seen so many different variations of the color green, as we flew above the landscape. It was truly beautiful. As our altitude continued to increase, we were soon into the clouds and our once inspiring view became completely obscured by a blanket of white. I occupied a window seat on the jet. Now with nothing more to see outside, I slid down the plastic shade. After a miserable night trying to sleep on the airport floor, I quickly fell asleep aboard the aircraft, with the hum of its four jet engines droning in my head. I occasionally woke briefly during the flight, and once I sat-up to eat my aviation dinner. But otherwise, I just tried to sleep as much as I could. I can't even remember now if there was anyone sitting beside me."

"After 6-hours in the air, it was still only 10:00 A.M. when we landed in Montreal. That time difference thing always confused me. After clearing Customs, collecting my bags, and another vicious battle in the airport bathroom, I was ready to get back on the road. After 7-months abroad, I was back in Canada and itching to make Vancouver. During my flight, I decided the cheapest way home would be by Greyhound Bus. When I enquired about cost and departure times at the Montreal Airport, they recommended that I take a bus from Sault Ste. Marie. I can't remember now what the logic was behind that, but I figured they knew what they were talking about."

"Anyway, it ended up taking me two full days to get from Montreal to Sault Ste. Marie. I couldn't understand why no one would pick me up. When I finally caught a ride from a young guy, who charged me 10 bucks for the privilege, he told me that some creep in Chicago had recently murdered 8 student nurses, and he was on the loose. That's why no one wanted to pick-up a strange hitch-hiker. Go figure. Where was my lucky tooth when I needed it? I finally caught the Greyhound Bus in Sault Ste. Marie. Four days and two layovers later, we pulled into the Kelowna Bus Depot. I had been writing Shirley during my travels and she asked me to stop over and visit her upon my return. Being a man of my word, I was now keeping my promise."

"When Shirley opened her apartment door, she didn't immediately recognize me. I must admit that my appearance had changed dramatically since she last saw me in November. Finally, she smiled and tugged me into her apartment, immediately heading toward the bedroom. Naturally, in good conscience, I stopped her, and admitted that I had contracted a venereal disease during my adventures. If I had sex with her, she would be stuck taking penicillin for the next 2-weeks. She thought about it for a moment, and then dragged me into the bedroom. What a trooper! After spending the night with Shirley, I returned to the Bus Depot the following day and purchased a ticket on the Greyhound to Vancouver. That same evening, we pulled into the main Vancouver Bus Station at Beatty and West Pender Streets. From there, I transferred to a City Hydro Bus and made the additional 40-minute ride eastbound to the Joyce Road Bus Loop. As I stepped off the bus, I was struck by a wave of nostalgia. I was once again on familiar territory. This was my turf. Where I grew-up. Where I played sports and attended school. Where me and my crew would battle it out with the Bobolink Park gang, whenever they had the audacity to breach our neighborhood."

"As much as I enjoyed my whirlwind, 7-month global excursion, this was where I truly lived. And I had to admit, it felt good. In ten minutes, I would finally be home. I started walking at a brisk pace and eventually rounded the corner onto East 44th Avenue. In a few additional heartbeats, I stood before my family's two-story house. I strode into the backyard and as I reached the foot of the wooden stairs leading up to the porch, there stood my mother. Her back was turned to me and she was unaware of my presence. She was removing the now dry, laundered clothes from the line and placing them into a wicker basket."

"Hey Mama Bear, I'm home," was my initial greeting. My mother quickly turned around and looked down at me standing on the walkway. I could tell that she wasn't immediately certain who I was. My appearance had obviously changed during the past 7-months. But then I could see the recognition in her eyes. Her first words were, "Robert Patrick McGuire, is that you?" I responded, "In the flesh." Without another word, "I dropped my bags and rushed up the steps. I gave her a tight squeeze and she held on to me for dear life. After breaking our hold, we both walked into the kitchen and spoke for a few minutes. As if just realizing something important, she immediately said that she needed to call my dad. He was working an afternoon shift and was about to call it a night. She wanted to let him know that I was home, and to prepare him for what he was going to see, when he clapped eyes on me. I just laughed at her serious tone, and was amazed that they still worried so much about my appearance. I was now 21, so I hoped they weren't going to demand that I cut my hair. However, that still seemed to be the case. She rushed off into her bedroom, and returned momentarily, holding a ten-dollar bill. The note was supposed to be for a haircut. I just shook my head and told her to put away her money."

"After speaking with Mom for a few more minutes, she could see that I was exhausted, and ordered me straight to bed. As I walked into my old bedroom and turned on the light, I could see that nothing had changed during my absence. As I lay down on top of the bed fully dressed, I made a point of remembering that I needed to make an appointment with Dr. Britten tomorrow. He was our family physician and I urgently required his expertise. I tried to rise in order to remove my clothes, but then fell back on to the bed. I was more tired than I realized. Just before entering my slumbers, I remember whispering, 'Welcome Home, Bobby'."

Chapter 6
Interlude
Poseidon's Bounty

Today's dialogue had been the longest session McGuire and Jesse had yet spent with the travel reminisces. As a result, it was now not only late into the afternoon, it was actually bordering on early evening. Both storyteller and listener had become so engrossed in this final installment, that each seemed unaware of the late hour. McGuire trekked back to the pickup truck in order to secure his travel journal, while Jesse began the task of folding the deck chairs and removing the makeshift table. He then gathered together the loose tools, along with the remaining nutrients and garden cart, and then returned them all to the greenhouse in preparation for the compound lock-up.

Securing the greenhouse and protective barrier surrounding it, was quickly accomplished. It was then decided that the Texada Island Inn would once again be their choice for dinner. Neither McGuire nor Jesse relished the thought of returning home at this advanced hour and having to prepare their evening meal. Besides, prior to them leaving for the greenhouse site that morning, nothing resembling a meat product had been previously removed from McGuire's freezer to thaw. Essentially, there was nothing noteworthy to eat at the house. The pair had now frequently visited the inn for meals and were already on a first name basis with most of the restaurant staff.

After consuming their meals and the relaxed draining of a couple bottles of beer, they returned home. McGuire headed toward his bedroom while Jesse had first dibs on the bathroom. When he returned to the living room sometime later, Jesse found McGuire already there, occupying his favorite chair. Stacked on the coffee table in front of him were five, dog-eared and somewhat similarly bound books, one on top of the other. Jesse immediately recognized the first volume as being the travel journal McGuire had been referring to this past week. He could only surmise that the remaining four books contained related content. His hunch was confirmed during their resulting conversation.

"Hey, Kid. Have a seat. I've got a proposition for you," was McGuire's opening statement. As Jesse complied, his host once again commenced speaking.

"This past week was kinda' special for me for a couple of different reasons. But what really turned me on, was being able to talk about my travels again. That really brought back a lot of great memories, and I gotta' say, you've been a great audience."

"Now that I've finished my first journal, I kinda' want to keep going. While you were in the can, I started digging around in my closet and located a few more volumes, of other trips, I made over the years. I haven't gone through them for ages but my juices are flowing now, and I figured it was high time that I reacquainted myself with them."

Jesse remained mute while McGuire spoke, and was anticipating that the crux of the conversation was soon to be divulged.

"You did a great job at the greenhouse this past week and really saved me a lot of work. Not to mention, saving more wear and tear on my aching back. Now that my sixteen young ladies are finally in the ground, I can handle the daily greenhouse duties myself. However, if you're not in any hurry to go home, I sure could use your help around the homestead. You've probably noticed that my yard looks like hell, and for one reason or another, I've been ignoring it. You're under no obligation to stay and I certainly don't want to pressure you. However, if you agree, I would really appreciate it. That way, I can also continue relating my other travels to you after work is finished. I can't pay you, but maybe there's something else I can do to make up for it. What do you think?"

These past several days, Jesse had been secretly contemplating what he was going to do once the work at the greenhouse was completed. Not having done any drugs since leaving the mainland, had definitely improved his sleep pattern. Plus, the daily physical work and regular meals he was receiving, had him feeling better than he could remember for a considerable time. Here on Texada, he wasn't completely on his own but Bob was allowing him quite a bit of freedom. Despite their age difference and aside from talking up a storm, whenever they were together, he actually found McGuire's company pleasurable. He knew that if he were at home, his parents would be constantly hovering over him and inundating him with their "good advice." In addition, he still hadn't communicated with his best friend since their unpleasant emails. That unnerved him somewhat. Although McGuire's property did indeed require a significant 'clean-up', Jesse was also intelligent enough to realize that the work was secondary. What McGuire really wanted from him was his solicited companionship, while he continued to recapture momentous elements of his misspent youth abroad. Those travels helped define the person he was today. And like most human beings, reliving good times, even just through reminisces, can often be tantamount to reliving them in real time. For Jesse, when it came right down to it, the decision whether to stay or to leave, was easy to make.

"Yah, no problem," was Jesse's eventual reply. "I've got nothing pressing back home, so I guess I could stay a while longer. I'll message my parents before bed and let them know."

"Great!" was McGuire's immediate and enthusiastic response. He then rose and walked toward the kitchen. "I'll get us some wine while you grab the remote and check-out the PVR library. You can pick something for us to watch tonight. It'll be your choice for a change."

As it turned-out, the program selection that night was not completely Jesse's decision. With a little prompting and subliminal suggestion from McGuire, they agreed upon the gritty, 1968 police-action thriller, 'Bullit', starring Steve McQueen, Robert Vaughn and Jacqueline Bisset. McGuire assured Jesse that he would enjoy the car chase sequence near the end of the film, and in fact he did. Apparently, McQueen, who was an amateur race car driver, performed the actual stunt driving of his green, Mustang GT during the entire chase. Jesse remarked that McQueen had nothing on McGuire's need for speed and driving prowess. He further suggested that the actor should visit Texada in order to garner a few driving tips from him. These final remarks from Jesse elicited a resounding belly laugh from McGuire, bordering on tears of mirth.

As the movie ended, and after having consumed several glasses of wine during its run, it became apparent that McGuire was once again itching to talk. "Did you know that I was down in California around the same time that movie was made?" he rhetorically questioned Jesse. Since they had never spoken of this before, there was no way that Jesse would have known, and he merely shrugged his shoulders in response.

"Yah, I was. Except, I was living in L.A. rather than San Francisco. I was there for about a year. A good friend of mine owned about a dozen food trucks, and I was operating one of them. We were working all over the Greater Los Angeles area from the beaches and business districts to a lot of the tourist hot-spots throughout the San Fernando Valley. We served the usual burgers, fries, hotdogs and tacos. Nothing fancy but business was good. Things got a little hairy in the late Summer of '69, when there were several high-profile murders in Beverly Hills, and citizens were beginning to panic. One of the victims was a real gorgeous actress named Sharon Tate. She was pregnant at the time and those bastards actually cut the baby right out of her womb. They not only killed nine people over a matter of weeks, they actually butchered them. At some of the crime scenes, they even wrote cryptic messages on the walls, in the victims' own blood. What a bunch of animals!"

"It turned out that a real sleazy creep named Charles Manson, and a bunch of his followers that thought he was God, were responsible for the murders. The cops finally arrested Manson and his cronies in October '69 in Death Valley at a place called Spahn Ranch. It used to be an old movie location where some of the Studios filmed westerns. All Summer long and into the Fall, me and the other food truck operators were always on the look-out for weirdos while we were driving around the city. We convinced ourselves that it was our vigilante actions that caused Manson and his 'family' to become nervous and finally retreat from L.A., to hold-up at Spahn Ranch." With this last sentence, McGuire chuckled to himself but somehow Jesse wasn't convinced that he was completely joking.

"I made some pretty good coin while I was down there, but eventually went back to Vancouver near Christmas of that same year. To tell you the truth, it had been a few years since I had done any serious traveling, and I was chomping at the bit to get out there again. However, in order to do that, I needed more money than I had at the moment."

Now Jesse piped-in. "That's what I wanted to ask you. I now know what you've been doing for work these past few years, but what other jobs did you have when you were younger?"

Since Jesse usually spoke sparingly during their interactions, this last question from him both inwardly surprised and pleased McGuire. It meant that he was paying attention and it also meant that these reminisces were becoming of interest to him. A thin smile now crept across McGuire's whiskered mug as he readjusted his position on the chair.

"I'm glad you asked that Kid!" exclaimed McGuire. "All this talk of the past has brought back a lot of memories for me, and not just those regarding my travels. Since you've asked, and if you don't mind, I'm going to put my journal off to the side for a bit and go way back in time."

Jesse was more than happy with that response and enthusiastically nodded his head. With that acknowledgement from his younger charge, McGuire continued.

"Yah, I guess you could say that my working life started early. Even before I finished my schooling. In 1957 when I was about 12 years old, my family moved away from where we were living in Vancouver, into a brand-new house about thirty minutes further east. This was going to be a fresh start for us and it meant that me and my three brothers were going to be attending new schools."

"My older brother Mike and I would be heading to the recently constructed Killarney High, while my younger brothers, Ken and John would be going to Carleton Elementary. Both were just a few minutes' walk from our home. Attending a new school and hopefully soon to be building new friendships, I wanted to make a good impression on my first day. I had previously seen a real cool pair of blue suede boots with white piping around the soles in a local shoe store, and realized that they were just what I needed to 'wow' the other kids. Unfortunately, my mom didn't agree with me, and quickly shot-down that dream faster than the Red Baron sent his opponents down to earth in a ball of flames while blasting them with the two front mounted MG08/15 8-millimeter machine guns on his Fokker Dr. 1 tri-plane. Since she was footing the bill for my school clothes, I had no say on what she bought me."

"I loved my parents, and they were good people, but they could sure be 'hard asses' at times. Instead of resembling a younger, red-haired version of Elvis Presley starting high school in a pair of blue suede boots, I ended-up looking like the comic book character Sad Sack, awkwardly shuffling his feet in an oversized pair of stiff, black leather Army-issue ankle boots. My hugely anticipated debut at Killarney High was now a bust, and I was relegated to the non-descript masses of other pimple-faced Grade 7 peons, standing at the bottom of the school's hierarchy ladder. Needless to say, I was both majorly devastated and pissed-off at the same time. It was then that I made the decision to somehow earn enough money to buy the things that I wanted, and not have to rely on my parents for everything."

"Within that same week, one afternoon after school, I went down to the nearby newspaper shack and applied for a paper route. For a kid of my age, this was the only paying job available to me. Well, a few days later I received word that my application had been accepted, so now I was not only a brand-new high school student, but also a part-time member of the local workforce. At the time there were two major newspaper publications in Vancouver being produced by the Pacific Press Corporation. The morning paper was the 'Province' and the afternoon/evening paper was the 'Sun'. For my sins, I was given a route delivering the 'Province' to about 40 residents in my general neighborhood. At 4:30 every morning, except Sundays, I would force myself out of bed, walk down to the newspaper shack to first receive and then fold my papers. After making my deliveries, I would return home, eat breakfast and then go to school."

McGuire ceased speaking momentarily to ensure that he still held Jesse's undivided attention. Once satisfied that Jesse was focused on his words, he recommenced.

"Now get this Kid. I kept that paper route for approximately four years. For six days a week at the ungodly hour of 4:30 A.M., without fail, I'd leave my house and would then faithfully deliver those newspapers to the front doorsteps of my customers, regardless of weather conditions or how I was feeling. In fact, there were many days when I felt like crap, but I always ensured that my newspapers were delivered efficiently and on time. I don't know if those readers ever wondered what a paper boy went through to get those 'rags' to them every morning, but it definitely wasn't easy. Through hard work and dedication, I eventually worked my way up to the enviable job of sub-manager. And I held that position until the time I decided to move on. Now here's the bombshell." McGuire hesitated for greater effect and then slowly continued.

"During those years, I almost walked my feet flat and nearly broke my hump for the unbelievable wage of $10.00 per month." As Jesse's jaw slackened and his eyes opened wider, McGuire repeated himself. "You heard me right, Kid. I said $10.00 **per month**! I know that doesn't sound like much now, but back then, it helped me to acquire many of the things I wanted, plus it provided me a huge sense of accomplishment and independence. That was extremely important to me."

"When I eventually left the newspaper business to start my new career as a 'gas jockey', it took the Pacific Press Corporation several years to recover their prestige. However, even without their star employee, they eventually regained their momentum." From the huge grin on McGuire's face, Jesse could tell that his leg was being pulled by that last comment, so he began laughing as a result. McGuire's sense of humor was beyond corny, but something about his mannerism, delivery and facial expressions, always made his attempts at jocularity quite hilarious and endearing.

"So like I said," continued McGuire after Jesse had regained his composure. "While in Grade 11, I started working at an Esso gas station not far from home, near the corner of Kingsway Avenue and Boundary Road. I had really hit the big time now, since I was earning $1.00 per hour plus a small commission on oil sales. I was literally on cloud-nine."

"That might not seem like much now, but once again put into context the standard of living back in those days. Then, for the cost of only $1.00, I could get a combination dinner of three separate tubs of Chinese food from the Asian restaurant in our neighborhood. Nowadays, you can't even buy a side of fries from McDonald's for a buck. Consider that!"

McGuire allowed Jesse to absorb these facts before returning to his dialogue. "After graduating from high school, I quit the service station and began working in a downtown warehouse on Water Street. This was a large pharmaceuticals distribution center and we shipped medicines and related supplies to Drug Stores all over the province. I was now making even better wages than before. As a result, I was eventually able to afford my dream machine, a gorgeous, 1955 Chevy Bel Air, 2-door hardtop. I loved that car! I was also saving every extra nickel I could toward that world-wide expedition that my buddy Jim Bridge and I were planning on making very soon."

"Life was plugging along just fine as I continued working and saving as much money as possible. By November of 1965, my travel fund had grown, but still not enough to set my plan in motion. I was becoming anxious to increase that total at a more rapid pace. The anticipation of taking that foreign trip was eating me alive and I was hoping to realize it sooner than later. I eventually made a fateful decision. One of my co-workers had previously offered to purchase my beloved '55 Chev, but then, I just couldn't let her go.

Later in time when he offered to buy it once again, my resolve had collapsed. It nearly broke my heart, but I finally decided to sell my baby in order to boost my travel fund. The sale of that car put me over the top and I was now ready to solidify my plans. Well, you know the rest. When I contacted Jim with the good news, he informed me that he had changed his mind. While just having purchased a new car, and now juggling a couple of ditsy girlfriends, world travel was no longer on his agenda. I was seriously choked by his betrayal, but that didn't alter my plans in the least. As you know, I started out on that first trip without Jim, but afterward, I never regretted it for a second."

Jesse's respect for McGuire increased exponentially as he digested this latest history lesson from him. The old adage of "you can't judge a book by its cover," definitely applied here. Looking at him now, Jesse would never have imagined that McGuire possessed such a strong work-ethic and sense of dedication. First as a juvenile, and then as a young adult. He was completely floored with what he had just learned. Sensing that he had made an impact on him, McGuire re-addressed Jesse and commenced the next installment of his rendition.

"Not including the small weekly allowance that I received from my parents' throughout my boyhood for performing daily chores around the house, the jobs I just mentioned were my introduction to wage earning. In addition to those, I've also worked in many other minor situations over the years, but never for very long. Having said that, the lengthiest and most rewarding time I ever spent working was within the West Coast fishing industry. I devoted the better part of twenty-five years to this occupation and I loved every aspect of it."

McGuire rose momentarily to refill their glasses with his patented golden elixir and then reseated himself. "Okay, Kid. You've got me on a roll now, so I'm going to keep talking." Jesse merely nodded his head in response as he raised a brimming wine glass to his lips.

"I guess my good buddy, Hermie Gruhn figured he owed me a favor for all of the free eats I fronted him from my food truck gig in L.A., so not long after we had returned to Vancouver from California, he sent me down to Granville Island, underneath the Burrard Street Bridge, to meet a friend of his that was looking to hire a cook for his commercial fishing boat. The skipper's name was Bob 'Porky' Patterson and I found his vessel tied-up at the wharf in Burrard Inlet. Porky's boat held a five-man crew and most were novice fishermen. When we met, he gave me a stern looking over and asked if I could cook. I figured, how hard could it be to cook for such a small group of men, so I fudged the truth a bit, and told him that I was a great cook. He then asked me how I handled myself on the open ocean. This time I didn't have to lie. From my experiences on board the S.S. Independence during my first overseas trip, I knew that rough or choppy waters didn't bother me. The next thing I knew, we were shaking hands and I had just been hired on as 'Cooky Robert'."

The nickname caused Jesse to chuckle as he began envisioning McGuire with a French accent, wearing a knee-length white, chef's apron and starched, white stovepipe culinary hat. McGuire ignored Jesse's snicker and continued orating. "I was happy to have gotten the job, but I wasn't thrilled with the boat's condition. It was an older Japanese-built vessel named the MV Purse Seiner. At 49-feet, it was short in length as well as on the beam, but at least it was afloat. It was originally built to troll for salmon, but was now refitted to resemble a reasonable facsimile of a purse seiner. A combination of hemp rope and steel chain maneuvered the antiquated steering assembly, and a 1948 Mercury truck transmission atop a gearbox, brought the net back on board after it had been set."

"Top speed for the old scow was only 6-knots, which was nearly 4-knots slower than its closest contemporaries. To give the vessel greater stability in the open 'chuck', the net drum was sunk lower into the deck. Unfortunately, as we later discovered this caused a variety of problems, not which of least was the difficult removal of numerous captured salmon that would always get trapped at the bottom of the drum well. However, that issue was yet to be experienced. In the meantime, I was now becoming a newly minted fisherman. Porky recorded my personal details and told me to expect a call from him as soon as the salmon run commenced. I threw him my best 'sailor's salute' and left the dock with a spring in my step. At my earliest opportunity, I called Hermie to give him the good news."

"Before long, I received the call from Skipper Porky and I headed back down to the dock. The wharf was abuzz with activity when I finally arrived. Provisions, equipment and the crew's belongings were being loaded on board as the mini-seiner's fuel tanks were being topped. I received an enthusiastic handshake and friendly smile from Porky as I boarded the boat, and was then introduced to the rest of the crew. Soon afterward, the skipper took me 'below' and showed me where to stow my gear. Stepping inside, I immediately bumped my head on the low cabin ceiling and was shocked by the cramped and claustrophobic living quarters. The vessel was definitely better suited to Japanese fishermen whose physical stature was generally much smaller than that of Western Europeans."

"Compact was an overly kind word for the galley where I would be preparing the crew's meals. This was also the case for the fo'c'sle, which contained four child-sized bunks and was separated by a soiled, threadbare Army blanket from the 'Screaming Jimmy', Detroit diesel 6-cylinder engine that powered the vessel. The skipper earned the minor luxury of sleeping alone in a single bunk above ours. To highlight the galley's limitations, it had a farm-style, fresh water hand pump that took at least fifty cranks just to fill a Kool-Aid container. We were so cramped inside the confined space, that during meal times, only four persons could be seated. This meant that the skipper and my three other shipmates ate their meals within, while I waited outside on deck until they finished. During those times, I definitely felt like a second-class citizen."

It was here that McGuire paused momentarily and allowed Jesse time to visualize what he had just been describing. It also took him back to that first day starting out as a novice fisherman and galley cook. Pondering the memory and the picture re-created in his mind, he just couldn't help breaking into a smile.

Within a brief interval, McGuire returned to the present and realized that Jesse was politely waiting for him to continue. "Sorry, Kid. I just got caught up in the moment. So, getting back to that first day. We eventually loaded our net, stowed away our gear and fueled-up in preparation for departure. As we slowly chugged out of the harbor and further into the inlet, we were being overtaken by newer and sleeker seiners also heading out for the 'run'. They were all about 60 to 90 feet in length, decked-out with impressive looking gear and equipment, and passing us like we were standing still. I had to admit that I was a bit embarrassed watching them gracefully gliding past us. However, motoring along at a less than impressive 6-knots had its benefits. At least it provided me the opportunity to take in all of the wondrous sights surrounding me. I was really stoked and looking forward to what lay ahead."

"As the initial first few days melted away, I quickly got into a routine. When I wasn't busy with my own duties, I was always asking the skipper or my crew-mates if they needed help with anything. I was eager to learn whatever I could regarding the fishing business, and no job was either too hard or too dirty for me to undertake. In addition, whatever I did, it was always with enthusiasm and with a smile on my face. It wasn't long before I had won the respect and gratitude of both Skipper Porky and the crew. That is, all except for our 'skiff-man'. I don't recall his name at the moment, but he was a real miserable dude. He was forever griping, disrespecting the skipper and complaining about my cooking. Now, Porky was a great guy but not a disciplinarian. As a result, the dorky skiff-man's behavior was not kept in check, and it was beginning to affect the crew's morale."

"One day after his usual bout of dissing and bitching, I was fed-up and called him out onto the deck. Before long, I had him lying on the boards, nursing a split lip and a bleeding nose. That afternoon he quit the boat and we sailed to nearby Kelsey Bay to drop him off. The skiff-man on a fishing boat holds a key position, and through my actions, we had just lost ours. As a result, I thought my days on board the MV Purse Seiner were numbered. How wrong I was! As we left the dock at Kelsey Bay, Skipper Porky approached me with a wide grin on his face and an extended right hand. As I shook his hand in disbelief, he announced that I was no longer 'Cooky Robert', but now 'Skiff-Man Bob'! I had dodged a bullet and I couldn't have been happier. After that, I still helped-out in the galley, but my elevation to skiff-man was a Godsend. I held that prestigious and rewarding position for the remainder of my fishing career."

Jesse could see that see that McGuire was very pleased to have relived that moment in time and raised his glass to him. Acknowledging the gesture, McGuire rose, leaned over, and in appreciation touched his glass to Jesse's. These two unlikely companions were slowly but unmistakably strengthening their bond. Reseating himself, McGuire continued.

"Having been watching the crew at work for a while now, it wasn't long before I became quite adept at the duties and responsibilities of my new role. Our fishing boundaries were between Campbell River to the south and Seymour Narrows to the north at Granite Bay. The area wasn't a large 'producer', but it was great for a novice crew. We also still managed to achieve quite a few 'jackpots'. Some of our catches were so large that we required the assistance of more capable seiners to take hold of our tow-line, in order to allow our net to be returned onboard. Standard operating hours for seiners were during the daylight, while daybreak and dusk was for gill-netters. The occasional flare-ups occurred whenever certain boats encroached upon the operating hours of other vessels. As a result, the chilling echo of ripe comments and ribald language could be heard floating across the wide and salty sea. I lasted the season on Porky's boat but the remaining crew rotated out several times before we finally returned to port in Burrard Inlet."

"After my inaugural fishing season on the MV Purse Seiner, I too left Skipper Porky's employ, and worked on several other boats over the next couple of years. I continued honing my skills as a skiff-man and also picked-up my Scuba Diving Certificate. I knew that this skill might come in handy 'on the job', as well as for recreational purposes. My hunch was correct when I beat out a bunch of other applicants for a chance to work on a 'highliner'. These fishing boats were the top of the line and were basically given license to print money. If there were fish to be had under the waves, a highliner would snag them. They were also operated by skilled fishermen and less prone to technical or mechanical problems. My opportunity came aboard the MV Western Commander, skippered by a former RCMP officer named Jack Fast. She was a west coast double decker and one of the B.C. Packers Fleet of nearly one-hundred seine boats. She was an 88-foot well-seasoned, wooden hulled beauty, capable of handling the treacherous waters of the Bering Sea. She was beamy, tough and luxurious to crew. I was one lucky son-of-a-gun to be working on her."

"As predicted, our catches were bountiful and my wallet began to swell. I even got to utilize my diving skills a couple of times when other fishing boats requested our assistance due to snagged lines or tangled props."

"On these occasions, the distressed boat would send one of their crew to replace me, while I donned my wetsuit. The first rule of diving is never dive alone. It's simply too dangerous. However, on these occasions, the extra money was too good to pass up. With a sharp knife in hand and a quick glance toward the heavens, I'd plunge into the cold, briny deep, locate the snag, and methodically saw through the offending coil. Before long, I was back on the surface and upward of $300.00 richer. This might seem like highway robbery for such a short job, but flying-in a licensed diver from the mainland to perform the same task would have cost that skipper thousands of dollars. Everyone agreed that me performing the job instead, was a pretty fair trade off."

Hearing this last statement, Jesse raised his eyebrows and let out a sustained, high-frequency whistle. "I see what you mean," was his verbal response.

"I continued working on the Western Commander throughout both the salmon and herring seasons, enjoying the thrill of the hunt and making good money. Thinking back now, I guess my most memorable fishing experience came during the salmon run of 1989. By now, I was into my 19th year of fishing. Then, I was working on the 85-foot, steel-hulled MV Mary Isle, skippered by Captain Teddy Assu. His wife Denise, also worked onboard as the cook, and man, she produced some pretty fantastic grub! We had started this season on a Sunday night, just south of Kelsey Bay at a spot named Robson Bight. This unique place was world famous, since this is where Killer Whales would come to rub their massive, blubbery bellies on the gravelly bottom of the bay."

"Anyway, the skipper came down from the wheelhouse onto the deck and announced that Fisheries had designated a 'clean-up' catch in Knights Inlet the following morning. We were chosen to be one of a few seiners to become involved. Fisheries had intimate knowledge of the Province's river systems, plus the numbers of spawning salmon that would be depositing their eggs in the sandy bottom of the river beds. Over-spawning was not welcomed, so extra fish needed to be removed from the equation. That's where we came in. After a brief pow-wow, we decided to sail to Campbell River and trade our shallow sein net for one that would reach lower depths. Before sunrise the following morning, we motored out to Knights Inlet. Arriving near 5:00 A.M., we prepared our equipment and awaited the signal from the Fisheries representatives. We could clearly see the 6-foot boundary markers that delineated our capture area and were then later informed that Fisheries would soon be increasing them even more."

"Before long we were all ready to rock-n-roll and I anxiously sat with my 'beachman', inside the skiff, atop the gentle swell of the surface. The beachman was the guy that would secure our lines to the designated markers. Bobbing around in that vast expanse of emerald green, bone-chilling water, several hundreds of feet deep, would have been foreboding to some. However, for me, I felt right at home. Exactly at 6:00 A.M., we received the signal to commence fishing. I began rowing the skiff out toward the target zone as quickly and evenly as possible, using deep, forceful oar strokes. I gotta' tell you Kid, in those days I was in pretty good shape, with biceps the thickness of full-grown pythons. My heart kept pumping faster the nearer we reached the target zone, since the surface of the water was literally pulsating. It resembled a giant pot of potatoes boiling on top of the stove. Scores of Pink Salmon were jumping from the sea into the air, in an attempt to escape the huge mass of other salmon, all sharing that tight space."

"We soon reached landfall, and my beachman jumped ashore to secure our one-and-a-half-inch poly line to a massive fir tree that held a boundary marker. I signaled to our crew that all was ready, and Captain Teddy began using the fishing boat to slowly tow the net into what eventually became a large circle. Once the main line was released, we received a loud blast from the boat's air horn, and I began quickly rowing the skiff back to the boat. Once arrived, the beachman and I climbed back aboard to help bring in the net. Within a few minutes, cheers of triumph and giddy laughter was heard echoing across the inlet as we witnessed what our net had actually captured. Lowering the hydraulic stern, and then using the main boom, we slowly dragged the massive, overflowing net of wet, shimmering and squirming salmon into the boat. As we released the catch into the 'hold', what I saw dropping inside wasn't fish, it was crisp, green dollar bills. We were all beside ourselves with joy."

"With his many years of fishing experience, Captain Assu estimated that our first haul was approximately 15,000 pounds. Having passed our initial moments of excitement, the skipper noticed that Fisheries had not yet taken down the boundary markers. That meant we could continue to fish. Everyone now jumped back to action stations, the net was reset, and I was rowing the skiff back toward the target zone. The beachman soon retied the line around the same massive fir tree as before, and the entire process repeated itself. The net was circled around the school, the line was released, the horn blew, and I then rowed like hell to get back to the boat. This time, Denise, the skipper's wife, had grabbed my video camera and raced to the top deck in order to film the action."

"On this occasion, I was required to remain inside the skiff and tie-off on the cork-line. This was a necessary step to take in order to avoid the net sinking to the bottom of the inlet. We had scooped so many fish into our net, that the sheer weight of their numbers was immense. Since the net was now too heavy to raise with the main boom, we broke out the 7-foot brail, that was essentially a heavy-duty mesh scoop. With each dip of the brail, we removed three to four-hundred writhing salmon from the net and lovingly placed them into the awaiting hold. This went on for several hours until finally every last 'pink' was brought onboard. By the end of the day, we were all exhausted but elated. That last net full of salmon was estimated at 80,000 pounds and we later learned that our total catch for the day had been nearly 110,000 pounds. We were just about 'shitting ourselves' with joy."

Jesse couldn't even imagine what 110,000 pounds of salmon would look like, but he could tell from the excited tone of McGuire's voice and the moist gleam in his eyes, that this had been a very momentous day for him and his crewmates. Once again, he raised his now empty wine glass toward his mentor in appreciation for a job well done.

McGuire observed the empty glass and paused his commentary in order to refill both glasses. He then returned to his reminisce. "A company packer cruised toward us later that afternoon and offered to take our catch to the processing plant. Captain Teddy had a much better idea. We would deliver our prize to the plant in person, and several hours later, our hold was being pumped-out at the processing plant in Steveston. Watching the action and the sheer tonnage of fish being removed from within our hold, we were all still floating a foot off the dock. When all was said and done, and reckoning time had arrived, my payday was nearly $10,000.00. It was the most money I had ever earned for a day's work and the largest one-day catch made by any purse seiner on the West Coast to date. Plus, on a final note, the video Denise filmed that day was 'killer' and I carried a copy of it with me on my later trips to both Thailand and Australia to show the people there what a real fishery looks like!"

On that note, Jesse stood up and gave McGuire a standing ovation. Whether the tale was actually truthful or just another exaggerated fishing story from an overly enthusiastic fisherman, didn't matter to Jesse. He thoroughly enjoyed the rendition. Now slightly embarrassed, McGuire playfully tossed a cushion in Jesse's direction and told him to sit down. With a final chuckle and clap of the hands, Jesse complied.

"Now where was I when I was so rudely interrupted?" scoffed McGuire. "Right, I was just nearing the end of my fishing adventures. By 1995, my old buddy Hermie Gruhn and I were together again on his boat, the MV Vera G. We were 'shaking gill-net herring' off the east coast of Vancouver Island near French Creek, and having a great time. One day, another fishing boat, named the MV Katlin, sidles up to us. Hermie shouts out a big 'hello' to the skipper of that vessel and soon introduces him to me as Captain Tom May. I soon discovered that Tom was contracted to pack Hermie's herring, and that of other fishermen, on to his boat and then transport them to the Gulf Herring Fishery processing plant at French Creek. This meant that boats could have their catches delivered to the processing plant without having to break-off their fishery. It was a win-win situation for all concerned. Tom would receive a fee for delivering the herring, while the active fishing boats could continue to fish."

"During our conversation, I also discovered that Tom owned and operated a salmon hatchery on the Sunshine Coast. Hermie and I had a great afternoon with Tom and talked about many different things beyond fishing. He was quite a bit older than me, but we really seemed to hit it off. We eventually transferred our herring over to the MV Katlin and said our good-byes to Tom. Later, I had a bit of an epiphany. Now Kid, you know that I'm not overly religious, but that night I dreamt I spoke with God."

Absorbing this last statement, Jesse nearly choked on his sip of wine, and looked at McGuire as though he had lost his mind. "Yah…yah, I know what you're thinking, but I'm not going nuts! Anyway, God is talking to me in this really deep, haunting voice and he says, 'Bob, for many years you've been slaughtering millions of my salmon and herring. Now, I think it's time you reversed that trend and begin raising fish instead'."

"The dream only lasted for a few seconds, but when I woke up in the morning, I was still thinking about it. Naturally, I didn't tell Hermie about the dream and just went about my daily routine. Over the next few days, we saw quite a bit of Tom May. One morning, Tom approached and complimented me on the way I handled myself. He said that he really liked my personality and work-ethic, and offered me a job at his salmon hatchery. I immediately thought back to my dream a few nights prior and got a chill up my spine. I immediately wanted to accept Tom's offer, but I also didn't want to leave Hermie in the lurch. When I later informed Hermie of Tom's offer he couldn't have been happier for me. What a super friend!"

"Anyway, Tom and I shook hands to finalize the deal and before long, I was dropped onto a little piece of heaven on earth near the top of Hotham Sound. That was Tom's property and where he and his wife Michelle lived in a beautiful house right on the water. My place was in a secondary home on the property, a short distance behind them. I lived alone upstairs while a couple of single guys, that also worked there, lived in rooms downstairs. There were thirty large, above-ground fish tanks situated on the property, containing upward of a million salmon smolts. Over the next 20-plus years, Tom taught me all there was to learn about fish farming, and I began to consider him and Michelle like surrogate parents to me. In early 2019, after working over 24-years at the hatchery, I slipped off a tank ladder with a 20-pound bag of fish food and nearly broke my arm. This kind of scared me and got me to thinking. I still had lots of living to do, and I wanted to ensure that I remained healthy and physically strong. I realized how easy it was to become seriously injured working at the hatchery, so I made a gut-wrenching decision. With a heavy heart, I later informed Tom and Michelle that I would be retiring that coming June."

Throwing Jesse one final meaningful glance, McGuire had reached the end of his dialogue. "Well, that's what ended up happening, and here I am telling you about it."

With this last sentence, McGuire abruptly stopped speaking and turned his head to one side as moisture filled his eyes. Jesse could see that this was an emotional time for him and thought it best to end the evening on that note. In any regard, it was already well into the early morning hours, and suggesting that they both 'hit the sack', wouldn't be unusual. As always, McGuire soon recovered from his reverie and concurred with Jesse that tomorrow was another day. Without formality, both men then rose, said their goodnights and retreated to their respective bedrooms.

Chapter 7
It's Better the Second Time Around

Before calling it a night, Jesse messaged his mother from his laptop computer informing her, and his father, that he would be remaining on Texada with Bob McGuire for a little while longer. Although she would not read the message until later that morning when she rose for the day, he was already confident that she would be pleased with his decision. After all, it was her and his aunt Pearl who orchestrated this subterfuge in the first place. He was certain that they probably felt the longer he remained there, the more likely they could re-program him. Jesse no longer resented their interference, but he still hadn't come to any decision, one way or the other, as to what his next moves would be.

Plans for commencing the property clean-up the following morning had to be temporarily halted. It was raining. That didn't bother Jesse and it also didn't seem to faze McGuire. In fact, Jesse formed the opinion that his mentor had completely forgotten that the project had even been discussed. This further affirmed Jesse's earlier belief that the proposed yard work was simply a deception concocted by McGuire and designed to keep his house guest static, so that he could continue plying him with past experiences. Either way, Jesse was in agreement with that arrangement. He was actually becoming quite intrigued with these tales of former glory.

After breakfast, Jesse accompanied McGuire to the greenhouse and observed while he performed a few rudimentary functions. The compound was inspected to ensure its continued security, the indoor plants were watered while wilted leaves were plucked and disposed. Finally, the outdoor plants were examined as the rain gently inundated them with its liquid nourishment. Overall, everything appeared to be in order, with all interior and exterior growth thriving to McGuire's satisfaction.

Before returning home, the pair drove to the Texada Market to pick-up a few more essentials, now that Jesse had elected to remain. Pearl was working that day and immediately ceased her shelf stocking duties when she observed the men entering the store. Displaying a beaming smile, her first words were directed at Jesse. "I hear you're staying with us a while longer. That's great!" He did not immediately respond but simply grinned and nodded his head. He was not surprised in the slightest that Aunt Pearl was already aware of the news. He was certain that his mother must have immediately jumped on the telephone to her, as soon as she read his email message.

McGuire left the relatives to their conversation while he grabbed a shopping cart and began selecting items for his larder. He took his time until Jesse eventually broke away from Pearl and joined him. He invited the youth to pick out whatever items he wished in addition to the things he had already chosen. With their shopping eventually completed and the financial reckoning transacted, Jesse began carrying the groceries to the truck. Once exiting the store, Pearl quickly grabbed McGuire by the arm and inquisitively asked him how things were going. McGuire informed her that everything was fine and reassured her that Jesse was in good hands. Moments before Jesse returned inside the store, Pearl quickly reiterated that she would leave the pair to themselves but that she was just a telephone call away if they needed anything. McGuire nodded his affirmation and then called out to Jesse, "Okay, Kid. Say goodbye to your aunt Pearl and we'll head home for a late lunch."

Once back home and with their meal now consumed, McGuire instructed Jesse to tidy the kitchen while he went for a short nap. With the inclement weather painting everything a gloomy, depressing hue, and his daily chores now completed, he felt the treat of an afternoon snooze was just what the doctor ordered. For his part, Jesse felt the same way but instead of napping, he later occupied himself with his computer. Several hours later, the pair reconvened in the living room and assumed their usual positions. McGuire was in possession of his next travel journal, and placed it open on his lap, as he began to speak.

"Okay Kid, so we fast forwarded from 1966 to 1970. When we last spoke, I was rockin' my fishing gig and enjoying every minute of it. Although, I had already made enough money to 'hit the road' by the end of Summer that year, I decided to stay on with my crew and finish out the fishing season. Those guys depended on me and I didn't want to let them down. In addition, I wanted to leave on good terms, so that when I returned home, I'd have a job waiting for me. Staying on a while longer, also allowed me to make even more money."

"It was now November 1970 and it had been four years since my last major trip. As you know, I had been in L.A. for a while, plus I made a few extra side trips here and there in Canada and other States. But my real ambition was to cross the ocean again. Like I said before, I developed my interest in globe-trekking while I was still in high school, but it was when I actually made that first real trip in '65, that I knew traveling was all I wanted to do. I can't explain it, but it was like I was infected by something and the only cure was to see the world. Not just once, but constantly."

"I'm sure that if I had lived a few hundred years ago, I would have hooked-up with some famous explorer like Jacques Cartier or Captain James Cook or Vasco De Gama and sailed with them. I can't think of anything more exciting or fulfilling than visiting different lands and meeting new people."

"When I made my first excursion in the mid-60s, young people were just beginning to cut loose by hitch-hiking and generally bumming around. By the time of my next trip, the flood gates were opened. Now, there were freaks traveling everywhere. Exotic places, tourist hot-spots and highways throughout the European Continent were now packed with free spirits all searching for independence, enlightenment, love or generally just a good time. Previously, I met quite a few people that I connected with. Some for a short time, while some others for a while longer. After four years, I was still in regular contact with some of those people that had now become good friends."

McGuire picked up the journal from his lap and began shaking it in Jesse's general direction. "During my next trip, I met even more people. I recently started re-reading this journal and I came across a bunch of names. A few were good friends but more were just people that came and went throughout my trip. I apologize in advance that my memory of them will either be vague or non-existent. So, if you hear a name and ask me who they are, I might or I might not be able to answer you adequately. Just so you know."

Having now set the stage for the next act, McGuire rose and entered the kitchen. Hearing the sound of glasses clinking, Jesse realized that their evening libation was about to be served. This might become another marathon session. McGuire returned moments later with two empty glasses and the entire jug of wine. This way he could refill their glasses without having to leave the room. This might indeed become a long-distance event. Once the glasses were filled and without standing on formality, McGuire opened his journal. After briefly glancing down for reflection, he commenced his oration.

"For us, that year's fishing season ended in early November. A few days later, I purchased my ticket for Europe. Last time, I took the leisurely route by crossing Canada and then taking a cruise ship out of New York City. This time, I wasn't going to fool around. From the same travel agency as before, I purchased a one-way charter airline ticket from Seattle, Washington to London, England."

"I would have preferred to fly out of the Vancouver Airport since it was closer to home. However, the American flight was much cheaper. I think you know me well enough by now, that if I can save a buck or two, I'm going to do it."

"I'd been living on my own for a while now but I always kept contact with my family. My parents had finally gotten off my back regarding my long hair, although I knew it still drove them crazy whenever they saw me. My older brother Mike was still climbing the corporate ladder with the international Safeway grocery chain, while younger brothers Kenny and Johnny were still in school. Kenny had just entered the University of British Columbia on a Phys Ed major, with Johnny commencing Grade 12 at Killarney High School. I was very proud of them all, although for the most part, I kept that sentiment to myself."

"Whereas before, I had received a big send-off from the relations in '65. This time when I broke the news to them that I was going abroad again, it was more or less, 'Okay, have a good time. Just be careful and don't forget to write.' Anyway, I don't remember the exact date but sometime around mid-November, a couple of my buddies drove me down to the Sea-Tac Airport and dropped me off. I then flew over the Polar Icecap to London, England, arriving ten hours later. This was a big difference compared to the two weeks I had previously spent on the cruise ship."

"From Heathrow Airport, I took a shuttle bus into the city and eventually made my way to Earl's Court where Peter, another buddy of mine, was living. It seemed that almost all of the 'heads' in London were living in that district. It was close to all of the action, plus had some neat shops and pubs close by. It was also cheaper than most other areas of the city. Peter was expecting me, and had organized a bit of a welcoming party. Several other people were inside the flat as well when I arrived and most were already three sheets to the wind.

It didn't take me long to make myself comfortable and join in. There was lots of beer to drink, some great hash being passed around and good tunes playing on a cassette recorder. I had only landed in the country a couple of hours ago but already, I felt like I was at home."

"The next few days were spent reacquainting myself with the city, being introduced to a bunch of new people and basically just partying. Being the frugal person that I am, it wasn't long before I realized that I was spending more money than I wanted to. That wasn't the plan. I wanted to stretch out my finances for as long as possible, since the more money I saved, the longer I could prolong my holiday."

"My immediate goal was making it to Spain where living costs were cheaper and the weather was much warmer. If I wanted to hang around in a cold, damp climate, I could have saved my money and just remained in Vancouver. At the moment, my only concern was the attitude of the fascist Spanish Government toward long haired hippies such as myself. I had heard that they weren't exactly rolling out the welcome mat for my sort. In fact, people were being turned away from the borders. In the last week of November, I visited the Spanish Embassy downtown to check-out my options. I was reassured that everything should be fine and that I could just carry-on with my plans. This response boosted my spirits somewhat, but the word 'should', left a lingering worm of doubt in my mind. It was all well and good for some 'suit' sitting in a comfortable office in downtown London to tell me that everything would be fine. However, it was another matter convincing some armed, narrow-minded Spanish Customs official at a remote border crossing in Spain, that he should permit me entry into his country and accessibility to all of those luscious Spanish virgins."

"Anyway, I didn't come all this way for nothing, so I decided to take my chances." McGuire now glanced down at his journal and briefly took a moment to apprise himself of some facts. "On November 24th, I visited a nearby Medical Clinic and obtained a Cholera vaccination. After Spain, I was planning to venture further south into Africa and this protective measure was strongly recommended. After leaving the clinic, I attended the International Youth Hostel Association at #29 John Adams Street and purchased a membership. I then decided to take a long walk. This activity always allowed me to clear my head and concentrate on future plans. After about an hour, I arrived at Hyde Park and noticed a young, black boy dribbling a soccer ball. I asked him if he was interested in a little 'one on one' action. He agreed. We were having a pretty good time when suddenly this older black dude shouted out and asked me what I was doing there. I thought that this was a stupid question since it was obvious what I was doing. I ignored the guy and kept dribbling the ball. The guy calls out again but this time, I can tell that his tone is becoming hostile. He's got a heavy West Indies accent and he starts adopting an aggressive stance."

"Now, he's got my full attention and he's beginning to piss me off. I've got half a mind to just deck him and continue with my game. However, I decided to ask him what his problem was. From the sound of my voice, he recognizes that I'm not British and immediately assumes that I'm American."

"When I correct him and tell him that I'm Canadian, he starts taunting me and saying that's nothing to be proud of. He then tells me to get a shave and a haircut."

"Now, I'm really starting to boil and I've completely forgotten about my soccer game with the boy. I throw a dig at the jerk and tell him that at least I can grow a beard, something he's unable to do. Well, this really sets him off and I'm thinking that the fists will be flying at any moment. Suddenly, as if they came out of the woodwork, there's a bunch of other 'blacks' out on the street and they're yelling at me as well. Mothers with their kids, other guys with big Afros and wearing Malcolm X badges and the like. It looked to me like I was just about to get involved in some type of race riot, so I called out a few more choice comments and started backing off. As I was walking away, I could still hear them jeering at me but I decided to ignore them. I'm not one to back away from a fight but I learned a long time ago to choose my battles wisely. That's what I was doing then. I just found it ironic that people who were formerly oppressed through bigotry and racism, had now become racist themselves."

"Returning to the flat, I was still steaming. But after a few pulls from the hash pipe and with a lovely little 21-year-old Winnipeg filly who had just arrived that day, tucked snuggly under my arms, I was soon feeling much better. Also, in the flat that night was a couple from California who were en route to India. They had a real sweet gig that I would love to have been a part of. They were hired by some large clothing store chain in the States to go to India and purchase all kinds of Indian arts and handicrafts. They were then to return to the States with them for some type of advertising campaign. Their trip was being entirely sponsored by the firm, with all expenses paid. Not only that, the company was also providing them two, brand new 4-wheel drive Land Cruisers for the trip. That's why they were now in London. They were waiting to take delivery of the Land Cruisers from the factory. Not bad, eh?"

Jesse digested this latest detail from McGuire and raised both eyebrows in response. "Yah, that sounded fantastic!"

"Anyway, they were 'in'," but I couldn't convince them to take me along. The next day, I ate my final greasy English breakfast and started getting ready to leave London. I said my goodbyes to everyone and purchased a sturdy pair of boots, for 3 Pounds, from Peter. After that, I was on the 11:00 A.M. train to Dover. Once there, I easily cleared Customs and boarded the ferry for Ostend, Belgium. By late afternoon, "I'm back on the European Continent and starting to get excited."

"I'm quickly getting good vibes about this trip, since within ten minutes of sticking out my thumb on the highway, I've already gotten my first ride. It was with an American named John, driving a Volkswagen van and he took me all the way into Antwerp. What a great ride! The icing on the cake was when he tells me that he's heading to Amsterdam, Holland the very next day and that I'm welcome to join him if I want. If I want? Is the Pope catholic? Of course, I want to go! We then head straight to the Youth Hostel in Antwerp and check-in. The next stop was to a nearby pub for dinner and a few beers. Even before downing my first beer, I was already flying three feet off the ground. What a great start to my trip!"

McGuire was now beaming like a leprechaun who has just found a pot of gold at the end of the rainbow. And Jesse can tell that he's been metaphysically transported back fifty years into his psyche. This is probably what he was hoping to achieve and Jesse was convinced that he had.

"Breakfast at the hostel the following morning was fresh bread with jam and tea. With one last hasty gulp of tea, we grabbed our belongings and piled into the van. Unfortunately, we only managed about 18 kilometers before the engine seized and we had to call for a tow truck. Now back where we started, John was performing some intense dickering with the mechanic of a repair shop where we had been towed. He finally struck a deal where for $50.00 USD, the mechanic would install a used, 1200cc engine into the VW. It was smaller than the former engine, but John didn't care. At least, it would get us back on the road."

"The job would take all day, so we just started going around the city checking out some of the sights and hitting a few bars. Antwerp appeared to be a very conservative city, since we didn't see anyone else that resembled us. In fact, from the way everyone was staring, I felt a little like a monkey in a zoo enclosure. Later that night, we checked out a skin-flick called 'Knockers McCall'. It was pretty cheesy but good for a few laughs. And the six or seven other guys in the theater wearing trench coats seemed to be enjoying themselves…if you know what I mean? Afterward, we headed back to the repair shop but work on the van wasn't completed until after 11:00 P.M. By then, it was too late to go back to the hostel, so we decided to sleep in the van. Early the next morning, we were back on the road and actually made it out of the city. However, the engine is still running a bit rough. We're hoping that it's just 'growing pains' and that the replacement motor will settle down after a while."

"After driving for another hour or so, we crossed the Dutch border. Unfortunately, the engine is still unsettled. Looking for a place to refuel and maybe locate a mechanic that can check-out the VW's motor, John decided to pull into the next town ahead of us. Before long we approach a small community called 'Goes' and John exited the highway. As we pulled into the main part of the town, there appears to be some type of fair taking place that day. The square is all done up with ribbons and streamers and there's some kind of horse show in progress. Man, they had some of the biggest horses there that I had ever seen. Goes was a farming community and most of the animals on display all had something to do with that kind of work."

"Since almost everything in town was closed due to the fair, we never found a mechanic, but we did manage to fill-up the VW's gas tank. John asked if I wanted to drive the rest of the way to Amsterdam and naturally, I jumped at the chance. I had never driven outside of North America before, so this was a first for me. I was a bit nervous at the beginning but at least the cars here drove on the right side of the road as compared to England where they drove on the left. I thought, how hard could it be? Holland is flat as a pancake and the roads are pretty decent, so it wasn't long before I felt pretty comfortable behind the wheel. I just asked John to navigate for me and keep an eye on the road signs so that I could concentrate solely on driving. After leaving Goes, it soon became quite foggy, so aside from the road ahead of us, there wasn't much else to see."

"Anyway, Kid. You are already familiar with my superior driving skills, so I don't need to tell you that we made it to Amsterdam safe and sound. After parking the van, we spend the next few hours just walking around the center of town and checking-out the sights. Afterward, we found the local Youth Hostel and registered. The cost per night, with breakfast in the morning, was 5 guilders. That translated into about $1.35 per night. The only downside was that the hostel was jam packed with travelers. Amsterdam was a favorite destination for back-packers, so I wasn't surprised."

"I don't know what dummy turned me on to vegetarianism, but at this time of my life, I wasn't eating much meat. I didn't stick to the diet too faithfully, but I was attempting to eat healthier if I could. As a result, John and I ate supper at Kosinos Restaurant that first night in the city. It was a micro-biotic locale, and to be honest with you, I wasn't turned-on by what they served. When I look back now, I wonder what the hell was I thinking? I've always been more partial to seafood than meat, but a good steak or roast chicken for dinner once in-while, sure hits the spot."

"After eating, John and I looked into a few of the 'weed bars' that were all over the place downtown. I thought these bars were a great idea and I hoped that more cities and countries around the world would eventually get on board with that concept as well. Outside one of these bars, we were approached by this dude who offered us 10 grams of Turkish hash for 35 Guilders. When you figured out that this was only about $10.00 dollars in our money, it was actually a steal. Naturally, we accepted his offer and skipped back to the hostel as giddy as a couple of school boys on Christmas morning."

"The next day we received the usual starvation rations that seemed prevalent at all Youth Hostels world-wide. It consisted of two thinly sliced, single pieces of rye bread with a bit meat and cheese on top and a cup of weak tea. Hardly enough for a growing boy like me. Naturally, we left the facility that morning still hungry. Also, in an effort to save even more money, we decided to sleep in the van from now on. As was par for the course during all of my travels, we spent most of the day just walking around and checking out everything. Later in the evening we went to see the new Mick Jagger movie called 'Performance'. It was kind of weird but it had a good soundtrack."

"After the movie and a few more beers, we returned to the Youth Hostel. Not to stay, but to post a notice on the corkboard that we were looking for a travel companion to share our expenses en route to Spain. While pinning up the notice to the board, we met Jason, another Canadian circumnavigating the globe. He turned us on to the 'Prince Bar' where beer was only 25 cents a glass. After drinking a snoot full, John and I eventually located the van and went to sleep. It was actually quite comfortable inside there but a little cooler than I would have preferred."

"November 29th was a Sunday, and as it so happened, almost everything was closed. I didn't realize the people in Holland were that religious. However, I later found out that this was the case over most of Europe. Sundays were considered family days and people should be at home with their families, not working. We ended up spending the better part of the day at the National Art Gallery and I was glad we did. The building housed some beautiful works. Everything from modern-day abstracts to canvasses hundreds of years old. We then headed back to the van and I started writing a few letters for the folks back home. As our body heat was beginning to warm the interior of the VW, I'm noticing a foul smell. It then dawned on me that I was the foul smell. I hadn't taken a bath in over a week and as a result, I was definitely over-ripe."

"The next day, I hit the public bath house. For the approximate cost of 14 cents, I kept myself under a stream of hot water with a bar of soap for over an hour. When I finally left, I felt like a new man. Near the bath house was an Automat where for 34 cents per load, I had all of my dirty clothes laundered by a friendly, non-English speaking Indonesian woman. John was an art student and the reason for him coming to Amsterdam in the first place, was that he had entered a competition with a few pieces of glasswork that he had blown. I accompanied him to the Art School later that afternoon and discovered that the academy also had a restaurant in the basement. For a few pennies, you could get a great meal prepared by the culinary students studying there. John and I ended up stuffing ourselves."

"The next few days seemed to blend together with not much exciting going on. We eventually went back to the Youth Hostel to see if there were any bites on our 'traveling companion' request. In fact, there were two. The first was from a single American dude named Irv. While the second was a young couple, Brad and Patricia. They were also both American, hailing from Minnesota."

"On the morning of December 4th, we planned on leaving Amsterdam. John and I started driving toward the hostel in order to pick-up our trio of new found friends. Wouldn't you know it, we got a flat tire, and it was pouring rain outside. Not only that, John couldn't find a tire iron in the back of the van.

There was nothing left for us to do but turn up our collars and start walking. After hoofing it for about four blocks, we were fully drenched. But a least we found a repair shop that sold John a used tire iron for a good price. After finally changing the tire and picking up our traveling companions, we were on our way."

"We made it into Brussels by late afternoon and stopped at the first bar we came to. Inside the bar, I was dancing with Patricia when this older fruitcake cuts-in and starts dancing with me. He then leans over and kisses me on the ear. Not once but twice! I guess that I must have turned about three shades of red, since everyone around us started laughing their asses off. That is, everyone except me. I angrily shook my fist at the old queen but he merely blew me another kiss and tripped off to try his luck elsewhere. We got well-oiled that night but the conversation always seemed to drift back to me and my unrequited lover. What an embarrassment!"

McGuire then stopped to face Jesse directly and resignedly admitted, "See, Kid. Not only am I irresistible to women, but the guys can't seem to keep their hands off of me either. It's a curse, but I guess I just need to learn to live with it."

It wasn't so much McGuire's words but the expression on his face after this last statement that caused Jesse to become unhinged with laughter. The wine might have also have contributed to his outburst, but whatever it was, before it was finally contained, Jesse was wiping the tears from his cheeks. He was now becoming quite adept at imagining McGuire's descriptive phrases and this last envisage was just too much for him to withhold.

"That night the guys slept in the van after we took Patricia to the Brussels Youth Hostel. The next morning, we were back on the highway en route to France. Frustratingly, the VW's engine began running rough again. However, we hoped for the best. At noon, we crossed the French border without any hassle. Before doing so, we changed our Dutch and Belgian money into French francs. In those days, one franc was worth about 20 cents to the U.S. dollar. And since all of my traveler's cheques were in American dollars, it was these little facts that were important to me. It was a beautiful sunny day that afternoon but quite cold. The French countryside was absolutely fantastic and everywhere you looked, you thought you were seeing a picture postcard."

"About mid-way between Brussels and Paris, we rested at a very quaint little town called Suzanne. It appeared to be lost in time. All of the houses and related buildings appeared to be at least 200 years old and the citizens were really laid back. It was a bit too quiet for my liking but a nice place to stop for a while. Since the van's motor was still stressing us out, John went looking for a mechanic. He finally found one and the guy started tightening this and adjusting that until it seemed to be running much better. He didn't even charge us for his services."

"Once back on the highway, creeping ever closer to Spain, we were feeling more confident in our van's ability to do its job. Unfortunately, once again a wrench was thrown into the works. Brad asked John to pull over to the side of the road. He then explained that he and Patricia had been counting their finances, and they were much shorter than originally thought. As a result, they wouldn't be continuing with us to Spain but needed to bail-out now. That was a real downer, since the whole idea of bringing them along was so that they could share the travel expenses."

"We were now going from a five-way split to a three-way split. Not cool. They were also a pretty nice couple so we would miss their company. Their plan was now to hitch-hike toward Luxembourg and re-think their itinerary. Well, out they went and now it was just us 'Three Musketeers'."

"From this point onward, I was doing a lot of the driving, and it was so cold inside the van, that we were nearly frozen. At times, I could barely feel my feet and I didn't know whether I was stepping on the gas pedal, the clutch or the brake pedal. Volkswagens are extremely reliable vehicles, but anyone who has ever owned one knows that their interior heating systems suck big time. Our van was no exception. We eventually reached the outskirts of Lyon and stopped at a lodge in order to warm up. While I was thawing out my nearly frostbitten feet and sipping on a headless beer, Irv and John tucked into a bottle of red wine."

"I pulled out my writing supplies from within my pack and began penning a few letters and postcards. I lost track of time and when I looked up again, John and Irv were absolutely blasted. I have no idea how many bottles of wine they polished off, but judging by the sight of them, it must have been quite a few. Irv could barely talk and John wasn't doing much better. While inside, we met a real friendly French-speaking Moroccan guy. After the lodge closed at 9:00 P.M., he invited us to another bar near-by, that stayed open later. Against my better judgement, we agreed and proceeded to the next drinking establishment. When we finally left there, John and Irv were very nearly incoherent and neither could walk unassisted. I was manhandling John while the Moroccan guy was attempting to support Irv. Well, I guess Irv was a bit too heavy for our escort, since he falls forward and does a face-plant onto the sidewalk. He really banged up his face when he hit the cement and it's bleeding like a professional wrestler that has just had his forehead driven into the ring's turnbuckle by his opponent."

"The Moroccan invites us up to his apartment to meet his family and to patch-up Irv. We finally managed to stagger to his flat and as we entered, there was his wife and about five or six kids to greet us. To say it was crowded inside the tiny flat was an understatement. It reminded me of all those tins of sardines I consumed during my last big trip. We were practically sitting on top of each other's laps. Irv ended up with a pretty good gash over his left eye and the Moroccan guy was attempting to bandage him while his wife started fixing us something to eat. The kids just sat there staring at us, as if we were from another planet."

"I don't know if it was too much wine, the sight of Irv's blood or the smell of raw onions that the Moroccan's wife was now beginning to fry. But before I could get to him, John started heaving and then barfed all over the Persian rug on the living room floor. The kids immediately screamed and jumped back, while lifting their feet, as if a torrent of molten lava had just flowed into the room. The adult Moroccans just stood frozen in their positions with their mouths gaping open in utter disbelief."

"I quickly attempted to steer John into the flat's only bathroom. But before he could reach the toilet bowl, another torrent of projectile vomit spewed from his mouth. This time it was all over the bathroom door, the sink, the bathtub and the floor. I think he hit everything but the toilet and the ceiling. The wife and kids immediately burst into tears while the husband just stared at us in shock, trying to say something, but not able to find the words. All I could do was continuously apologize, while attempting to gather up Irv and John for a swift departure. I was worried that if the Moroccan regained his composure before we could leave, that he'd probably try to kill us!"

"It seemed that John and Irv also realized that we were in a bad situation, since they seemed to sober up slightly. At least enough for us to get away from there. As we eventually exited the flat, the wife began shrieking hysterically and I kept looking over my shoulder as we stumbled down the sidewalk to ensure that the Moroccan wasn't coming after us with a butcher knife. We finally made it back to the VW van and I chucked both Irv and John inside. I then locked all of the doors for security reasons. As I gradually headed off to dreamland, in my mind, I could still hear the woman screaming. I can honestly say that this was one of the most embarrassing moments of my life. It was now the morning of December 7th. For the Americans in 1941, that was the date forever remembered as the day the Japanese armed forces attacked Pearl Harbor in Hawaii. For that Moroccan couple, I'm certain that they will remember it as the day the Americans defiled their flat. Although, I wasn't American, I guess to them, we all looked the same."

Jesse wasn't even present that night the event occurred, but even he sat there with a look of shock on his face. It was obvious that he was visualizing the scene. What originally began as an attempt of camaraderie and hospitality for the Moroccan couple, turned out to be absolutely disastrous. Afterward, Jesse could only imagine the woman warning her husband that if he ever came home again, unannounced, with a pack of drunken strangers, she would probably castrate him.

McGuire once again commenced speaking. "The next morning, Irv and John were obviously both hung-over. They had been so bombed, they now had no memory of the debacle. I had to inform them of what occurred, but I'm not certain if they believed me. I think the gash over Irv's eye and the smelly, dry crusted puke all over the front of John's shirt, finally convinced them. As a result, I had to drive the van. To be honest with you, I couldn't get out of Lyon fast enough. I kept thinking about that poor family and the mess we had caused them. We didn't even stay behind to clean it up. At that moment, I told myself that from now on, I was going to have to keep an eye on both John and Irv when they drank. By the end of that day, I ended up driving over 300 kilometers."

"We arrived in Marseilles shortly after 3:00 P.M. It must have been the start of their rush-hour, since the traffic was bumper to bumper. I was tired from driving all the way here and now instead of being able to rest, I was still stuck behind the wheel. John was still feeling sick and Irv had lost his glasses last night somewhere in Lyon. As a result, he was nearly half-blind without them. As I slowly snaked my way through the winding streets, pedestrians were giving us the evil eyeball. I felt like a gazelle on the savannah, surrounded by ravenous hyenas. Most of the people I saw were truly shifty looking. I got the impression that half of them had just committed a crime, while the other half were just about to commit one. I needed to get out of there, so I started driving up a steep hill. Traffic was thinning out nicely but I had absolutely no idea where I was going."

"We finally reached the top of the hill. The view of the city and inner harbor below was quite impressive and the homes appeared to be more affluent. I noticed that many of the houses had platforms constructed on portions of their roofs with railings surrounding them. These additions seemed out of place and I wondered as to their purpose. As we sat parked watching the activity below, I observed a pair of cute, younger girls walking toward us. They looked a bit friendlier than most of the other faces I saw earlier on the streets, so I rolled down the van's window and called out to them. Thankfully, they spoke English. I asked them what the story was regarding the platforms on the rooftops and surprisingly, they had an answer for me."

"Marseilles being a port city, naturally, the sea played an important part in its commerce. This meant that many of its men-folk were either in the French Navy, were merchant sailors or fishermen. With the unpredictable nature of the oceans or during times of conflict and storms, these were all potentially dangerous professions."

"Mothers, wives and secret lovers of those sea-farers would often pace the rooftop platforms gazing into the harbor below. Hopefully to observe their loved ones returning to port. Frequently, ships and other boats were lost at sea for various reasons, and as a result, some men did not return. Over time, these platforms became known as 'Widows Walks' or 'Widows Watches', depending on whom you spoke with. I thought that this was an interesting tidbit to tell the grandchildren sometime in the future, and I was glad that I had asked these young historians. We decided not to remain in Marseille but rather just pass through. However, I wasn't going anywhere until I had a bit of a rest and something to eat. We left the hilltop and drove back down into the heart of the city."

"After a good rest and a couple of sardine sandwiches, we filled-up the VW's gas tank and continued on our way. I drove well into the night but by the time we reached Perpignan, I was spent. There was no way that I was going to move another inch. I pulled the van well off the road and hit the mattress like a ton of lead. Irv and John were already asleep and had actually slept away most of the day. My upper body muscles were stiff from hanging on to the steering wheel all day and my lower back was killing me from hours of sitting.

Gradually, all of the aches and pains were beginning to disappear as I fully stretched out and pushed John over to his side of the mattress. It started raining as I drifted off to sleep. The water droplets hitting the metal roof of the van and the sound it created within the hollow vehicle, reminded me of someone gently tapping on a snare drum. The pleasing rhythm sent me off nicely."

"The next morning, my traveling companions were in better condition after their extended rest and I was getting excited as we inched ever closer to Spain. We were just reaching the base of the Pyrenees Mountains when the VW's engine started acting up again. It was still running, but hesitating and sputtering quite a bit. As we began ascending the mountains in earnest, the van was really struggling. At some points, we were barely making 15 mph and I was worried that we were going to get stuck there. That was the last thing we needed. Thankfully, the old buggy eventually managed to reach the summit, and we all breathed a sigh of relief as the subsequent downward slide afterward went much smoother."

"We were now only a couple of kilometers away from the French/Spanish border and I was beginning to sweat bullets. Were we going to be permitted entry into Spain or not? The moment of truth was now at hand as we approached the check-point near the town of Le Perthus."

"The Spanish border guard requested our identification and started peeking into the back of the van through the side windows. The look on his face told me that he was having a bad day and I started getting that sinking feeling. He spent what I thought was a lifetime checking our papers and comparing our faces to the pictures on the passports. But in the end, he just returned our documents without saying a word and waved us under the barrier. I couldn't believe it. After all that time spent worrying since leaving England, we just breezed right through. I waited until we put some distance between us and the check-point before letting-out a loud cheer. After four years, I was finally back in Spain and anticipating good times to come. Along with my much-inflated demeanor, the weather was now improving as well."

"The sun had broken through the scattered clouds and the further south we traveled, the warmer it became. Driving along the Costa Brava was a little slower going, but worth it, since it was very scenic. The invitingly turquois-blue water was a much-anticipated familiar sight for me. And the waves thunderously crashing on the shore conjured memories of many fun-filled former hours spent swimming and body surfing. One noticeable difference was that now there was quite a few more resort developments and condo buildings along the beaches as we motored past. That usually meant more tourists, and increased tourists, meant higher prices. I wondered how much this might impact my bottom line. We were now only about 60 kilometers away from Barcelona and eventually reached its outskirts within a couple of hours."

"It felt as though I was getting reacquainted with an old friend. I was beginning to recognize a few familiar landmarks the closer we drove toward the city center. We eventually parked the van and started walking around. We quickly located a reasonably priced Pension and checked-in. I was also relieved to find that most other costs such as food and drink were still fairly cheap. Happily, I would be able to stretch my budget quite a bit further."

"Looking around, it appeared as though the U.S. Navy was in town. Everywhere were sailors in their white uniforms and gob hats, souvenir shopping and generally checking-out the sights. This would be a big boost to the city's economy and would also keep the 'hookers' busy until they left again. I wouldn't be surprised if they had to recruit extra girls from neighboring towns in order to keep up with the demand. We visited a few bars throughout the day and on each occasion, I would challenge the odd 'sucker' to an arm-wrestling match."

"I don't mind admitting that I finished the day undefeated, and thoroughly impressed more than one drunken sailor, that underestimated the strength of an over-fed, long-haired, Canadian leaping gnome."

"Buoyed by my successful physical exertions and reinforced with a belly full of beer, I was ready to take on the world. As we continued through the city on foot, we arrived at the Arreo Funicular cable car station. We figured that this would be a great way to view the city from above and we could also rest our sore feet for a while. The ride in total was less than ten minutes but its highest point was nearly 300 feet and that definitely sobered me up a bit. In fact, I don't mind admitting that it was actually kind of scary. The ride ended near the inner harbor. In order to re-establish our composure after that hair-raising experience, the necessity of another drink…or two, was required. And speaking of hair. Wherever we went, there were quite a few people staring and giggling at my magnificently flowing red locks. Men with long hair was still a bit of an anomaly in Spain, and I got a kick from their attentions."

"We soon found a suitable bar and spent the rest of the night drinking beer and stuffing our faces with calamari. During our stay, I kept half an eye on both Irv and John to ensure they behaved themselves. If I could help it, I wasn't going to relive another night like we recently experienced in Lyon."

"We remained in Barcelona for a few days just bumming around and enjoying ourselves. I was getting anxious to 'hit' the Canary Islands so, I kept prodding John and Irv that we needed to move on. But first stop would be the Balearic Island chain for a few days. Finally, on December 13th, we piled into the van and drove it to the Barcelona Airport. We could legally park it there for as long as we wanted at a cost of only 5 Pesetas. At that time, 69 Pesetas equaled $1.00 U.S. dollar, so basically, we parked for free. After removing our belongings and securing the van, we all took a shuttle bus back into the city and headed for the harbor. From there, we boarded a boat bound for Eivissa Island. Total cost was 108 Pesetas."

"Within a couple of hours, we docked at Sant Antoni on the tiny island of Eivissa. What a gorgeous location! I immediately fell in love. Between the temperate weather, the azure-blue water with golden sand beaches and incomparable scenery, it was probably the most beautiful place that I had ever visited. Now that I was finally there and seeing what it was like, I never wanted to leave again. Our first challenge was finding a suitable place to stay and then somewhere to party."

"We eventually checked into a place near the beach. While registering, I started chatting-up a cute chick named Carole, who was also staying there. She had already been there for a few days with some friends but they had since returned to the mainland. She was going to rejoin them in a couple of days. Since we were new to the scene, Carole recommended a nearby cantina that served great seafood and reasonable drink prices. That sounded right up my alley, so I invited her to join us there."

"We had a blast that night, and by closing time we were all pretty gooned, as we stumbled back to our lodgings. John had purchased a big bottle of cheap red wine before leaving the cantina and we all repaired to our room in order to enjoy it. As I expected, Irv and John were murdering the bottle of wine while Carole and I became a bit more acquainted with one another. Before long, the bottle was empty and Irv and John had passed-out. I thought it was high-time that I introduced Carole to my 'middle leg', so I then treated her to a hot beef injection. I found her to be a very receptive client."

"The next morning, I awoke to find Carole asleep in my arms. Before Irv and John could rouse themselves, I took that opportunity to slip in my dipstick to better check her oil level. I found that she was down a quart, so I decided to top her up. Later, Carole told me that she had appreciated my service provision."

"It was already noon before we all managed to motivate ourselves enough to leave the room. Carole suggested that we should rent mopeds and tour the island. She had done it once already and really enjoyed it. The cost was minimal and since the island was so small, we could probably circumnavigate its entirety before nightfall. We all agreed and headed out. I had never driven a moped before, but essentially, it was a motorized bicycle. We had a lot of fun throughout the day, stopping now and again to check-out an exceptionally pretty vantage point, or to grab a snack at the occasional roadside stand along the route. It was already dark before we returned to the rental shop. The owner kept the business open after hours and wasn't impressed when we finally arrived. However, all was right with the world when we slid him a few extra pesetas for his trouble."

"Later, we all visited the same cantina as previously and once again, closed the place. Since, Carole was leaving the following day, I spent the night in her room while Irv and John kept themselves company in the alternate room. Carole checked-out early the next morning as I tip-toed back to my room. I didn't sleep much that night, so I was hoping to grab a few winks before Irv and John awoke."

At this point, McGuire threw Jesse another one of those looks wishing to emphasize his unspoken meaning. By now, Jesse was completely familiar with McGuire's near addiction to the female species and he took it all in stride. He had long since lost count of the many sexual exploits attributed to his aged friend. The notorious Renaissance Venetian womanizer, Giacomo Casanova had nothing on this guy.

The next couple of days for the trio were spent sleeping, eating, drinking, sunbathing and swimming. In essence, everything attractive for those with nothing else to do but enjoy themselves. McGuire then spent a moment to review his travel journal before returning his attention to Jesse.

"I was quite happy to spend the rest of my life in the town of Ibiza on Eivissa Island, but John and Irv wanted to check-out Las Palmas de Gran Canaria. That group of islands was approximately 500 miles away from where we were, off the northwest coast of Africa. This would mean another lengthy boat trip and we would need to return to the mainland in order to catch that vessel."

"On December 16th, we sailed back to the coast and landed in Valencia. It had been a horrible trip, since the sea was very rough. Kitchenware and deck furniture was flying all over the place, while those less acclimated with the ocean, were leaning over the railings puking out their guts. Once again, I felt fine."

"We only remained in Valencia overnight and then got ready to leave for Las Palmas via Alicante, the next day. Unfortunately, we arrived at the Ferry Terminal a bit too late that morning, and as a result, all of the 3rd Class bookings had been sold. This meant that we needed to purchase the last three remaining 2nd Class tickets. If we didn't like that, we could stay behind. That was our only choice. The kicker was that a 2nd Class ticket cost $42.00 USD. That was way more than I wanted to spend! Man, was I pissed off but we were between a rock and a hard place. Reluctantly and while swearing under our breath, we purchased the tickets. Waiting to board the boat, we walked over to a nearby park and started throwing around the Frisbee. Before we knew it, there was a huge crowd of Spaniards watching us and cheering whenever one of us made a great throw or an impossible catch. They were unfamiliar with this form of recreation and we suddenly became minor celebrities. I thought an astute entrepreneur could probably make a million bucks marketing those plastic discs over there."

"Once aboard the boat, we were up to our usual hijinks of drinking, sneaking the occasional toke of hash and trying to chat-up any and all loose chicks. In Alicante, we picked up some more passengers. Most were Canadians or Americans. In one of the groups boarding, I noticed a real cute tomato with a killer body and I sidled up to her. She told me her name was Bobbie and I informed her that my name was Bob. This elicited a huge laugh from us both and the ice had now been broken. We hung around together on the vessel during the trip to the Canaries and quickly hit it off. After a relatively boring trip, we eventually made it over to Las Palmas and docked in Arrecife."

"As we watched the stevedores unloading the ferry, we remarked that they were moving slower than molasses. They were even more useless than the longshoremen back in Vancouver and if they could work any slower, they'd be going backward. After an hour, most of the passenger's bags were still on the boat. I soon found out that Bobbie was a real mischievous imp when she leaned over and whispered in my ear. I was a bit surprised by her suggestion but game to play along. Following her lead, we slipped behind one of the large baggage carts and both removed our tee-shirts. Bobbie owned a luscious pair of tubes and obviously, she wasn't shy to show them off. We came back around from the baggage cart and started parading back and forth in front of the stevedores. You should have seen their faces. They had all stopped what they were doing and just stared at Bobbie. I don't think they even noticed me. Although, with my long hair, I could have been a flat-chested girl. Some of them were whistling or pointing and giggling, while others just silently gaped."

"We left them a couple of minutes later with their tongues dragging on the ground while we slipped behind the baggage cart once more. Replacing our tee-shirts, we spent a long time afterward roaring with laughter. I don't think our little topless show made the stevedores work any faster but it definitely left them with something to talk about at the end of their day. I hadn't known her very long, but Bobbie was definitely someone I wanted to get to know better."

"Over the next two-plus months, the island of Las Palmas de Gran Canaria was my home and little piece of heaven on earth. Along with Irv, John and now Bobbie for company, there was also a multitude of different characters that blew into and out of our lives like winter snowflakes descending from the sky. Some stuck to us and spent time sharing our experiences, while others quickly melted away, leaving us little to no recollection of them. The island itself was considerably larger than its related landmass, El Hierro."

"It boasted twice the population and its denser inhabited communities nearly doubled that of El Hierro. It also possessed far more attractions and amenities for visitors. Most of our days and nights were spent chasing the seven deadly sins and we successfully achieved many of them on a regular basis. Of the seven, the three I was most guilty of were…Lust, Gluttony and Sloth."

Lust, because I couldn't stop myself from chasing anyone possessing a vagina. So long as she was moderately attractive, over eighteen and under fifty years old, with a relatively strong pulse, then she was fair game for me. Gluttony, because I spent every waking moment either eating, drinking or smoking drugs to excess. And when I wasn't engaged in any of those activities, I was thinking of ways to correct that absence. My motto was 'too much is never enough'. And lastly, Sloth. "During my time in the Canary Islands, the most strenuous activities I engaged in, were getting-up in the morning, reading books while sunbathing, and frequently splashing around in the ocean. Occasionally, I would really exert myself and throw the Frisbee around or kick a soccer ball. Overall, I was living the life-style of an English Country Gentleman without actually possessing any of the wealth or honorific title that usually accompanied such personages."

Jesse wanted to dispel any feelings of guilt McGuire may possess as a result of this last admission, so he offered his own comments. "Well, it was a tough job but someone had to do it." McGuire acknowledged the youth's remark with a brief but appreciative nod. He actually wasn't feeling guilty in the least. That was a great time in his life.

"Just like New York was a city in the state of New York in America. So was Las Palmas a city on the island of Las Palmas in the Canary Islands. While on the island, we spent most of our time either in the city of Las Palmas or else we would bounce over to towns like Galdar, Sardinia or San Mateo. Each community had something special to offer and if we couldn't catch a ride to any of them from a local citizen or some other visitor with a car, then we took a transit bus."

"As we neared Christmas, I was still trying to wrap my head around the fact that I would be spending it on a warm, sandy beach in paradise while my Bear Family back home were going to spend theirs in the cold and damp of Vancouver. A few days before the holiday, I headed into the city and attended the Main Post Office. Lo and behold, there were two letters for me. One was a Christmas card from my parents containing a $15.00 CDN money order, while the other one was a letter from my brother John. It was like hands across the ocean. I was glad to hear from them."

"In the main square that day was a large pro-Franco rally with marching bands, military ordinance and a big parade. There were banners and flags flying from many of the buildings and thousands of cheering people packed the streets. There were uniformed police and soldiers everywhere and I suddenly started feeling a bit paranoid. I got the impression that people were beginning to notice this fair-skinned, long-haired Gringo in their midst and their mood toward me wasn't exactly friendly. After all, the fascist government led by its dictator, Generalissimo Francisco Franco, was on record of disliking decadent western governments and their tolerance of dissidents and free-spirited, anti-establishment types. I guess, I could have been lumped into the latter category. Anyway, rather than overstay my welcome, I made a hasty exit, stage right, and got the heck out of there."

"Christmas and New Year's Eve were a blast with lots of good food, liquor and drugs. Unfortunately, I was still trying to stay faithful to my vegetarian diet, so while everyone else was stuffing themselves with turkey, beef roast and other delectable viands, I stuck to salad and baked potatoes. Somewhere along the way, Bobbie had evaporated and now I was hooked-up with a sweet little maiden named, Wendy. Our main lodgings were at the Hotel Diana in Las Palmas. Six of us had rented the penthouse. This might sound extravagant, but in fact the rooms were reasonably priced. Split between the entire group, it was quite affordable. However, having said this, depending on where we partied on any specific night, that's where we would probably remain. Sleeping in a stairwell or on the beach was not out of the norm."

"One night in early January, I found myself in Sardinia with a girl named Sandy. We had missed the last bus back to Las Palmas, so we decided to spend the night there. We started looking around, but the cheapest room in town that we could find, was 580 Pesetas per night. There was no way that I was going to spend that kind of money, so we started walking toward the beach. En route, we came across a couple of Guardia Civil. That was what they called the Spanish National Police. Anyway, I guess they figured out that we were going to sleep on the beach and stopped us. One of them spoke decent English and he addressed Sandy. He warned her against sleeping on the beach. He said that if the Guardia Civil was to find anyone sleeping on the beach, first they would shoot them and then they would throw them in jail. I thought that was a pretty ignorant thing to say, but I guess they wanted to scare us."

"We waited for them to leave and then continued walking toward the beach. Along the way, we met a younger Spanish dude and I asked him if there was a safe place on the beach where we could sleep and not get hassled by the cops. He volunteered to show us a spot so we followed him. It was quite a drop from the top of the road to the beach below but it could be accessed by a series of wooden stairs. It was pretty dark and looked dangerous, but we trusted the kid to get us down safely. At one point, he stopped and showed us a location where apparently the night before, a drunken German tourist had fallen to his death over the railing. Picturing the event in my mind, I thought that mishap would most probably have ruined the fool's holiday."

Jesse laughed at McGuire's cryptic remark and agreed that indeed, such an event, would definitely put a permanent damper on things.

"The local brought us safely to the beach and took us down to where several wooden fishing skiffs, containing nets and related gear, rested on blocks. He told us that the police never come down this far and we would be safe there. He then started back from whence we came. Sandy and I went to the boat furthest away from the rest and moved to its opposite side. From that angle, we wouldn't be seen by anyone approaching on foot. I started rummaging through the boat and found a couple of old seat cushions. We wouldn't have a blanket but at least we'd have pillows. It was still pretty warm outside, so if we snuggled up a bit, we should be fine."

"It was out of character for me, but that night, I declined attempting any hanky-panky with Sandy and we both fell asleep relatively soon. It felt like we hadn't been sleeping very long when I was suddenly awoken by a loud racket. The sky was just beginning to lighten, but mostly, our surroundings were still in relative darkness. Being a seiner-hand myself, it now dawned on me that the noise was being created by the local fishermen arriving for work. Naturally, the best fishing occurred early in the morning, so they were beginning to prepare their craft and head-out to sea. Just my luck! By now, Sandy was also awake. We rose and shook the sand from our clothing. I tossed the seat cushions back into the skiff and we then sleepily started walking back toward the staircase leading up to street level. The fishermen were busy with their duties and didn't even notice us strolling past them. We had to wait for a few hours afterward, but we eventually caught the bus back to Las Palmas. Not having slept much that night, I was beat, but made up for it once I returned to the hotel. After grabbing something to eat, I spent most of that day in the sack."

"One day, I didn't feel like leaving the hotel but I also didn't want to just lay in bed staring at the ceiling. I actually forgot which girl from our group taught me embroidery, but for a lark, I started stitching designs on my jeans. It was slow going at first, and a few times I had to pull out some false stitches and start over. But after a while, I was getting the hang of it and basically doing a pretty good job. On my first attempt, I started stitching at 11:00 A.M. and by 6:00 P.M., I had created a flower, two fish and a star. I was pleased with the final results and had to admit that they all looked pretty good. I decided that I'd attempt a few different patterns some other day, but at the moment, that was enough."

"As the days and weeks past, I was really beginning to vegetate. And not just because I was eating mostly vegetables. By now, my motivation level was at about a minus-10 and I was beginning to put on weight. That's not good for a Love God like me, so I thought it was high-time to start doing something about it. Firstly, I cut back to eating only one large meal per day. This should help me lose some weight and I'd be saving some money, to boot. Secondly, I thought that it might be cool to organize a soccer match. I was thinking maybe North America against Spain. I started throwing out the idea to a bunch of the 'heads', and to a few of the local guys we had come to know. Everyone thought it was a great idea, so we started training."

"Although our hearts were in the right place, in actuality we didn't train very hard or very often. Every time someone would show-up with booze or some dope, we would get hammered. As a result, the training went out the window. As we neared game-day, our resident pharmacist had been a no-show for a while, and everyone started getting the 'heebie-jeebies'. As a result, we were relegated to soothing our frayed nerves with only liquor. And although that was a fairly good substitute, it didn't compare with a good old-fashioned 'carrot-chillum'."

"After all," McGuire continued, "According to those television commercials of the 1950s and 1960s, 'Eight out of ten doctors prefer the **high** received from a potent Afghani joint as compared to that of a 4-finger glass of Scotch on the rocks.'"

This last sentence went right over Jesse's head but McGuire thought that his attempt at humor was at least worth a try. I guess the kid couldn't be expected to understand television commercials of the mid-twentieth century, when he hadn't even been born until the twenty-first century.

"We were all pretty bummed over this situation, so I thought that I would try to lighten things up a bit. Although, I've never considered myself a literary genius, I pulled out my travel journal and put pen to paper. The following poem resulted:

No dope today – Our contact ran away,
Liquor was a substitute – In our one room institute,
But we're so very versatile – Either way, we laugh and smile.

As McGuire completed reciting his poem, Jesse began clapping and raucously shouting-out for more. In mock appreciation, McGuire rose from his chair and took a few deep bows. As both men then broke-out in hysterical laughter, the storytelling had to be temporarily suspended until they could finally regain their composure. Before continuing, McGuire briefly referred to his journal once again in order to obtain accurate facts."

"February 3rd was my dad's birthday. It was also the day of the big soccer game. That morning, we spent several hours tidying-up the playing field by removing large rocks and related debris such as broken branches, palm leaves and general garbage. By early afternoon we were ready to play but it turned out that the Spaniards were only able to field three players, aside from their goalie. We ended up donating several of our North American players to their side in order to even the rosters. The Spanish players were pretty solid while their defense was nearly impregnable. Every time we attacked, either their defenders or goalie would stymie our attempts. Going the other way, the North Americans were less effective and by half-time, we were down five to zip. In the second half, I traded my sandals for our goalie's running shoes and my traction increased dramatically. In fact, I ended up scoring two goals but it still wasn't enough to make a difference. Begrudgingly, we lost the match. After getting blasted with the gang later on and licking our wounds, I completely forgot about my vegetarian diet and inadvertently ate a couple of hotdogs. Afterward, I wasn't struck by lightning, so I guess Ceres, the Roman goddess responsible for the growth of food plants, either hadn't seen me or otherwise, didn't care."

"As time continued ticking closer to March, I realize that our stay in Las Palmas is winding down. Near the end of the month, me, Irv and John went into the city and visited the Spanish Consulate Office. We were hoping to acquire a travel Visa for Western Sahara, since we had to move through there while en route to The Gambia, Africa."

"It turned out that we could actually grab a ship from Las Palmas and navigate to Africa from there, rather than first sailing back to the mainland. The only delay would be having to stop at Santa Cruz de Tenerife Island, in order to pick-up extra passengers. That was fine with me. By now, I've become an old salty dog, and there's nothing I like better than a good boat-ride."

"On March 6th, after saying goodbye to the many friends we had made in Las Palmas during our memorable stay, I was back at sea and gradually steaming toward yet another unfamiliar destination. Our new found friends, guitar-player Dave and master freak Rusho, were coming along as well. They were both great guys!"

"Our leisurely voyage from the Canaries to the west coast of Africa had its moments. Some were good and some were bad. But as always, there was something important to be gained from them. As we reached La Guera, where Western Sahara meets Mauritania, Irv, John, Dave, Rusho, me and a few other passengers needed to transfer from our mother ship to a rusty old barge, in order to take us to the peninsula. There were no proper docking facilities in La Guera, so our ship dropped anchor off shore and we sat on deck in the blistering hot sun for 2-hours, while the inept crew on land got their act together. A small tugboat finally sputtered its way out to us trailing the barge and we precariously transferred from the ship on to it. The huge tidal swells kept sucking the barge into the larger ship and then bouncing it backward. I thought the old tub was going to sink as a result. The tug then proceeded to laboriously tow the barge approximately 500 yards to shore. Travelling as far as it could in shallow water, a thick, long rope was then tossed toward the shoreline and about a dozen scrawny black laborers ran from the strand into the water's edge. They grabbed a hold of the rope and began pulling as though their lives depended on it, while their Caucasian overseer shouted a constant stream of abuse at them. Eventually, we landed. As we gathered our belongings, we were approached by a large, evil looking thug wearing a full military uniform, complete with a riding crop, a chest full of campaign medals and gold-braided epaulets. He was accompanied by a couple of armed goons, also in uniform, and their arrival seemed foreboding. This dude figured that he was West Africa's answer to Idi Amin, the notorious Ugandan dictator, and he immediately began barking orders at us. We were forced to empty the contents of every bag on to the sand, so that he could conduct an inspection. Since he was the only person with the authority to stamp our entry documents, we all reluctantly complied. This power-tripping pissant seemed to relish the additional time spent examining the female traveler's undergarments."

"Having arrogantly exerted his power over us, we were eventually allowed to reclaim our goods and leave. Our immediate group then took a taxi into Port Etienne. Today, the formerly French named town has reverted to its Arabic title of Nouadhibou. However, not to confuse matters, I'll keep referring to it as Port Etienne."

"I had never been in this part of the world before, and although it takes a lot to shock me, observing the state of this place, I was truly unnerved. Everything was covered in dust and the air was dryer than a camel's fart. There were swarms of large, black flies everywhere and the conditions of poverty the inhabitants endured, was unbelievable. People seemed listless and just aimlessly shuffled about. Even stall owners or other business persons seemed not to care whether they were making any trade. They were like zombies. Everything around us was in a terrible state of disrepair and it looked like the buildings were ready to collapse. The old nursery rhyme of the 'Big Bad Wolf and Three Little Pigs' came to mind. From where I now stood, the wolf in that ditty could probably have huffed and puffed and blown down this place."

"Customer service for anything was nearly non-existent. This included the Customs Agents and Border Guards. They were all nonchalant and essentially useless. They practically ignored us. Before they finally examined our documents and authorized our travel papers, we had missed the last train toward Nouakchott. That was Mauritania's capital city and also its largest. The next train wasn't due until the following morning. Thoroughly pissed-off, we grabbed our bags and started walking through the town. There was absolutely nothing to see there, so we parked ourselves and started throwing around the Frisbee. Once again, its pull was magnetic, and soon there a small crowd of Mauritanians watching us."

"One man in the group was an educated fellow who had spent a couple of years in England. He invited us into his house and his gorgeous, school teacher wife served us mint tea. During conversation, I politely raised the issue of the city's apparent neglect. Our host explained that most Mauritanians were extremely lazy and had no interest in either working or making anything of themselves. As a result, a regular stream of citizens from neighboring Senegal arrived in Mauritania to work. It was mostly their efforts that was keeping the country's economy going and much of Mauritania's disposable income was being absorbed by the Senegalese as a result. I found this to be a very interesting fact, but I couldn't understand the Mauritanian's attitude."

"After a pleasant visit with the couple and their children, we headed to a restaurant where all the other former passengers from the ship were congregating. It was decided that everyone would remain there overnight and await the morning train to Nouakchott. A financial arrangement with the restaurant owner had been concluded, allowing everyone to remain there until morning. By now, my lower back had been bothering me for several weeks. The thought of sleeping on the restaurant's floor or semi-reclined in a wooden chair, wasn't very appealing. Unfortunately, I really didn't have another choice."

"The following morning, the interior of the restaurant appeared as though it had suffered a Sarin Gas attack. There were bodies strewn all over the place. The only difference was that these bodies were still alive. There were people sleeping on top of the service counter, propped-up behind and in front of it, laying on tables tops or with chairs pulled together creating makeshift beds. The place stunk of overused cooking oil and stale perspiration. Swarms of flies had already reappeared inside the premises, or perhaps they had never left in the first place. Needing to relieve my aching back and stretch my legs, I exited the so-called eatery and went out on to the street. It was still early in the morning but the air was already stifling, and the day to come was shaping up to be another scorcher."

"Being bored to tears, but still wishing to leave my mark on this dump, I pulled-out a piece of purple chalk from within my pack and adorned the building's wall with a picture of a monster driving a taxi cab. I was quite pleased with the result, since it reminded me of our cab driver from yesterday who had driven us from La Guera to Port Etienne."

"As time passed, we were uncertain if our train would ever arrive, so we all started walking into the desert toward the railhead. We reached that point about an hour later, and noticed a massively long freight train that was ready to depart and heading more or less in the same direction as us. The entirety of the train was nearly a mile and a half in length and carried raw ore from the Atlas Mountains to port cities, from where it was then loaded onto ships and transported globally. We all jumped into one of the open and empty freight cars just as the rusted, protesting wheels began to slowly rotate. Due to the sheer length of the train, it seemed to take forever before we actually began making any noticeable speed. Unfortunately, we had made the mistake of choosing a freight car near the end of this gargantuan snake on wheels. As a result, all of the dust and sand straddling either side of the tracks, was being kicked-up from the train's forward movement, and directed straight into our open-air coach."

"Before long, we were all coated in dust and could barely breath. We used whatever articles of clothing we could to cover our heads and faces, but the dust still managed to infiltrate. My hair and beard had never been so filthy and my lips and eyes were caked in grit and grime. The train rattled on through the night and it soon became freezing cold inside our coach. That was one of the problems with desert climes. It was unbearably hot during the day but freezing cold at night. It was one of the worst nights of my life and there was no way that we could even relax, let alone fall asleep. This living hell on wheels continued for almost twelve-hours. Mercifully, near 5:00 A.M., the following morning, we reached our jump-off point. We left the train just as it was beginning to depart once again, but this time in a different direction. We were happy to be out of that tin can, but we were still a long way from Nouakchott. Because of the desert terrain, it wasn't easy getting there."

"Inside our freight carriage, we had met a Mauritanian named Ali. He also left the train with us and volunteered to help with acquiring a ride for the next leg of our journey. Still covered in the filth from our horrific overnight ride, we all walked into the marketplace and within a few minutes, Ali was able to barter a ride with the driver of a large truck. He would take us to his destination which was the town of Atar, in the northwestern portion of the country. The cost to us was 700 Mauritanian Francs, which sounded like a lot of money. However, when you realize that 100 Mauritanian Francs equaled only 37 or 38 cents in our money, he was practically paying us for the pleasure of our company."

"The man drove well and it was an extremely interesting ride all the way to Atar. The landscape turned semi-arid the further east we drove, with patches of shrubs and stunted trees along with the interminable sand. We observed Bedouin tents and encampments dotted here and there throughout the region with their meagre goat herds and the occasional dog. I even noticed a few gazelles, grazing on tufts of grass and a couple of Albino camels. The unrelenting brutality of the place really hit home when we passed the occasional dead camel at the side of the road. If these resilient and hearty animals could not survive that environment, we probably wouldn't even last a few hours if we ever got stuck out there."

"At one point, the driver stopped the truck so that we could all have a drink of water. I was parched and grateful for the opportunity to wet the inside of my mouth and throat. At this point, it was difficult swallowing, or even speaking for that matter. As we stood in the blistering heat, the driver reached under the front bumper and removed a bloated goat skin water bag."

"He then reached into the front seat of truck and removed an empty oil can. Filling the can with water from the goat skin, he first took a big gulp and then passed it over to us. When I eventually received the can, I looked inside to see remnants of motor oil still floating on top of the murky water. In addition, the liquid was very hot after being tucked under the front bumper all day in the desert heat. I did my best to take a long drink, but could only manage a miserable sip. So much for that. I was still thirsty and still dirty, so for me, reaching Atar couldn't come soon enough. Once back inside the rear of the truck, a brief interlude of entertainment was on hand. Inside the truck with us was a wizened old man. We were amazed to watch as the codger produced from underneath his clothing, a small empty tin, a flask of water and a sponge. With what appeared to only be a few ounces of water, he poured that into the tin, dipped the sponge inside, and then methodically and meticulously began washing every inch of his body. This went on for several minutes, and although hardly clean, he did appear refreshed afterward."

"Atar was even more miserable than Port Etienne. I had seen needy people before in North Africa, Turkey and Beirut during my first travels. But this was taking poverty to a whole other level. I couldn't believe what I was seeing. The multitude of beggars and other severely deformed people was horrifically pathetic."

"Most of them were literally starving to death and could barely function. The bloated, undernourished children were so used to having the relentless swarms of flies buzzing around them, that they no longer even bothered trying to chase them away. Those filthy insects would sit on the faces of those poor waifs and attempt to enter their mouths, nostrils and even their eye sockets, all in search of moisture. I'm no softy by any stretch of the imagination, but my heart was breaking to see the state those poor bastards were in. I just wanted to get the hell out of there as soon as possible."

"Ali managed to barter another ride for us soon after, but this next driver was a real prick! He owned a minibus that held about 14 or 15 people max. However, he ended-up stuffing 27 passengers in there. When we found out, we were really pissed-off but we had already paid our money, so there was nothing we could do about it. You didn't get refunds in a place like that. We could barely breath inside the bus all crammed together and it was hotter than Hades inside. At least, he stopped every once in a while, so that we could stretch our legs and take a few deep breaths."

"When we finally reached the small town of Tachot near the dinner hour, the driver stopped so that we could all grab something to eat at a roadside restaurant. Most of the passengers were skinned, so they just sat outside of the bus. Since we were better off, we entered the restaurant. To our surprise, Ali carried a few record albums inside his case and he convinced the proprietor to put them on the restaurant's record player."

"After that brutal bus ride, we were now sipping on a cool Pepsi Cola, eating finger food, and all the while grooving to John Mayall, Jimi Hendrix and Ten Years After. Unfortunately, we had to get back into the minibus after our tune-filled break and it wasn't until after midnight before we eventually reached Nouakchott. Ali suggested that we go to a district police office to see if they had room for us to sleep. He said that they often allowed that, and it would be 'no charge'. I didn't like the idea of spending the night in jail, but these would be non-criminal circumstances. As we reached the police station, Dave remembered that he left his guitar on the minibus. He headed for the main square in order to grab a taxi and hopefully catch-up with our former ride. It turned out that Dave was really having a bad night, since not only did he not catch-up with the minibus, he also left his backpack in the square when he jumped back into the cab. He later rushed back to the square but his backpack was gone. Inside the pack was 12,000 Mauritanian Francs, $30.00 dollars in American cash and $380.00 dollars in American Express traveler's cheques."

"Now, he was not only without his guitar, but all his money was also gone. As he checked the pockets of his jeans, all he managed to find was a miniscule 100 Pesetas left over from the Canary Islands. That was less than $2.00 dollars American. At the time, there was nothing more for us to do other than report the incident to the police. However, from their attitude and lack of response, it appeared as though they weren't going to be doing anything to assist Dave. He was just going to have to suck it up and carry-on as best he could. Hopefully, the American Express office would at least be able to reimburse him the $380.00 dollars in uncashed traveler's cheques."

"It was a bit unnerving sleeping in the police station that night, but overall, I had a great rest. The best part was the next morning when I was able to shower away the previous day's grime. Stepping out into the street afterward was like walking into blast furnace. Within a few minutes, I was sweating like a galley slave and my body had quickly forgotten all about the great scrubbing it had just received. However, everywhere you looked, the local people were dressed in long robes, as if for winter."

"I was surprised that Dave was actually in pretty good spirits considering his losses. Ali guided us to a large marketplace near the city's center and we looked around. Despite the large-scale poverty everywhere around us, the market's prices were quite expensive. I ended up only purchasing an elephant hair ring for 38 cents and a blue, faux gemstone for my necklace. One thing I really did enjoy there was the coffee. It was strong and satisfying, as well as being affordable."

"Everywhere we went, little kids were following us or hanging around our legs. This seemed to impress many of the locals, since we received quite a few handshakes from them accompanied by friendly greetings of, 'Bonjour' and 'Ca va'? To entertain the kids, we pulled out the Frisbee and once again, drew a large crowd. People loved it and we gave as many of them as we could, a turn at throwing it to one of us. We ended up sleeping at the police station again that night but this time, I slept poorly since they kept the lights on all night."

"The next morning, we quickly determined that there wasn't much shaking for us in Nouakchott, so we decided to split. We found our way to the Senegalese Embassy in order to acquire a travel Visa for that country. I didn't have any more legal sized photos of myself for the Visa, so I cut-out the picture from my expired California driver's license and handed it to the puzzled clerk. At first, he didn't want to accept the picture, since it no longer resembled my current appearance. However, after a brief hesitation, he took it from me and attached it to my Visa."

"Ali afforded us one last favor by bartering another ride for us. This time it was in a truck filled with giant sacks of rice. The driver's destination was the town of Rosso, right at the border of Senegal and Mauritania and separated by the Senegal River. As we traveled further southbound, I came to the full realization that I was actually in Africa. The desert sand was beginning to recede and replaced by pockets of vegetation. Soon we were passing groups of Nomads herding small flocks of goats or lean, long-horned cattle trekking over the savannah. Further south, we observed the occasional indigenous villages surrounded by stick walls. These enclosures contained clusters of conical huts with wisps of smoke rising from openings in their thatched roofs. The smell of manure was heavy in the air and we later learned that the natives burned dried dung from their livestock, as fuel for cooking and heating."

"We eventually arrived in Rosso and waved goodbye to our driver. We could see the border checkpoint across the river but couldn't figure out how to get there. A local tribesman in a large dug-out canoe noticed us standing on the river bank and waved us down. We couldn't communicate with him verbally. However, we soon managed to complete a financial transaction with him, whereas he would take us, and our things, across the river. I felt like I was starring in a Tarzan movie. It was quite a bit different than hitch-hiking in Europe. We managed to get through the check-point without any hassle and bartered a ride to St. Louis with a truck driver who was waiting at the Customs shack to have his papers processed. We were now officially in Senegal."

"As we arrived in St. Louis a couple of hours later, we needed to make the obligatory visit to a police station. Here, our money was counted and our travel documents would be authorized and stamped. Once this was done, we were given permission to leave our bags inside the station so that we could do a recon of the area. As we made our way into the town, we noticed that most people were very friendly toward us. We received a lot of smiles and waves from passersby and that helped to put us at ease. I wouldn't say that I had been worried about traveling to Africa, but I also didn't know how we would be received. After all, for hundreds of years, Western European countries such as Britain, France, Germany, Portugal, Belgium and others had been exploiting the African nations and their people. They imposed their own rules, often brutally, and tried to shove their religion down the throats of the natives. They raped them of their natural resources and treated them like third-class citizens in their own countries. It would be completely understandable if they resented us for all that, but I wasn't catching that vibe from them."

"As we ventured further into town, I noticed a distinct difference between Senegal and Mauritania. Where the latter was barren, dry, impoverished and depressing, Senegal was lush, developing and more optimistic. Sure, we saw evidence of poverty if we really looked for it, but it wasn't overwhelming. The people even appeared to be livelier. Almost everyone we saw was dressed in colorful clothing, especially the women. This was opposite to most North African and Arabic countries where females usually only wore drab colors such as black or other dark tinted shades and were forever covered from head to foot. Many of these Senegalese women wore full-length, brightly patterned sarongs with bare shoulders and large, turban-like head coverings. Most appeared very well endowed and when they walked in front of us, their plump and ample butt cheeks would jiggle back and forth, almost as if they were winking."

"I've got to admit that despite the essence draining heat of the African sun, I was getting slightly turned-on by these chocolate-colored goddesses. After a couple hours of walking, we were knackered. The heat and pitiless sun were really doing a number on us. Thankfully, we found a café that looked like it served something we could digest, so we went inside. The food actually turned-out to be very tasty and we stuffed our faces. One thing I discovered since arriving in Africa, is that water is like liquid gold there. Soda pop, beer and hard liquor were all cheaper to purchase than plain old H_2O. Afterward, we all went back to the police station and crashed in the open courtyard behind the building."

"I was so exhausted that I just dropped off to sleep without even considering what types of creepy-crawlers or wild animals might try to ravage us, while we were out there unprotected. The next morning, we awoke to find about fifty, young Senegalese men all standing against the wall of the police station and inquisitively staring at us. Once again, I felt like a caged monkey at the zoo with everyone wondering what I was, and where I came from. I was intrigued by this sight, so I took my camera out of my pack and snapped their picture. They loved it and all started hamming it up for me. After drinking a cup of coffee with the 'fuzz' and thanking them for their hospitality, we made our way to the main road and waited for a possible ride. Many of the larger roads in Africa were paved but most others are just hard packed dirt. As a result, it's always dusty. And on one those rare occasions when it does rain, they then turn into mud tracks. Right now, a rain shower appears to be the last thing we're going to be seeing."

"The undeniable reality of the African climate reinforced to us that 'hitching' in Africa was going to be much different than anything we had ever experienced before. Traffic, was also much less frequent. Here, in the smaller towns and villages, ox-carts or camels were still the preferred mode of transportation. It also didn't help that there were five of us with our thumbs out. We would need a truck or large automobile in order to accommodate us all. We had only been on the side of the road for a few minutes and already we were bitching and complaining. It was already hotter than hell and there wasn't a cloud in the sky. It actually hurt your eyes to look up. Not just because the glare of the sun would instantly melt the cornea of your eyeballs, but also, because the brilliant blue of the sky was so intense. I had never seen that unusual shade of blue before and wondered if it could be duplicated on a painter's mixing pallet. I would love to use that color in one of my drawings. At least, the wind was constantly blowing, so this helped ease our suffering somewhat."

"Also, there were quite a few large trees here and there and we would utilize the shade they provided. After about three hours without a ride, we're beginning to dry-out like Sun Maid raisins and the locals were coming by occasionally, to make sure we're still alive. Finally, a uniformed chauffeur driving a big, black limousine pulled over. He was heading into Dakar and offered to take us with him for 300 Francs each. This was more than we wanted to spend, so we started bartering with him. That didn't work out very well when he suddenly tromped on the gas pedal and sprayed us with dust. We wondered if we had made a mistake."

"Not long afterward, a small bus came by and this time we weren't going to play around. We all paid our 100 Francs and climbed aboard. Unfortunately, the bus didn't take us all the way to Dakar, so we still weren't out of the woods yet."

"In a move of desperation, I tucked my long hair under my hat and we moved toward the road again. The other guys looked a bit more presentable than I did, so we were hoping that someone driving by might mistake us for college kids. This seemed to work, since with 15-minutes, a French couple and their teenaged daughter, driving a large, 4-door sedan pulled over and offered us a ride. We couldn't believe it, but we weren't going to argue. We all managed to squeeze in, with our bags safely tucked away inside the vehicle's massive trunk. It was music to our ears when we learned that they were going to Dakar. The daughter spoke passable English, so I had a good conversation with her during the entire ride. In addition, the landscape was becoming more interesting and attractive the further we drove."

"The density of the greenery increased as did the number of thatched hut villages and colorfully dressed natives populating the area. Even though it was evident that the indigenous peoples were far from affluent, they all appeared to be satisfied with their lot in life and everyone smiled and waved as we passed them. I took this as a positive sign. We arrived in Dakar shortly after 8:00 P.M.

and we thought that this was the end of the line for us and our French philanthropists. However, to our surprise and gratitude, the patriarch invited us all to join them for dinner. We ended up eating steak with all the trimmings at a pretty fancy joint. The other well-to-do diners were giving us the once over, and wondering what a bunch of bums like us, had done to deserve what we were receiving. We were actually served fresh water along with our dinner wine and it was the best tasting drink I had savored in an awfully long time."

"After supper, the father removed a very detailed map from the glove compartment of his car and gave it to us. We pointed out where we had already been and where we were still hoping to go. Afterward, they drove us out to the coastal fishing grounds where we found a thatch-roofed hut on the beach. The Frenchman informed us that it was a sculpturing sales office during the day but it was disused at night. The owner was his business partner. They wouldn't mind if we slept there for one night. He produced a key and unlocked the door. It would lock again automatically the next day when we pulled it shut. Once again, the kindness of strangers was absolutely amazing and had I not been made of sterner stuff, I might have shed a tear or two out of gratitude."

"As they drove off, I couldn't help but remark that what had started out as a pretty crumby day, actually ended up as being one of our best. It was shortly after 11:00 P.M. when I finally crawled into my sleeping bag and I was out like a light the minute my head struck the rolled-up sweater I employed as a pillow."

Once again, Jesse was dumbfounded by the unbridled benevolence of certain strangers toward McGuire and his friends. By now, he no longer doubted Bob's claims of other's generosity toward him. However, he just couldn't understand why someone would do this. He was certain that this would never occur during this time and age, but he considered how nice it would be if it could. He wondered if one day he would ever be the recipient of such kindness or if he were ever placed in a position where he could help a random stranger to that degree. It was an interesting thought and one that he would need to ruminate on. McGuire allowed Jesse a moment to absorb his last few sentences and after taking a healthy gulp of his wine, continued speaking.

"The closeness of the rapidly warming air inside the small hut and the bright sun beams filtering through the windows, caused us to wake fairly early the following morning. We made certain to clean-up after ourselves and then left the sculpting sales office before its owner arrived. We eventually made our way into the city proper and were pleased to discover that Dakar was a sophisticated and cosmopolitan community, modelled in European fashion. There were office towers, high-rise apartments and all the shops that we were used to seeing, including the designer stores where only a select few could afford to buy. There were neatly groomed boulevards, numerous trees and shrubs and flower baskets everywhere. If I didn't know better, we could have been in Paris or London or even Madrid."

"There is strong French influence in Dakar and most of the professional and business people we saw were 'white'. There were a few affluent Africans as well, and they immediately stood out as compared to their less successful counterparts. The former individuals were all dressed 'to the nines' in Armani silk suits or very colorful and intricately designed traditional outfits. They also wore mirrored aviator sunglasses, heavy gold chains around their necks, solid gold rings on their fingers and expensive watches on their wrists. This was quite a contrast to the majority of their countrymen who were poorly dressed and held jobs such as gutter sweepers, rickshaw drivers, trash collectors or street vendors. As we walked around, it wasn't long before the vibrant portion of the city was gradually replaced by a shanty town."

"Still in distant view of the business district skyline, but attempted to be tucked discreetly out of sight, were thousands of scrap wood and cardboard shacks supporting corrugated tin roofs and packed very closely together. These shacks seemed to go on for miles ahead of us and spread across the horizon as far as the eye could see. I had no idea how many people existed in these conditions, but judging by the amount of humanity milling about the slums, it must have been considerable. What a difference from the Dakar we had been experiencing just a few minutes earlier. Before long, there was a hoard of half-naked kids in ragged clothes and snotty noses all bouncing around us with their hands extended and jabbering in a mixture of French and some other language we didn't understand. I don't blame these people for trying to get something out of us, but it soon became a real pain, when we couldn't go anywhere or do anything without a bunch of kids, and even adults, continuously asking for a hand-out or attempting to sell us some cheap and cheesy souvenir."

McGuire perused his journal briefly and then addressed Jesse directly. "Okay, Kid. There's a pop quiz coming up now. I want to see if you've been paying attention to me during our conversations, or if I've just been wasting my breath." Jesse sat-up straighter with a somewhat puzzled expression, not knowing exactly what to expect. McGuire continued.

"March 17th...what's the significance of that date?" Jesse hesitated before answering and racked his brain, attempting to think of an answer. Then it struck him. "March 17th is St. Patrick's Day and he's the patron saint of Ireland. Your family heritage is Irish and your middle name is Patrick!" Jesse was certain that he answered the question correctly and awaited a response.

"Way to go! You got that one right. Now here's the second part of the question. On my first trip, where was I on March 17, 1966?"

This next question was much easier for Jesse to answer since he remembered quite vividly that this was the day that McGuire and his friend had met a couple of American sailors in Beirut and then later snuck them into to their Youth Hostel. "You were in Beirut!" Jesse quickly blurted out. "Bingo! You're two for two," exclaimed McGuire. "You have been listening. Straight A's, for you!" Jesse actually surprised himself for remembering that fact. He was pleased to have passed the test but was hoping that there weren't going to be any more. However, just in case there was, he decided to listen even more acutely from here on in.

"Well, now its St. Patrick's Day again, five years later and I'm in West Africa. It's after dinner and I'm sitting at the curb on Route de la Corniche Ouest, staring out at the jade-green, North Atlantic Ocean watching the perfectly round sun, now the color of a ripe orange, slowly sinking behind the horizon. The heat of the day is gone and replaced by a comfortably warm tropical breeze that's wrapping itself around me like a child's security blanket. There's a Senegalese musical ensemble just down the road playing a blend of African-Euro jazz that's softly caressing my ear drums, while the smell of spicy street-food from a dozen different sidewalk stalls, is wafting through the air. Ask me if I was feeling pretty good right about then, and I would have answered that I was actually in heaven."

"At night, we slept on the beach with a small army of other young travelers from all over the world. No one seemed to bother us there. We would usually sit around a camp fire, have a few drinks and sing songs. Dave had some money wired to him from home and he now has another guitar. It sure helped in the sing-alongs. I was getting eaten alive by sand ants while trying to sleep, so I picked-up a makeshift, single person tent from another 'hitcher' that was hitting the road. It seemed to help a bit. There wasn't much dope around Dakar and from talking to others who had been there longer, it wasn't worth the hassle, what with the law and all. There was a rumor going around that if you were busted here for dope, they would take you to a mental hospital, instead of jail. We didn't want any part of that, so we decided to wait for our next carrot chillums somewhere else."

"In the mornings, I would watch the local fishermen go out in their large, dug-out canoes equipped with a light horse-powered outboard motor, or else they would take about four or five additional men that would paddle the canoe. They would then throw out the net in a fanning pattern in order to catch fish. It kind of reminded me of purse seining. Being interested in the customs and habits of people in their own environments, I began noticing certain things. I observed that people not gainfully employed slept a lot. They would be splayed all over the place and would remain there for hours. The flies, sun and the heat didn't seem to bother them. I think they might have done that to conserve energy or maybe to forget about the fact that they were hungry. Women's hairstyles were also unusual. Many Senegalese city women wore their hair braided into tight strands and then made them into cornrow or checkerboard patterns. Mauritanian women wore their hair pulled upward and fashioned into a large bun near the top of their forehead, while most countryside women completely shaved their heads."

"I also noticed that everyone, men and women, chewed on a stick derived from a particular species of wood. This was used to clean their teeth. It must have been pretty effective since everyone I saw had beautiful choppers and they were forever smiling. The locals seemed to be enchanted with my beard. Especially the kids. They kept wanting to touch it. I don't know if it was the reddish color or what, but they couldn't get enough of it. Maybe they thought it would bring them luck to feel it."

"It was also very frustrating shopping there. You had to barter for nearly everything you wanted to buy and even something as simple as a bottle of pop or loaf of bread was priced differently, regardless of where you went. Even if the shops were right across the street from one another, prices for the exact same articles would vary. Since it was such a hot climate there, you would think that swimming in the ocean would be a good way to beat the heat. If you thought that, you'd be wrong. In fact, the water temperature was so cold that you couldn't remain in there for more than just a few minutes. It was also quite ominous sitting on the beach, while up in the sky was the constant presence of huge vultures circling around. I don't know if they were attracted by the large amounts of garbage heaps in the shanty town, or if they would soon be feasting on something more substantial, like maybe a dead body or something similar. I didn't even want to think about it."

"It wasn't long before the sheen of Dakar was beginning to lose its luster, so one morning we attended the Gambian Embassy in order to acquire a travel Visa. The group decided to split-up hitch-hiking for practicality reasons. Don, Rusho and I left Dakar first and headed toward the highway. We managed to catch a couple of short rides relatively soon but we were still a long way from our next destination, which was Kaolack, Senegal. This was the last significant Senegalese city before reaching the Gambian border. We attempted hitch-hiking a while longer but it was slim pickins. Once again, we weren't going to be playing Russian Roulette with Ra, the Egyptian Sun God, so we caught the next bus going by. We paid our 300 Franc fair and enjoyed a relatively uneventful trip into Kaolack. The only notable event was when Rusho caught a glimpse of a group of topless, female natives walking at the side of the road. Before, I could get into position at the open window with my camera, they were already gone. So much for my National Geographic moment. It was so hot in Kaolack when we arrived that we decided not to stay and ended-up taking a taxi the rest of the way to the frontier town of Amdallai."

"We had just paid our 400 Franc entry fee into The Gambia at the border check-point and had our documents stamped, when a British couple were coming back into Senegal from The Gambia. They had only been there for two days and couldn't hack it any longer. It seems that the Gambian Immigration High Commissioner was on a 'permanent downer' regarding people with beards and long hair. In other words, people like me! It didn't matter how much money you carried or what influence you might have. He just didn't want you in his country. And to make matters worse, his word was final. There was no appeal."

"We were handed a bunch of forms by one of the border guards and instructed to complete them. We were then supposed to present ourselves to the Gambian Immigration Office the following day with the forms. After speaking with the British couple just a few minutes prior, we knew where this was heading, and promptly left the check-point, after agreeing to their demands. We had no intention of walking into that Immigration Office the next morning only to be kicked out of the country by the afternoon. We decided to take our chances and tossed our immigration applications on to the nearest garbage heap."

"From Amdallai, we grabbed a bus into the town of Essau that sits on the edge of the River Gambie. Here, the only way of reaching The Gambia's capitol city, Bathurst, is by ferry, since the river separates the two communities. The Gambia was originally a British Crown Colony in the late eighteen and early nineteen-hundreds."

"Since the British are no longer there, the name of the capitol has since been changed to Banjul. But for the purposes of my story, I'll keep referring to it as Bathurst. As we reached Bathurst, we were constantly checking for police, or anyone in a uniform, that might be in the military or government. Since many like-minded travelers as us, spend much of their time on the beach, that's where we headed. Sure enough, already from a distance we could see groups of pale-skinned hippies prancing around on the sand."

"We introduced ourselves to a group of Swedes and before we knew it, the joints were afloatin'. Praise the Gods of hippies and aimless wanderers! They told us if we kept a low profile and didn't bother anyone, then the police would more or less turn a blind eye to our presence and our activities. After about six hours on the beach of smoking and tossing around the Frisbee, we started walking toward the city. The ganja had calmed our nerves, and thoughts of police, immigration officials or expulsion from the country, had completely evaporated."

"As we entered the city, we were happy to see that it resembled Dakar for modernity and amenities. The only negative thing we noticed right off the bat was the stench of sewage. There were a lot of open manholes on the streets in order for the methane gas to escape. This really ruined what had initially been our good first impression of the city. Once again, people were mostly friendly toward us and since this had been a British colony in the past, most of them spoke English. I don't know if it was the dope messing with my head, but almost every native girl I saw was gorgeous. Especially, their big brown eyes. They looked exceptionally round and the whites surrounding the iris' made them pop-out like jewels. I was mesmerized by them."

"We managed to score some cheap accommodation and decided to remain in the city for a while. Like the other 'heads' suggested, we're laying low and spending most of our time at the beach. One afternoon, me and Rusho bumped into a couple of Swedish girls from the group we met upon our arrival. We took them to a secluded area of the beach and did some skinny-dipping. Trying to impress the chicks, I stayed out longer than I should have and ended-up with a serious sunburn. As we were walking in the sand back to the main beach, we noticed that some type of fenced enclosure had been erected in a field not far from us. Peeking through the slats, we saw a small group of native wrestlers, training. They were all different ages and weight classes and wore a simple loin-cloth to cover their privates. Once opponents were paired, they would go into a kind of ritual dance and then get on their hands and knees."

"They then assumed a crouching position and rushed each other. They always commenced bouts by grabbing each other's forearms, but after that, anything went. We watched a couple pretty intense matches before a policeman approached us and said we would need to pay if we wanted to watch. That was that."

"That night, my sunburn hurt like hell and I couldn't sleep. The next day, I ignored the discomfort and got on with life. Attempting to play it safer, Rusho and I each gave our landlord a couple of extra Pounds. He said he knew someone working at the Immigration Office and would attempt to bribe him so that we would be granted an interim Visa, allowing us to stay in the country. We were keeping our fingers crossed. I was actually really beginning to enjoy it here and it would be a shame if we had to leave."

"Everywhere we go, the boys call us a word that sounds like 'tubop'. It means 'white man', but they're saying it in a friendly manner, not to be rude or disrespectful. They grow-up fast around these parts and all have that 'rough and ready' look. You always see groups of them gambling for pennies in the alleyways and smoking cigarettes."

"They're almost always barefoot and they can walk over anything without it bothering them. It doesn't seem to matter if they're standing on the blistering hot pavement or walking over sharp rocks. They take it all in stride. Our landlord didn't return home that night and we wondered if he was arrested for attempting to bribe an official. We later found out that he was a bit of a nut-case as well as a scammer. We were told that he didn't have a friend working in the Immigration Office and he was probably enjoying a mini-holiday with our money. Our source also mentioned, that he wouldn't be surprised if the landlord 'ratted us out' to the authorities, just to be rid of us."

"Well, that settled it. We weren't going to be able to remain in the city. And for all we knew, there was a team of immigration police on their way right now, to arrest us. Heading further south was also not an option since most of the countries near or around the Horn of Africa were in a constant state of upheaval. The last thing we needed was to be caught in the middle of parliamentary forces and rebel fighters battling it out through cities, villages and jungles. These troops were poorly trained and brutal, so the Geneva Convention meant nothing to them. Collateral damage in the form of murdered civilians or obliterated buildings was all part of the game. Rusho and I spent the morning writing letters home and contemplated our next moves. While posting our letters, we met a couple of Gambian boys, and struck up a conversation with them."

"We spilled our guts out to them regarding our troubles in the country and that we worried about our immigration status. They appeared sympathetic to our situation and told us of a possible solution. Apparently, they were both from a village near a small town approximately 100 miles east of Bathurst on the River Gambie called Mansa Konko. The place was really laid-back and we shouldn't be bothered there. In addition, transit buses went there a couple of times per day. The boys also turned us on to a cool guy that lived in the same village compound near Mansa Konko. His name was Ego. They were certain that he would probably let us crash there super-cheap. Back into the city we went. First, to cash a couple of traveler's cheques and then to find the bus depot."

"Me and the gang eventually located what passed off as the bus depot, and after paying the exorbitant fare of 75-cents each, we boarded a rickety, wreck of a vehicle loaded down with all of our belongings, plus a cache of foodstuffs and a quarter pound of Gambian pot. Needless to say, after several hours of choking on the constant clouds of dust wafting into the bus through unglazed windows and bouncing through a million potholes, we were finally nearing our destination."

"The sun had just set when we arrived in Mansa Konko and we peeled ourselves out of the sweaty bus with our clothing sticking to us like flypaper. We had to ask around a bit but we finally discovered where Ego's place was. It was only a short taxi ride, but already dark when we arrived at the village compound. We were greeted by a bow-legged ancient sole holding a lantern that barely lit more than a foot in any direction. We asked to speak with Ego and received an unintelligible response from the geriatric sentry. We soon realized that Waliff was all he spoke, however, he did seem to know what we wanted. It was quite spooky as we entered the gloomy, silent village with the canopy of a billion stars and the eerie glow of a silvery moon above our heads. The old man knew where to go and could probably have found his way around there with his eyes closed. We soon found ourselves standing before Ego, and at that same moment, he was rolling a nice fat joint with one of his friends. I immediately knew that we were in the right place at the right time. Ego was happy to offer us some space in the compound and it barely cost us anything."

"Our elderly guide then escorted us to a raised, thatched hut displaying three unglazed windows and containing two large straw beds. We had just unrolled our sleeping bags and settled ourselves, when supper was served. It was nothing fancy, just rice and fish-balls infused with hot sauce, but it sure hit the spot. My raw nerves were slowly beginning to calm."

"By 9:00 P.M., it was still uncomfortably warm outside and very humid. Whereas, in Bathurst there's always a cool breeze coming in from the Atlantic Ocean, out here in the interior of the country, there was none of that. As we were digesting our meal, some inquisitive villagers started inching toward us, and we attempted to introduce ourselves. Most couldn't speak more than a word or two of English while others couldn't speak it at all. Here it was either the local dialect of Waliff or Mandinka. But using the international communication system of hand gestures, facial expressions and body language, we made do. After the impromptu conflab and a few puffs on a water pipe, it was time for bed. I couldn't believe that I was actually going to be sleeping in a traditional African hut, complete with a reed mat floor. But that's exactly what I did. The only thing that was missing was something soft and warm to snuggle with. I hadn't lain with a woman since arriving in Western Africa and I was beginning to develop a permanent limp trying to keep my third leg under control. If I didn't find someone soon, I don't know what I would do."

"The following morning, Ego had gone to his job in the town while the village women ensured that we were fed and our clothes washed. I dug that big time. As I took in my surroundings, I noticed that all of the huts were raised at least 4-feet off the ground. I later learned that this was to prevent venomous snakes, scorpions and other harmful creatures from entering the buildings, as they sought shelter during the monsoon season. Each hut also held a small bamboo enclosure that protected their meagre vegetable gardens from the village goats. After writing a few letters for the folks back home, Rusho and I ventured further afoot and stumbled across several village women carrying large clay gourds filled with water, on top of their heads. As they noticed us, they all screamed, dropped their precious water gourds, and ran off in the opposite direction. I looked at Rusho and just shrugged. As far as I could tell he didn't look any uglier than usual."

"Thinking that it would be best to avoid trouble, we doubled back to the village proper to bide our time. Not long afterward, our two young Gambian friends from Bathurst stood at our doorstep with the local chief. I asked the one boy, named Lumpy, if we were going to be on the dinner menu that night. He just laughed while I offered the chief a pair of my near-new striped jeans. Rusho gifted one of his better shirts to the chief's principal wife. Both were very pleased with our gesture, and we realized that it had been a smart move on our part. The chief eventually left, but throughout the day, native women would bring their children by our hut to take a gawk at us. I guess they had never seen Canadian royalty before."

"Becoming a little bored with hanging around the village watching the women pounding their laundry on large boulders and listening to the scores of screaming kids running about the place, we all went walkabout and followed the sound of drums off in the distance. As we neared a clearing shaded by several large trees, we observed a group of native women dancing. As we came closer, several of the participants noticed us and ran away. All of a sudden from behind a tree, a big black dude dressed from head to foot in reeds and foliage jumped in front of us waving a wicked looking machete. As he's motioning it around in our direction, he's also extending his other hand. It appeared that he was asking us for money but I didn't know for certain. We just stood there staring at each other in a state of frozen shock. I wasn't digging the vibe."

"After what seemed like a life-time, the drumming tempo increased and the dude backed away from us still holding the machete. He then went into this real intense dance of stomping around, grunting and swiping the machete from side to side. We couldn't help but move to the beat of the drums, but that was our first mistake. Some of the other dancers thought this meant we wanted to join in, so they grabbed a couple of my buddies and pulled them into the circle. The guys tried their best to follow the dancer's moves but their sense of rhythm wasn't quite the same. I felt embarrassed for them and prayed they wouldn't try to get me over there. We ended up staying there for nearly two hours and had a great time."

"Afterward, we walked into Mansa Konko and met Ego at his workplace. While we were in town, we saw a young boy whose hand had been bitten by a scorpion. It was all swollen and turning ugly colors. It looked pretty scary. That put a different spin on things for us and we starting thinking that hanging around this part of the world for too long probably wasn't a great idea."

"The next day, Lumpy and his friend were back at the village and they asked if we wouldn't mind helping the villagers make bricks from straw and mud. They were intending to build a small schoolhouse. That was fine with us. At one point, I also noticed a petite native girl making rice flour in a large hollow stump by hammering the rice grains with a long wooden pole. It was similar to the pestle and mortars we used back home, but on a much larger scale. I asked the young lady if I could have a go at what she was doing, and she agreed. Let me tell you, it was much harder than it looked, and after a few minutes, my shoulders and arms were beginning to ache. Feeling a bit embarrassed, I handed the thumper back to the girl, only to receive a resounding chorus of laughter from all the other women."

"We spent another night in the village and had the usual dinner. However, this time we were joined by the chief and his wives, plus Lumpy with his friend, and numerous other villagers we hadn't yet met. We would be leaving the next day. By now, my financial situation was becoming a concern and I was thinking it was time to start slowly making my way back to England. Instead of taking the bus back to Bathurst, we took an old paddle-wheeler steam boat down the River Gambie, belching thick, black smoke all along the way from her twin stacks. We sailed through the night and slept on deck since the stink of animals and humanity below deck was overpowering. We nearly puked after only spending a few minutes down there. We arrived in Bathurst at about 11:30 A.M. and I mailed my letters home. We then immediately set-out on making plans to head back northward. Rusho was making our travel inquiries and came across an Italian freighter that would take us to Italy for only eight and a half English Pounds."

"Jumping on that opportunity, we booked our tickets. The other guys were remaining in Bathurst, so it would only be Rusho and me leaving. Unfortunately, we put our horse before the cart, and now we were in a bit of a pickle. We required approved travel papers, and at the moment we didn't have any. In an attempt to remedy that shortfall, we visited the British High Commission, the Immigration Superintendent, the Bathurst Chief of Police and a lawyer, all in hopes of receiving the appropriate accreditation to sail with the freighter. We also needed the signature of a 'guarantor' on our papers, but no one was willing to provide it. The only thing left to do was go down to the docks and speak directly to the ship's captain. Maybe we could convince him to take us along. We arrived at the docks and located the Italian freighter."

"However, the gangplank was raised, so we couldn't board the ship. After waiting around for several hours, a crew member finally came out on deck and we shouted up to him. It turned-out that the captain wasn't aboard, and in fact, he wouldn't be returning until tomorrow when the ship was expected to sail. Now we were really crapping ourselves. After all we had been through, I thought I was going to come down with an ulcer. Disheartened, we returned to the city. At least, we caught a free taxi ride with the laundry man that was heading back as well. We spent the night on the beach chewing our finger nails and then went back to the docks first thing in the morning. When we arrived, the captain was already on board, and when summoned, he came on deck. He introduced himself as Captain Panas, of Greek origin. We explained our situation regarding the travel forms and that we were unable to find a guarantor."

"While he listened, I couldn't read his face. Now, it was do or die. He asked to see our tickets, our passports and our vaccination cards. After carefully examining each document, he eventually looked-up and said that we were good to go. I couldn't believe it!"

"I was so excited that I wanted to kiss him. But then thought that might not be such a good idea. Anyway, the weight of the world was now off of my shoulders and the next stage of my journey could begin. I hadn't thought about it for years, but suddenly the memory of my lucky Palermo tooth came to mind. Although, it was no longer in my possession, perhaps it was still working its magic in absentia. One of the crew showed us to our cabin and I was pleasantly surprised to see that it was much more luxurious than I had imagined. As I'm stowing my gear, I believe I heard the anchor being raised. Now, I'm really getting excited. Today is March 27th and it's my older brother, Mike's birthday. I just hoped that he was receiving a present as fantastic as the one that I was getting right now!"

"As we're slowly steaming out of Bathurst harbor, lunch is served, and we are the captain's guests. My eyes are bugging out as I watch the steward begin laying several different plates of food on the table. There's a mountain of spaghetti with tomato sauce, beef steak with peas, a large stack of buttered bread, salad, ice-cold filtered water and oranges. And this is only lunch! The captain is a real friendly guy with a great sense of humor. The ship is carrying 200 tons of ground nuts that will become peanut butter and peanut oil after processing in Italy. I was embarrassed when Rusho started munching on his fourth blood orange, while the captain still hadn't eaten a single one. However, the captain didn't seem to mind."

"There were very few other passengers on board aside from us. However, there was one fellow near our age that we befriended. He was Scottish and his name was Bill. True to our hippy, social protocol, a pipe was soon produced and filled with herb that we managed to smuggle on board. It was over 2500 miles to Italy and our voyage would take approximately eleven days to complete. At the rate we were now eating, drinking, and smoking, they would need to roll us off the ship when we finally reached port."

"Unfortunately, the seas were mostly rough throughout the entire time and everything was rocking from side to side. I considered myself a pretty hearty sailor, but now the constant buoying was starting to get on my nerves as well. There was nothing much to do except read, so I spent most of the time inside my bunk skimming through my library."

"In short order, I read 'The Voyeur, Force 10 from Navarone and Chairman Mao's Manifesto.' During one of our suppers, the captain mentioned that the ship would be docking near Portugal in order to take on more fresh water. Knowing that my funds were nearly depleted, I wondered if I should jump ship then, rather than sailing further on to Italy. After all, Portugal was much closer to England than was Italy, and it might work out better for me in the long run."

"During supper of April 2nd, I informed the captain of my plan and asked if I could disembark when the ship docked for fresh water. He seemed disappointed that I wasn't continuing all the way to Italy but assured me that it wasn't a problem. That night my beers were free and the captain even gifted me three tins of luncheon meat and three tins of sardines for my trip into Portugal. What a tremendous person he was and once again I was thanking my lucky stars. The next morning, we were approaching land and I was packing up my gear. The captain asked if I wanted to attend the bridge and observe the docking of the ship. I jumped at the opportunity. The captain was all done up in his formal white uniform as he expertly piloted the ship into port. We had just arrived in the Spanish autonomous city of Ceuta on the tip of northern Morocco. Not quite Portugal, but close enough. It was raining to beat the band. Rusho and Bill were continuing with the ship the rest of the way to Italy but accompanied me on to the wharf."

"Using the dollar that Rusho gifted me, I purchased a ferry ticket to the mainland and said my final goodbyes. It was still raining and the wind was really howling when I reached the Spanish mainland. I got right down to some serious hitch-hiking, but it took about an hour and a half before I caught my first ride. A couple of additional short rides throughout the afternoon brought me to the city of Seville."

"When I arrived, I bumped into a chick named Tina and her boyfriend John. I had previously met Tina when I was in Sardinia, and John was also planning on heading to Lisbon. Small world! We agreed to pool our money and then checked into a cheap room for the night. They were both drenched as I was and we all tried to find an empty space in the room to hang out our wet duds. The next morning, John and I caught the 8:35 sleeper train to Lisbon. The trip seemed to take forever, since we were stopping every few kilometers to either pick-up or drop-off passengers. It was raining in Biblical proportions and at that rate, I was going to need stilts for walking, to stay above the water, when we finally reached our destination."

"At least it was a scenic ride with lots of idyllic countryside to view. I made a mental note to myself that I'd like to come back some day and tour that country in a van. We finally made it to Lisbon and the first item on the agenda was to find a place to crash and get out of the rain. The city is not as cheap as I expected it to be, and in reality, it's actually quite expensive. I'm pretty bummed-out since I wanted to pick-up a few souvenirs and gifts while I was here. I'm also getting pissy, because everywhere I go, I'm getting all of the usual childish comments from people about my hair and appearance."

"John and I are truckin' all over the place but we can't find a decent room at a reasonable price. To make matters worse, John is lugging around a couple of surf boards with his gear and nobody wants them in their houses. We finally located a cheap flop-house but it sucked. There were seven beds in my room and most were occupied. I'm also not having any luck getting through on the telephone to various shipping lines regarding passage to England. I'm beginning to wonder if my luck is running out."

"Finally, we couldn't stand the fleabag we were staying in, any longer, and hit the sidewalks again. I wanted to head near the docks so that I could check-out some of the shipping offices. I figured that it was also probably cheaper in that area of town, since most other similar places in the world usually are. John checked-in his surf boards at the train station, since it was too much of a hassle humping them around. We finally found a room that we could agree on and that we could afford. As we returned to the train station to retrieve John's surf boards, we met a couple of cute chicks from Houston, Texas. We spent the rest of the day drinking beer with them and getting thoroughly sloshed. Afterward, we went back to their room but that was a bit of a bust. No luck there. If I don't have a woman soon, I might as well take my vows and join a monastery."

"We spent April 9th having a nice Good Friday, cum Easter Sunday dinner prepared by the Texan chicks. The next day, I wished John well as he boarded the train with his two surf boards, for the first leg of his trip to Durban, South Africa. I then went to a nearby hospital to donate blood. For one pint of lovingly aged Canadian blood, they were willing to pay me 250 Escudos. The only kicker was that I could donate blood today but it would be another four days before they would pay me. That wasn't going to work, since I wanted to leave for Bilbao, Spain right away and that money was going to go toward the train fare. Well, it was their loss and not mine, so I left the hospital with the same amount of blood as when I arrived."

"I decided to put a further dent in my meagre funds and purchased a one-way ticket to Balboa. It was another scenic train ride through Portugal and into Spain but somewhat spoiled by the smelly feet of my cabin-mate, Raymond, from Bombay, India. I didn't realize that a pair of dogs could stink that bad! One thing that made me chuckle during the trip was at nearly every railroad crossing we came to, there was a peasant woman wearing a reflective vest, sporting outlandish eye goggles, and proudly holding a tin horn while carrying a red flag attached to a stick. The women were there to stop traffic from crossing the tracks as the train approached. What a hoot!"

"At some point, I needed to transfer trains for the last stretch to Balboa, but I didn't know where that was. It was near 3:00 P.M. and I knew that we must be getting close. When the train made another one of its many short stops, I asked a fat little conductor in uniform standing on the platform, where I needed to transfer trains. He suddenly gets all excited and he's pointing down the tracks yelling, 'Miranda…Miranda!' I saw a different, stationary train sitting on the tracks and I quickly understood that I needed to get off of my train right now and head for the other one. I started running back to my train just as its starting to pull way again. Raymond sees me beetling his way and grabs my gear for me. As he hands me my things from the moving train, I thank him and say that I'll look him up when I get to India. Unfortunately, by the time I ran back to the secondary platform, the other train was also pulling away and is too far gone for me to catch it. It wasn't his fault whatsoever, but I laid into the fat little conductor with a barrage of abuse that would have made a sailor blush."

"I finally made it to Balboa, but much later than I wished. When I eventually found the ferry terminal, all the offices were closed. I would now have to wait out the night. I entered a small restaurant nearby, only to be accosted with the usual cat-calls, whistles and negative comments about my appearance. One idiot even started making whooping sounds like an Apache Indian."

"I was so angry with those assholes that I ended up drinking five Rum & Cokes, and if anyone of those pricks would have tried to start something with me, I would have killed them! With a massive headache the next day from no sleep and a bit hung-over from the liquor, I was back standing on the platform of the ferry terminal before sunrise. There was no way that I was going to miss-out again from leaving here. By 10:00 A.M. I had purchased my 2000 Peseta, second-class ticket for Southampton. It was an overnight trip but we wouldn't be sailing until evening. I didn't want to hang around the terminal all day, so I took a bus into the city. A couple of other likeminded Brits also boarded the bus and we had a nice chat during the drive."

"One of the guys owned a second-hand shop in London and while in Guelmim, Morocco, he purchased 800 hash pipes at what he considered was a very reasonable price. He figured that he would make a killing in his shop, selling them for triple what he paid for them. Ah, capitalism, you gotta' love it!"

"I didn't have any luck finding reasonably priced gifts in Balboa but I did manage to receive the usual stares, giggles and verbal abuse. On the bus back to the ferry terminal, a cute Spanish chick had been staring at me. As I exited the bus, I blew her a kiss. Judging by her smile and the glint in her eyes, I also knew that I blew her mind. That made my day."

"That night on the ferry, I met a couple of South African kids attending Leeds University in England. At their table as well was a very pretty girl named Lori. She lived and worked in Manchester as a hand and face model, and she was just returning to England with her father after a 10-day tour of the Algarve. We all enjoyed a philosophical conversation while sipping on a few cocktails and listening to the blaring jukebox inside the ship's lounge. I was getting strong vibes from Lori, and I could tell that she wanted more from me than just conversation. The only problem was that Lori shared a cabin with her father, and I was sharing my cabin with some junior monk-in-training. As a result, neither of those two spots were ideal for any form of intimacy. Suddenly, I had a great idea and nodded to Lori that it was time to go. She took the hint, and soon we were both inside the corridor giggling like a couple of kindergarten kids. Grabbing her hand, we raced down the corridor and into a second-class passenger bathroom. After locking the door, our clothes soon hit the floor. For lack of space, we ended up inside the bathtub and made mad, passionate love. I felt reprieved. After weeks of celibacy, I had finally gotten my 'rocks off' again, and boy did it feel good. I thought the top of my head was going to explode. I didn't realize that young, British gals were that liberated!"

"I finally got to bed shortly after 1:00 A.M., but couldn't sleep. Instead, I just lay there until around 11:00 that morning. Afterward, I went on deck for a swim in the heated pool. While I was there, they opened the 'skeet shooting' area of the ship and people started lining up for a crack at hitting the clay targets that were being catapulted out toward the open sea. The cost was 50 New Pence for six shots, and although I was tempted to give it a go, I needed to save what little money I still had left."

McGuire was now loaded for bear and the memory of his rendition was spilling out of his mouth like the open floodgates of a dam. He quickly looked into his journal for reference, and with hardly taking another breath, he continued speaking.

"When we arrived in Southampton later that day on April 14th, I received another kick in the nuts. As I was attempting to pass through Customs, the excise official refused to allow me entry, complaining that I had insufficient funds. He wanted to return me to the European mainland. I couldn't believe what I was hearing and started to lose it on the moron. Once again, the Travel

Gods were watching out for me, since at that exact moment, Lori and her father arrived. Lori pleaded with her dad to help me, so he spoke with the idiotic official to work out a deal. He agreed to vouch for me and signed a few papers to that effect. Well, that was that."

"Once out of the Customs Office, I thanked Lori's father for saving my neck and then gave Lori an innocent peck on the cheek. As we parted company, I thought to myself, that the old man would probably be wringing my neck instead of saving it, if he only knew what I had done to his daughter the night before. Soon the reality of my circumstances set in. I was hungry and nearly flat broke. Needing to change that situation in a hurry, I made my way to the nearest motorway and started hitch-hiking. I was trying to make my way down to Truro, Cornwall, in the southwest of England, to visit my aunt Mary. I hadn't seen her in a long while, and I also knew that she would take care of me. Thankfully, I lucked-out on the road, and before long, I was standing at Aunt Mary's front door receiving a huge bear-hug from her. I remained at Mary's for a couple of days, and during that time, she spent most of the waking day stuffing food down my throat. I actually gained a couple of pounds while I was there. As I left, she tucked a little money into the palm of my hand and tearfully wished me well."

"I started walking toward the motorway and trod about four miles before a freak driving an Oxford Minivan pulled over and asked if I needed a ride. Well, not only was this 'brother of the cause' offering to give me a ride, he was actually heading all the way to London. Could life get any better? Without another word, I jumped into his sardine tin on wheels and we immediately 'burned down' a tasty number. And then a second. By then I was feeling no pain but still a bit nervous. No matter how many times I travelled to England, I just couldn't get used to flying down the wrong side of a skinny road, they had the nerve to call a highway, doing 70 MPH in what a North American child would consider a kiddy-car. However, I shouldn't complain, since the 'head' eventually dropped me within six blocks of my final destination."

"I arrived back in Earl's Court and met a distant acquaintance named Barb Farris at my buddy Ian's flat. Ian wasn't home at that moment, so we settled down to a quiet evening of television. I watched the last episode of the last season of Monty Python's Flying Circus, plus an exciting soccer game between Leeds and Liverpool. Leeds prevailed one to nil. In between programs, I flipped through the newspaper checking out charter flights home. There was also a letter waiting for me there from Judy, my sweetheart back in Vancouver."

"I had been writing to her quite a bit from different locales during my travels, but hadn't received any replies until now. I greedily tore into the envelop and after reading the first few opening sentences, my heart dropped into my stomach. Talk about a 'Dear John Letter'. I guess I had been away too long and was having too much fun. I shouldn't have been so honest and mentioned all of the girls I was meeting and having good times with."

"As they say, 'all's fair in love and war', but I wasn't going to lose this fight so easily. I decided that I would confront Judy as soon as I returned home."

"The next day, Ian took me to a travel agency, where for 75 English Pounds, I booked a charter flight to Seattle, Washington on April 23rd, with a connecting bus to Vancouver. That was absolutely the last of my remaining money and now I was truly living on the charity of friends. After a couple more days, I started getting that uncomfortable feeling again in my nether regions. Not playing games this time, I located a Free Clinic and took a blood test. When the doctor returned with the results, he simply instructed me to turn around and drop my drawers. As commanded, I rotated, stripped off my jeans and touched my toes. He rapidly approached me clutching a gigantic syringe tipped with a needle that looked at least six inches long. After unceremoniously jabbing my butt and squeezing into me every drop of whatever was inside that hypodermic, he bluntly said, 'No sex or alcohol for the next three weeks.' Without another word, he then strode back toward the door and walked out of the examining room. Attempting to comprehend what had just occurred, I momentarily remained in that compromising position, bare ass to the wind, and with a stupefied look on my face. Reality finally set in when my memory drifted back to that recent moment of ecstasy inside the bathtub of the second-class passenger washroom on the ferry to Southampton. Alone with my thoughts, I then whispered, 'Oh, Lori…Lori…why me? You broke my heart. Talk about deja vu.'"

"As a result of my contracting the 'Big G' once again, the remainder of my time spent in London was pretty subdued. Sex and alcohol were a big part of my life, and saying that I couldn't have either for three weeks was like telling me that I wasn't allowed to breath. Anyway, I sucked it up as best I could and spent most of my time just walking around the city during the days and watching 'telly' at night. At least, I wasn't barred from smoking dope, and I took full advantage of that."

"The 23rd quickly rolled around and before I knew it, I was heading to Prestwick Airport. I boarded the plane, located my assigned seat and stowed my back pack in the overhead storage compartment. I sat down, fastened my lap belt and slid downward for added comfort. My lower back was bothering me again. I closed my eyes in order to relax, and soon a variety of images that I had encountered over the past six months began rapidly flashing within my conscious mind. Despite the ache in my back and the pain in my groin, I just had to smile."

Chapter 8
Twice Is Nice but Three's a Charm

Once again, the entire night had been completely devoured by McGuire's spell-binding recitation. As always, he was fully drained upon finishing this latest verse. Without fail, Jesse continued to be enthralled with the level of tenacity McGuire displayed toward his travel passion and his unswerving desire for a good time. There were also moments when Jesse felt that McGuire's need for having fun would occasionally even trump the regard for his safety or health concerns. After some additional discourse regarding this latest installment, with some minor clarification and embellishment imparted, both men were ready for bed. What was left of the evening sped by quickly and soon the pair found themselves once again seated at the breakfast table, contemplating the day ahead. The former inclement weather had given way to what had now become a pleasant Summer morning. Jesse wondered if further procrastination regarding the yard clean-up would ensue, or, whether the work would actually commence this day. As if reading Jesse's mind, McGuire immediately dispersed his uncertainty.

"Okay, Kid. I'm going to be heading up to the greenhouse for a while, if you wouldn't mind starting on the yard." As Jesse nodded his assent, McGuire continued. "Great! I'll explain what I want you to do before I go." After sufficient instructions were provided, McGuire drove off in his white pick-up truck while Jesse began his assigned chores. By now, he had been wearing his borrowed working clothes so often, that he now considered them his own. With little to no distractions afflicting him, and only pausing for the occasional bathroom or water break, Jesse was making good progress in the yard. For his part, McGuire had long since completed his tasks at the greenhouse. However, rather than return directly home, he decided to take this opportunity to visit his girlfriend, Linda. Since Jesse's arrival, he had only briefly spoken with her, a few times, over the telephone. From where he was now, she lived just a few minutes' drive away in Van Anda, and it just so happened that this was one of her rare days-off from work. Linda was an EMT and regularly 'on call'. During the island's off-season, and with few permanent residents, there was usually limited necessity for ambulance needs. This normally allowed her sufficient free-time. However, with more and more tourists and other visitors steadily arriving daily, requests for her services had increased existentially. McGuire was confident that Jesse could fend for himself for a few more hours. This was an unexpected bonus for him and he was going to take full advantage of it.

Their visit was brief but rewarding, and as McGuire returned to his ride, he was laden with the usual home baking and garden vegetables that Linda always provided him whenever he left her house. Knowing that he was entertaining Jesse back at his house, on this occasion, she even cooked them a large casserole dish of homemade macaroni and cheese, that he was now securing on the floor, in the crew-cab portion of his truck. As he drove away while honking the truck's horn and blowing one last kiss to Linda, he considered himself extremely fortunate to have this petite, talented and selfless lady as his special friend.

As he breezed up the driveway toward his Quonset Hut garage, he was impressed with the progress Jesse had made in the yard. There were several large piles of debris deposited throughout the property and much of what he asked Jesse to clean, appeared to have already been completed. "I'll make a man of him yet," he jokingly thought to himself. Jesse observed McGuire returning home, but kept working. Jesse surprised himself that he was still willing to continue, although he was covered in dirt and possessed a variety of scrapes and bruises that came from handling the never-ending, twisted strands of brambles and from falling down on the hard ground after tripping over hidden obstacles.

"Okay, Kid. Shut 'er down," McGuire called-out as he exited the garage. "Dinner is served." Jesse didn't need to be called twice, and trotted back toward the house while simultaneously removing his work gloves. "I'll keep the food warm in the oven while you get cleaned up." continued McGuire while he commenced preparing a salad with some of the items gifted to him by Linda. For some reason, he didn't wish to praise Jesse for the work he had done in the yard. Rather, he would just leave him with the impression that hard work was an expectation of all responsible people, whether they were rewarded for it or not. Linda's homemade macaroni and cheese was absolutely delicious and both men barely came up for air, while gorging themselves. Afterward, McGuire was pleased to observe that there was still enough left-over for the next day's supper. Before long, the pair were back in the living room with their brimming wine glasses and McGuire assumed his alter-ego as time-traveler and orator. As he raised and opened his third journal, he now set the stage.

"Okay, Kid. This will be my third trip to Europe, North Africa and the Mediterranean. I'll try not to bore you with a bunch of re-hash, but I will try to include some of the juicier and more interesting tidbits."

"If memory serves me well, I flew home from London in late April 1971. During the entire flight, it was driving me nuts that Judy had given me the "'bum's rush'. I wanted to get to the bottom of it, but since I was currently in the grips of another 'dose', I would have to wait until that cleared-up before calling her. The last thing I needed was to potentially get back into her good books again and then have to tell her that I couldn't touch her for another two weeks because I had the clap. I might as well just slit my throat instead. Those two weeks dragged on for an eternity but after getting a clean bill of health from my doctor, I beetled over to Judy's place. Thankfully, she invited me in and allowed me to plead my case. I could see by talking to her that there was still a spark between us, but she was still mighty pissed that I was continuously taking off and having a good time while she remained behind, keeping the home fires burning, so to speak. Well, after a lot of sweet talking and a bit of pleading, I was back in her good graces. Oh, and I also promised to take her along with me on my next adventure."

"She didn't have to wait long, since I was already raring to go in the Fall of that same year. I made a big score once again during that season's fishing cycle and the money was burning a hole in my pocket. I had also been in correspondence with some of my old buddies, and as a result, the travel bug had bitten me once again. On November 16th, Judy and I were in the air, soaring toward London, England. Through a stroke of luck, my good friend Farge, who lived in London at that time, had gained a new lady friend while drinking in an east-end pub the previous week. Vicki, as she was called, just happened to work the reception desk of the Hotel de France at 92 Queen's Gate in Earl's Court. She was able to garner us a 4-bed suite inside the hotel for the unbelievable price of 50 Pence each, per night. All we needed to do was keep the gas meter primed with shilling pieces in an attempt to stave off the cold and damp. Vicki even served us breakfast in bed the following two mornings. Such a deal!"

"We bummed around London for a few more days enjoying the benefits of our Afghani black hashish plus several gallons of 'scrumpy', an alcoholic apple cider from England's west country. On the 23rd of November, we said our goodbyes to Farge and Vicki. That morning, Judy, Rosco, Gee-Guy and me just managed to catch the 10:00 A.M., 'two-track special train' to Dover. It was cold and wet on the coast as we stood for over an hour in an outside queue, waiting to board the ferry for Ostend, Belgium. Just as we were about to board, a cancellation announcement was made over the Terminal's public address system. Apparently, Ostend port was temporarily off limits due to Force 12 gale winds. We would need to re-route to Dunkirk, France."

"This was a real drag for us, since Dunkirk was much further away from Amsterdam than was Ostend, and that's actually where we wanted to go. Reluctantly, and with cold-numbed hands, we picked-up our gear and trudged the quarter mile or so to the next ferry slip designated for France. Once again, we waited in a queue, kicking our wet feet together in order to stimulate circulation and blowing into our hands in an effort to keep them warm. Wouldn't you know it, the weather warning for Ostend was eventually lifted and sailings there could resume at 4:00 P.M. That was still a few hours away, but it was better for us than sailing all the way to France. Once again, we grabbed our belongings and shuffled back to the former departure slip. If someone would have tied a string to us, we could have become human yo-yos."

"Although the gale force winds had diminished, the sea was still very choppy and our trip to Ostend was accented with the sights, sounds and smells of passengers puking all over the ferry. Within a few hours, and while wading through ankle-deep barf, we stepped off of the H.M.S. Vomitus and onto the terra firma of Ostend, Belgium. Leaving the group at the terminal, Gee-Guy and I went off in search for lodgings. Before long, we scored an acceptable place and sent for the rest of the crew. To celebrate our arrival, I removed the remainder of our Afghani black hashish that I had secreted inside a hollowed-out loaf of bread and we commenced smoking our faces off."

"The following day, we fixed ourselves on the highway with thumbs thrust out proudly. However, after four hours in the wind and rain without so much as a single ride, we cashed-in our chips and bee-lined for the nearest Avis Car Rental outlet. We settled on a newer model Volkswagen Beetle, and with the aid of a shoehorn, managed to cram ourselves and all of our gear inside. By three-thirty that afternoon, we rolled into Amsterdam looking for action. The plan was to immediately rent a room for the night and then hopefully the following day, purchase a 'used' converted van or bus. This would double as transport and accommodation for us. It looked like every long-haired freak in Europe decided to converge on Amsterdam that week, and try as we may, we couldn't find any decent rooms. Just as things were beginning to look like a bust, we managed to secure rooms at the Hans Brinker Huis. It cost more than I wanted to spend, but at least Judy and I had our own room with a toilet and shower."

"The next day, after returning the VW Beetle to Avis Rentals, we headed to Dam Square in search of transport. This is where everyone congregated if they were looking to purchase anything their little hearts desired, or if they just wanted to hang-out."

"If you couldn't find it there, it probably wasn't worth buying. We checked out about a half-dozen different models before finally deciding on an early 60s VW van for three-hundred dollars. We initially split the cost four-ways but after second thoughts, I bought them all out, and owned it outright. You know I don't like sharing my girls. It had relatively low mileage plus the rear had been extended and partially camperized. I accompanied its former owner to an insurance agent and officially transferred the van into my name. I then purchased 3-months insurance for fifty-five dollars. While heading back to the Dam, I bumped into a fellow who was selling his camping gear. For twenty dollars, I acquired a 2-burner gas stove and accompanying gas cylinder, a lantern, pots, pans and utensils, and two super warm blankets in plastic containers. This was a real bonus, since I was expecting to have to visit the Flea Market in order to get these things. This saved me all that time plus the hassle of bartering for the items."

"This was my third trip to Europe and now, I finally owned my own ride. No longer was I going to be stranded for hours on the side of some highway in all kinds of nasty weather or sleeping in a stairwell or vacant building. I was on top of the world! There was still a couple of modifications I wanted done to the van plus a bit of servicing to ensure that it was roadworthy. I was directed to a nearby shop whose owner was said to be both affordable and reliable. I left the van with him and we all returned to the Hans Brinker Huis. As we arrived, I was pleasantly surprised to see my old friend, Jim Lowrey and his

Dutch girlfriend, sitting on the steps waiting for us. Jim had also purchased a VW van the day before for two-hundred and fifty dollars, and he was going to join us for at least a portion of our journey south."

"We all motored out of Amsterdam on November 25th and straight south to the Holland/Belgium border. Upon arrival, we cut cards to see who would need to hold the 'stash' and Gee-Guy lost the cut. Everyone from my van as well as all occupants of Jim's behind us, were directed into the Customs 'secondary office'. Here we were asked the usual questions regarding contraband, and naturally, we all denied possessing anything illegal. Unfortunately, our responses weren't good enough for them and we were instructed to empty our pockets. A couple of guys from Jim's party were even taken off into a corner and told to undress. While the first few of us were being searched, Gee-Guy managed to place his stuff on a shelf behind him, while Jim was able to drop his bundle unnoticed, behind a cabinet. After the physical search, they started in on our luggage. The Customs Agents found a starter pistol in Jim's bag and all hell broke loose. While they were distracted by this momentous discovery, Rosco managed to recover both stashes and quickly shoved them into his jacket pocket."

"Jim's starter pistol was confiscated, but since it wasn't an actual firearm, he wasn't charged. We were then given the green light and told to get lost. Without needing to be told twice, we split that scene as quick as humanly possible and didn't start breathing normally again until we were well away from the checkpoint. That was a little too close for comfort. By the time we reached Perpignan near the France/Spanish border, we hid our stash behind the front dash panel and crossed the frontier unhindered. We reached Barcelona on the 28th and were immediately plugged into its rush-hour traffic. After finally breaking free of the vehicular crush, we headed straight to the Caracoles Restaurant for a seafood feast. This was one of my most favorite restaurants in all of Europe and I never tired of going there."

"I've really been impressed with the van so far. It's so roomy inside that you almost forget that it's a Volkswagen. It rides real smooth and didn't burn even a drop of oil the entire trip from Holland to Spain. The one thing that continues to remind me that it's a VW, is the heater. It barely works. This seems to be the case with every other VW I've ever ridden in. They just inherently have lousy heating systems. The German engineers and car designers need to get off their butts and correct this problem. I've dubbed the van, 'Pokey'. I won't be winning any Gran Prix races with her anytime soon, but it keeps poking along and getting me where I want to go."

"After a few days in Barcelona, the weather deteriorated and heavy rains began. We decided to leave the city and head further south, but it's the same story everywhere we went. The roads in Spain aren't very good and often they're only one-lane in either direction. If you get stuck behind a tractor-trailer or heavy transport truck moving at 40 or 50 kilometers per hour, it's quite a while before you can pass. At night, the glare from oncoming headlights is nearly blinding, and so far, I've seen at least three large trucks tipped-over on their sides in ditches. For our part, Pokey is doggedly truckin' along and running like a top. Whenever, I get stressed or tired while driving, I just burn-down a joint and it sets me right. I can't believe that our single ounce of Afghani black is still with us since Amsterdam. That's mighty good stuff."

"The first week of December found us in Torremolinos, Spain on the Costa del Sol. This is a resort town with lots to see and do. One night in a bar, we met a couple of American sailors that were in port with the S.S. Concord, a military supply ship. There was Russ from Texas and Jerry from Illinois. Russ was a bit of a shy, quiet type but Jerry was an extravert. Over the next few nights, we drank with the pair and brokered a good friendship."

"One day, they took me aboard their ship and gave me a guided tour. I've got to admit that it was really interesting. There were even a few disassembled helicopters in the cargo hold that were worth a million dollars apiece. You've got to remember that the Vietnam War was still raging at that time. That thought made me glad that I was Canadian. The S.S. Concord was due to weigh anchor on December 10th, so the night before, we all went down to the ship after supper and watched a Spaghetti Western movie in their theatre. The movie sucked but we all had mucho laughs. Before we split for the night, Jerry pulled me aside and informed me that he was giving me an early Christmas present. I wasn't expecting anything from him but I thought that it was a very nice gesture. Well, an early Christmas present was a major understatement. I couldn't believe the amount of stuff he was laying on me. There was two, 6-gallon plastic containers for fresh water, several U.S. Navy pocket knives inside sheaves, a flashlight with 12 replacement batteries, 2 strobe light distress beacons, bottles of liquor, Navy ball caps, tea, cookies, sugar, a beautiful blanket and a bunch of other things that I can't remember right now. He did this strictly out of friendship. He didn't ask for anything in return. I was really touched by his generosity and he was someone that I wouldn't soon forget."

"The weather was pretty nice in Torremolinos, so we stuck around for a while enjoying the beach during the days, the dinners and drinks at Harry's Bar at night, and regularly moving Pokey from place to place so as to not attract any unwanted the attention from the Guardia Civil. One morning after breakfast, I was squeezing myself into my cut-offs and realized that I was becoming 'El Gordo'. If you're unfamiliar with the Spanish language, that means 'fat'. I decided that for the next while at least, it was going to be soup, salads and vino-tinto for old Bobby, until I was able to shed some of those unwanted pounds. On December 10th, Jim and his group parted ways with us and continued to follow their own travel plans. Rosco and I attended a soccer match in Malaga with about 30,000 other fans and watched as Celta FC and Malaga FC played to a 1–1 draw. On the way there and back, we checked the side streets for an empty bottle of Tia Maria. Jim said that he would leave us messages in this manner. Unfortunately, we didn't find one."

"It appeared that Pokey's interior and exterior lights were becoming dimmer and she wasn't too quick to start. I took her to a nearby VW specialist and he determined that she needed a new battery. She was also low on transmission fluid. He also recommended a lube job and a powertrain check. Normally, I always lubricate my girls without anyone's assistance, but after all, he was a professional."

"It was 1100 Pesetas for the battery and another 165 Pesetas for the remaining service. As usual, it was more than I wanted to spend, but never let it be said that Bob McGuire doesn't know how to treat a lady. We hit the smoothest Spanish road yet to date on the afternoon of December 15th and arrived in Algeciras near the Strait of Gibraltar, later that evening. Pokey purred like a kitten the entire way. En route, I passed stretches of highway that I had hitch-hiked almost a year earlier. The memory was like a sudden rush to the psyche. We hung around the docks all day and spoke with people returning from Morocco. It looked like long-hairs were being hassled there at the Customs Stations, and some were being turned back. A couple of guys with hair just over their ears said that they had to bribe a guard with money in order to be allowed entry. I wasn't digging the sounds of that too much. I later cruised a few of Algeciras' streets, but still no bottles of Tia Maria. I was hoping that Jim was alright."

"The next few days were very cloudy and windy. Not exactly beach weather, so we decided to tour some of the surrounding cities. We visited Seville, Granada and Motril. I wasn't really impressed with any of them and most were dirty and overcrowded. Half the time, I couldn't even find a parking space for Pokey, so that we could stretch our legs. The countryside between each city was at least a bit more appealing and picturesque, with acre upon acre of olive and orange groves. I managed to get a few decent photos. We eventually returned to Torremolinos and I started making plans for a jaunt over to Tangier. I figured that I had a better chance of entering the country with a tour group, so I purchased a Tour Afrique package on a hydrofoil for 1300 Pesetas. Rosco drove me to the docks on the morning of the 19th. Judy and the rest of the gang were going to follow later on the ferry with Pokey. As expected, I got held up for a bit in Customs and just managed to blitz out of there in time to literally jump on to the deck of the hydrofoil as it was pulling away from the wharf."

"Once onboard the vessel, I received the bad news that the skipper requested the passports of all eleven persons on the tour. He said that they would be returned to us when we sailed back to Spain. That wasn't going to work for me, since I wasn't planning on returning. The only reason I booked the tour in the first place was to side-step the Moroccan Customs Agents. During the entire 80-minute ride over to Tangier, my brain was feverishly attempting to devise a believable scenario that would have my passport returned to me without actually leaving North Africa."

"The tour was actually pleasurable and we saw some interesting sites. I even had an opportunity to ride a camel but that whole experience looked so commercial, that I decided not to take them up on their offer. The lunch included in the ticket price was at a high-end restaurant and the food was delicious. There were a couple of cheesy spots along the way where I declined to enter but overall, I had a good time. As the tour wound down, it was the moment of truth. I had a 'heart to heart' with the guide and he agreed to see if he could obtain my passport from the hydrofoil's skipper. In the mean-time, I was chewing my fingernails. The guide was able to score my passport from the boat and we then took a taxi to the nearest police station where I could fill out the necessary entry forms. I waited outside the station while the guide dealt with the police on my behalf. I couldn't believe it when he returned with my forms bearing that much coveted official stamp. I was in! I was so grateful, that I slipped him a couple of bucks and thanked him for all of his assistance."

"As the guide made his way back toward the docks and the remnants of the tour group, I started walking toward the town. On the way, I met an English-speaking Arab and he directed me to the Café Central. He agreed to accompany me into the town and we spent the majority of the afternoon discussing the mysteries of Islam and Arabic peoples. I eventually located the café and waited around for a while to see if Judy and the gang would arrive. After a while, I was getting bored, so I started walking toward the Souk. I hadn't gone very far when I noticed Jim's VW van sitting inside a repair shop. Knowing that he must be nearby, I waited for him to return. Within a few minutes, I observed Jim and his girlfriend, Marietta, walking down the road. I hid behind a wall and jumped out at them just as they were passing. Needless to say, they just about wet themselves, but then we all started laughing and patting each other on the backs."

"By 7:00 P.M., there was still no sign of my group, so, Jim booked me a room at the Hotel Istanbul where he and his pals were crashing. The next morning, we made our way to the Souk to do a little browsing and maybe some shopping as well. After a while, we started walking back to check on Jim's engine repair. To my surprise, there was Pokey parked at the side of the road in all her glory. I didn't see anyone around, so I left a note where to find me on the windshield. I was immediately stoked and looking forward to seeing everyone again. Jim and I continued toward the repair shop and were approached by a 30-something English dude coming out of the bus depot. He spoke with Jim and mentioned that he was returning home that day. He had some 'smoke' that he wasn't planning on taking along and asked Jim if he wanted it."

"Naturally, Jim agreed, and as the dude whipped-out over an ounce of top-grade weed from his pocket, our eyes nearly bugged-out. Another early Christmas present! We practically sprinted back to the hotel in order to give this unexpected gift a taste. A few minutes later, we blasted off to another galaxy and both agreed that we had hit the jackpot with the Englishman's generosity. Sometime later, we reunited with our errant pals and the party then began in earnest."

"We hung around Tangier for another day or so and then hit the trail in search of Nirvana. Soon afterward, we also began referring to ourselves as the

'Afghani Bunch'. So named after our favorite herb. On December 21st, my group split for Rabat, while Jim, Marietta and his crew headed back to Algeciras to hopefully 'make a deal'. If that proved successful, then they would be spending the Christmas Holidays in Madrid. We cruised into Rabat not long afterward, but then moseyed right back out again, since there was nothing shaking there. Since this was all new territory for my friends, I decided to show them Marrakesh. Driving through the countryside, I was once again reminded of the huge lifestyle disparity between Europe and North Africa. Judy and the others couldn't believe how grubby most places were, and appalled at how poor the general population was. I'd seen it all before, so for me it was no surprise."

"We blew into Marrakesh on Christmas Eve and checked into the Hotel de France for a much-needed shower and a bit of a rest. The next day, we headed straight for the 'covered market' and decided to splurge. For Christmas, I purchased a nice necklace for Judy and a wrist bracelet for myself. I then started checking out the leather goods. The remainder of our group were also picking up some great bargains. After leaving our hotel, we found a reasonably priced campground and quickly befriended several pods of liberated travelers. We all decided to put on a big Christmas Feast on the 25th and about sixty of us pooled our money and went grocery shopping in town. We ended up acquiring about five turkeys, half a lamb and all the necessary vegetable sides that would make the meal complete. We also bought about thirty gallons of wine. Unfortunately, cooking large quantities of meat using rudimentary means, left us eating most of the salads and other vegetables long before the meats were done. No worries, we all enjoyed ourselves anyway and I have to admit that my soup and salad diet was temporarily suspended."

"Approximately three days later, missing the seaside, and with no hot showers at the campsite, we pulled-up sticks and hit the road once again. We reached Essaouira on the coast a few hours later, but it was a bit too crowded there for our liking."

"We decided to drive a little further to the southwest and finally settled on Taghazout, a smaller but much more appealing fishing village. As one day blended into the next, we divided our time between crashing at a campground part of the time and directly on the beach for the rest. I actually dug a pit and built myself a bit of a lean-to on the beach where Judy and I could share some privacy. This turned out to be a wise move. The area surrounding the village was truly gorgeous and super peaceful. One of my favorite past-times was to occasionally rise early in the morning and just walk along the beach. The peaks of the sand dunes would sizzle and sing as the ocean breeze gently raised the upper-most grains and threw them together to create a sound similar to unearthly maracas. The brilliant-blue, cloudless sky had not yet absorbed the fiery heat of the sun, so it was still relatively cool on the skin. And with every stride taken, the perpetually breaking surf, would send forth a fluid carpet of foam-laden sea water straight up the strand to tickle the soles of one's feet. In my opinion, it was about as close to heaven on earth as one could find."

"Just as the mornings alone on the beach inspired me for the coming day, so did the glorious sunsets in the evening, soothe my soul and prepare me for the night's repose. As an artist, I appreciate the blending of colors to create a certain effect or to emphasize a point of reference on the canvas. However, the supernatural hues of red, orange, yellow and the deepest of purples that became our eventide, were completely out of this world and certainly beyond re-creation, to even the most gifted of painters. If it was within my power to gift each of the earth's 7-billion people just one thing, it would have been for them to spend at least a single day and night on the beach in Taghazout."

"What should have been for us a couple of days in Taghazout, actually became several weeks. New Year's Eve wasn't the major 'blow out', I expected it to be. Rather, it was fairly subdued and spent in the company of good friends, good food, and good dope. Our days involved mostly swimming and surfing in the ocean, throwing the Frisbee on the sand or playing impromptu soccer games. I was also becoming quite the accomplished surfer. So much so, that I began instructing others on the intricacies of remaining upright on the board for as long as possible. We cooked most of our own meals, and although I had the best of intentions, I just couldn't lose my jelly-belly. My brain would tell me to play it smart, but then I'd find my hands shoving food into my mouth. Sometimes, I would eat for hours at a time. It was almost like having a dual personality."

"It also didn't help when some of the local merchants would regularly come around and tempt us with their wares. There was the Banana Man, the Fish Salesman and even Doughnut Man, as we affectionately called them. I was like an alcoholic alone inside a liquor store. There was just no way that I was going to deny myself."

"Although most of our days were trouble free, we did encounter a few negative issues as well. On one occasion, I noticed that my scalp was beginning to itch, and when Judy went through my hair, she was disgusted to discover that I had head lice. When I asked to see her scalp, I noticed that she had lice as well. We tried to think where we might have picked up the vermin. The only thing that came to mind was near the beginning of our stay. A couple of dogs from neighboring campers spent a few nights sleeping in our van. Maybe the pair inadvertently infested us. Either way, we weren't going to let that slide and immediately drove into the town of Agadir to visit the local pharmacy. We purchased a couple tins of lice powder and returned to our camp. Although our problem shouldn't have been comical, we couldn't help but laugh at one another with our hair full of the white powder. We looked like a couple of gray-haired senior citizens. After a few days of faithfully maintaining a routine of applying and later washing out the medicated powder, our uninvited guests were eradicated."

"Another disaster occurred one day when a bunch of us were playing soccer. However, in the manner the game commenced, I wasn't surprised that something would go awry. Prior to the game, we had chosen an isolated strip of beach and cleaned away the majority of obstructions and debris. We managed to field seven players per team, including the goalies. This was far short of the usual eleven players per team, but we were confident that it would be enough for a rousing match. We took to the field, or should I say, the sand, and I adopted one of the forward positions. One of the less athletic 'heads' in our nomadic community, was acting as referee. Not possessing a whistle, he stood on the sideline, holding a woman's paisley headscarf in his raised left hand. It was a somewhat cloudy morning, and being not quite the noon-hour, the full heat of the day had not yet materialized."

"We all stood there barefoot in the sand, wearing only our bathing suits, and waiting for the scarf to drop. I was in a semi-crouched position, fully primed to charge the ball and take possession. In anticipation, we focused on our challengers, all the while attempting to stare them down and making cat-calls. What happened next, floored me."

"While we attempted a little psychological intimidation with our antics, I guess the goalie on the opposing team thought he would reciprocate. Suddenly, he pulled down his swim trunks and kicked them off to the side. He was now stark naked and displaying a facial expression of distain. Naturally, this quickly drew everyone's attention and a general hub-bub arose. After a few laughs and more than a couple lewd remarks, the remainder of his team dropped their trunks as well. What the hell? Momentarily confused, I turned to face my teammates. Within seconds, we all started laughing afresh, and as if of one mind, also pulled-off our suits. There were now fourteen naked freaks, in plain view, standing on a North African beach, about to play 'the beautiful game', as the Europeans call it. We were probably committing a venal sin in this conservative, Islamic country of Morocco, but we didn't give a damn about what anyone thought of us, or what consequences might ensue, if we were caught."

"Standing there in my birthday suit felt a little strange at first. However, I then thought back to ancient times, and realized that Greek and Roman athletes regularly competed in the nude. Seeing the rest of the guys all wearing a similar uniform, I soon dismissed my garmentless condition and waited for the game to begin. Once, the referee recovered from his initial shock, and both teams were alert, he dropped his scarf. I took off like a flash and reached the ball before the opposing forward. Now the game began in earnest. For over a half hour, we battled from one end of the beach to the other, but hadn't managed to break the 1–1 tie."

"The forward on the opposing team was a dude named Raphael from Caracas, Venezuela. We were both very competitive and playing our hearts out. Trying to turn the tide of the game, at one point, he and I were both charging toward a loose ball in an attempt to take possession of it. We arrived simultaneously, and in our desire to prevail, tangled our legs together fighting for the ball. The momentum knocked us off balance and we then both ended up falling backward. I hit the ground comfortably, but Raphael was twisted sideways. As he landed, I heard the sickening sound of a snapping bone, immediately followed by an intense cry of pain. I didn't need to be a psychic to know that Raphael had broken his leg."

"He was writhing in pain as we fashioned together a makeshift stretcher, and then carried Raphael to my van. We then drove him into Agadir and found the hospital. The doctor confirmed that Raphael's leg was indeed broken and he would need to set it with a plaster cast. We all felt bad for him, especially me, and wondered how we were going to pay the medical costs."

"None of us was rich, and when I last checked my traveler's cheques, I only had $400.00 dollars remaining. I was still expecting to be abroad for some time yet, so I needed to keep that intact. It turned out that Raphael's fibula received a clean break, so he didn't require surgery. The cast was all he needed. As a result, the hospital fee wasn't as dear as expected, so we managed to cover it by passing around the hat. However, this definitely put a crimp into our soccer matches. Raphael was a good player and his contribution would be missed. The game had been short-lived, but extremely memorable. We called the match a draw and vowed that one day we would all get together again to play, and then decide a winner. Of course, that secondary match never happened, but I'll never forget the day we played naked soccer on the beach in Taghazout."

"One afternoon, during the first week of January 1972, we were all higher than kites and just chillin' on the beach looking out into the vast expanse of shimmering, turquoise-blue ocean. Shifting my gaze from side to side and eventually focusing on Pokey, I determined that she looked a little sad. I figured that she could use a bit of a spruce-up and opted to perform a vehicular tattooing of her. I drove into a nearby village named Banan, and purchased seven basic oil-based colors from a variety store. Over the next week or so, I spent numerous hours creating my masterpiece. On the passenger side of her metallic skin was a large depiction of Mr. Natural. On the driver's side was a giant smoldering Afghani joint and her nose bore a multi-colored butterfly in flight, with a beaming, happy-face. I had to admit that Pokey's 'new look' was pretty sharp, and that opinion was ratified by the many compliments I received from fellow campers and passersby alike."

"One day, we caught word that there was going to be a souk in Agadir, so we all jumped into Pokey and started heading there. Unfortunately, there was now approximately eight-hundred hipsters living in and around Taghazout, so a bunch of them asked if they could come with us. I agreed, but then later regretted my decision. There were about twenty freaks inside the old girl, and she was really struggling to keep moving forward. After arriving in Agadir, and everyone piled-out, I was determined not to let that happen again. Agadir was a resort town near the foothills of the Anti-Atlas Mountains and popular with European tourists. The town had a festive atmosphere with musicians, snake charmers and all forms of performers wowing the attendees. However, once again, commercial tourism ruined our day. The souk was filled with French, German and Swedish visitors, all intent on taking home a little piece of Morocco, regardless of the cost."

"As a result, most of the merchants and vendors had dramatically raised the prices of their goods in order to cash-in on this windfall. They also weren't in the mood to barter, and kept firm on their prices, since they knew the tourists weren't going to argue with them. All we received was a bunch of dirty looks and the cold shoulder. I did have an opportunity of purchasing a beautiful ten-foot by four-foot hand-woven tapestry for twelve dollars, but I didn't have enough cash with me. Souk merchants dealt only in folding-green, so my traveler's cheques to them, were no different than toilet paper. Disappointed and bored, we left the souk after only a couple of hours and returned to Taghazout."

"By the end of January, I had still not received a single letter or card from home and it was beginning to rile me. I always took the time to write family and friends during my travels and this trip was no exception. It might have been selfish on my part, but I decided that I wasn't going to write anymore letters until I first received a reply from one of them. However, I made a slight compromise to that decision on February 3rd. That was my father's birthday and I wasn't going to be so ignorant as to ignore it. He was going to be receiving the card belatedly, but at least he would be receiving one, and would then realize that I was thinking of him."

"During our entire stay in Taghazout the weather had been fantastic. However, that all changed in early February. It started raining at times during the nights and then even during the days. This was our queue to 'up sticks' and begin searching for sunnier climes. Taghazout had been great, but as they say, 'all good things must come to an end.' None of us were trained geologists or accomplished meteorologists, but we figured the further south one went, the better the weather should be. Running on this logic, we hit the highway bound for Guelmim, approximately 225 kilometers away. As anticipated, the closer we neared Guelmim, the clearer the sky became. However, now that we were in and out of the valley depressions at the Atlas Mountain foothills, the wind was really beginning to blow. Three hours later it was still howling, so we continued through Guelmim and traveled 15-kilometers further to the northeast, where the thermal spa of Moulay Yacoub was situated. We all enjoyed a relaxing dip inside the hot, Sulphur pools and managed to scrub-off a week's worth of grime. Not to mention our first layer of skin as well."

"We wanted to head up the coast to Sidi Ifni, but the main road was closed and the only alternative was an 85-kilometer gravel track. There was no way that I was going to subject Pokey to that. En route back to Guelmim, we observed a Berber encampment and decided to stop."

"The Berbers are the indigenous peoples of North Africa and often include strains of the mysterious 'Blue Men' tribe. They were the most nomadic of the indigenous population and although their true identity was Tuareg, they were often referred to as the Blue Men of the Sahara. They derived their name from the indigo blue robes they constantly wore in an attempt at reducing the harmful UV rays of the sun. They also always wore a cloth covering their mouths and noses, in an attempt at keeping a moist flow of air for them to breath. I could see that the Afghani Bunch was very intrigued with the scene before us, so I encouraged everyone to jump out and stretch their legs."

"I saw a group of women leading a string of camels and gestured that I wanted to take a few photographs of them. When they nodded and smiled, I snapped off about a half-dozen pictures of them and a few of the encampment ensconced in the sand. I then waved goodbye. Well, all of a sudden, all hell broke loose and I was getting reamed-out by the old broads. They wanted payment for their super-model posing, but I wasn't up for that. I'm glad they weren't armed, since they probably would have gutted me. Man, were they ever pissed-off. I figured that it might be wise to put a bit of distance between us and the Berbers, so we all jumped back into the van and put rubber to asphalt."

"We hadn't traveled very far when I began to notice that telltale wobble in Pokey's front-end, which usually signaled that we were losing air in one of the tires. Sure enough, within a few more eye-blinks, we were beginning to lean toward the right-front of the van and then the unmistakable sound and sensation of metal rims on the road. I immediately pulled-over and confirmed that our right-front boot was as flat as the pancakes Judy served me at breakfast. I went into the tool compartment and retrieved the jack. However, when I placed it under the frame and started cranking on the handle, it began bending like a strand of saltwater taffy. Upon closer examination, the metal on the notched, lift portion of the jack, was completely corroded, and it literally disintegrated under the weight of the VW. So much for that."

"There was only one thing for it. I needed to 'hitch' to the nearest village and hopefully find someone with a jack. Luckily, before long, a taxi pulled up and stopped. Rather than pay for a ride into a town that may or may not provide me a jack, I politely asked the cab driver if I could borrow his. He agreed to let me use his jack but then said he needed to leave again right away. We ended-up piling some large stones under Pokey's frame and removed the flat tire. I then returned the jack to its owner and he split. Between Gee-Guy and Rosco, I decided to take Gee with me, and out went our thumbs."

"Motorists in distress always seem to attract generous people, so before long, a truck driver pulled over and we jumped in the back with our useless tire. Approximately 12-kilometers later, he dropped us off at a service station in a small village. The mechanic located the source of the flat and repaired it. However, as soon as the rubber was back on the rim, he found a second hole and had to start the process all over again. A half-hour later, with our wheel finally repaired and a borrowed jack from the mechanic, we grabbed a taxi and returned to the stranded Afghani Bunch."

"After making Pokey whole again under near squall conditions, we headed back toward the village in order to return the mechanic's jack. We hadn't traveled even a single kilometer, when 'bang', the tire went flat again. I couldn't believe it! By now it was getting late, and with the wind kicking more sand into our eyes than a body-building bully at Muscle Beach in California, I was in no mood to try to remove the wheel again. It would have to wait until morning. I was so upset, that I barked at Judy for some trivial reason and she ended-up giving me the proverbial 'deep freeze' for the rest of the night. I didn't like that, but actually, I couldn't blame her. With the wind whistling into every crack and crevice of old Pokey's weathered body, it soon became very cold inside the van. I could actually see the vapor exiting my mouth as I breathed and the reality of being near the mountains quickly set-in. Judy refused to let me use her as a blanket to help keep me warm and to make matters worse, I began feeling nauseous. During the night, I jumped out of the van several times in order to relieve myself both above and below. Half the time my pants were around my ankles so that I wouldn't fill them, and the other half, I was holding my hair and beard aside in order to avoid them being soaked in puke. On top of that, I was only wearing a thin tee-shirt and shivering like a new-born puppy away from the protective warmth of its mother. Needless to say, it was my worst night of the entire trip."

"The following morning, I was still feeling like death warmed-over and chilled to the core. I definitely didn't have the energy to remove the flat tire, and my friends were giving me a wide berth. In the late morning, we were spotted by Colin as he was heading toward Tiznit. He was a friend from

Taghazout, and stopped to assist us. He let us borrow his spare tire and then followed us the 85-kilometers or so, to Tiznit, so that we could have our flat repaired. Once this was accomplished, we returned his spare to him and limped our way back toward Agadir. You won't believe this Kid, but just as I started feeling comfortable again, the same tire blew for a third time. Now it had a 2-inch hole in the sidewall and there was no way anyone in the world would be able to fix that. I was ready to lose my mind! Luckily, Colin drove past us again an hour later and stopped."

"We repeated the entire process as before, but this time I had to buy a cheap used tire to replace the mangled one. I also bought a roadworthy spare just in case of emergency. Unfortunately, they didn't have any used jacks for sale. I decided that this area of Morocco was bad luck for us, so I convinced everyone that we should head back to Taghazout. When we arrived, the place didn't quite look the same as before. The winds had done a number on our primitive lean-to's and everything else looked disheveled."

"It was as if some colossal being had picked-up the entire village by its lapels and shook it so hard that it had become wrinkled and misshapen. I was still feeling sick and attributed that to nervous tension and the stresses of money shortages, worrying if Pokey would break-down and the recent flat tire episodes. Eventually, I finally managed to keep down a couple dishes of yogurt, and smoking the odd number helped to keep me calm. We remained in the region a while longer and bounced back and forth between Agadir and Taghazout. I regularly checked the post office in Agadir however, nearing the end of February, I had still not received any mail from home. I was beginning to think that I had been disowned by my family."

"Soon the weather began deteriorating at our beachside paradise again, so on February 21st, we packed-up our belongings and slowly started making our way back northbound. We were aiming for Casablanca. We hadn't driven very far, when I noticed that the van's generator light began occasionally flickering. That's all I needed! Between the rain, unreliable tires and now the generator, I was really beginning to stress. I drove for seven hours straight and eventually stopped near the outskirts of Casablanca. We spent the night parked near the massive stone walls of Al Adeer Prison, but I didn't get much sleep. Under torrential rain, we floated into Casablanca the next morning, only to be impeded by a crush of traffic. It was a very rude awakening to be back in bumper-to-bumper gridlock after weeks of little to no traffic from where we had just been."

"To compound my misery, it seemed like every idiot in North Africa with a death-wish was either driving a car, peddling their bike, or simply traveling on foot and continuously cutting me off, cycling without any concern for their safety, or walking into the road without even first having a look to see if any cars were coming. I was forever slamming-on my brakes, honking my horn and swearing a blue streak. If it wasn't for my superior driving skills, there would definitely have been a few 'kill notches' scratched into Pokey's dashboard. I wondered what the traffic accident statistics and road fatality numbers were like there. They must have been considerable."

"We spent the night in Casablanca but the weather definitely placed a 'damper' on our activities, no pun intended. It was here that Pokey needed to drink her first liter of oil. Not bad considering all the miles I put on her. The next morning, we headed to Fes and arrived that same afternoon. Fes has the largest Medina in Morocco and lots of merchandise to buy. While parked, a couple of young boys approached us and began speaking English. Both were named Mohammad. Considering that they were only about nine or ten years old, they were pretty savvy, and offered to show us around the city. We agreed, and they hopped into our van."

"They showed us areas of town not normally seen by outsiders and we even went to a district where all the local artisans plied their trade. We observed craftsmen making silver and brass objects, textile makers dying wool and creating rugs and fabrics and jewelers producing rings, bracelets and necklaces. This was all really interesting stuff and we would never have been introduced to it had we just shooed the kids away. One of the boy's family owned a curio shop nearby, so we agreed to have a look. Unfortunately, the boy's family didn't possess his personality and their prices were way too high. We soon got the heck out of there and headed back to where I'd parked the van. As we returned, there was a miserable looking dude standing beside Pokey. He said that I had to pay him 10 Moroccan Dirhams for parking on his property. I told him to shove off, but he persisted. Just to shut him up, I offered him 1 Dirham but he refused. I wasn't in any mood for his crap, so we all just boarded the van and I fired-up the engine. The goof was still standing in my way demanding money. When he refused my final offer for the second time, I just floored the accelerator and surged forward. He just managed to move out of the way before I hit him, and we flew past, spitting gravel all along the way. He should have accepted my offer of the 1 Dirham. Now he had nothing."

"On February 24th, we were approaching the port city of Oran in the northwest of Algeria. It was still raining most of the time and that had significantly lowered the outdoor temperature. The countryside was also much less arid and considerably greener than western Morocco. Before actually crossing the frontier from Morocco into Algeria, we gave Pokey a good interior cleaning. Used pipes were discarded, seeds, sticks and other remnants of weed usage were swept away and our remaining acid blotters were carefully secreted. With all countries' customs agents now forcefully checking for illicit drugs entering their respective lands, the last thing we needed was to be busted, or have Pokey ripped apart as a result of some carelessness attributed to us. Our vigilance paid-off, as we easily breezed through the border crossing and into Algeria."

"As we arrived in Oran, I cashed a traveler's cheque at the American Express Office and inquired if I had received any mail. Rosco was also expecting money waiting for him at the office, but when he was informed that nothing had been delivered, he called his mother in Vancouver. When he left the payphone and returned to where I was standing, his face was ashen and his lips were trembling. Concerned, I asked him what was wrong, but he just kept answering, 'I gotta' go home…I gotta' go home.' He wasn't making any sense but I finally managed to calm him."

"He said that he had just spoken with his mother and she told him that his wife was having a nervous breakdown. She either couldn't or wouldn't give him any more details but simply demanded that he return home immediately. She said that she would 'wire' money to him in Athens and he could then begin to depart Europe from there. What a kick in the head to all of us! Rosco needed to get to Athens as soon as possible, so this meant we had to reach there quickly and bypass all of the other spots I wanted to visit. I felt really bad for him, but also a little selfishly for me and Judy as well. To make matters worse, since there was no money waiting for him in Oran, Rosco was broke. We also learned that Gee-Guy was now out of cash as well. That meant me and Judy needed to float them both, until they could score some funds."

"The further east we drove, the worse the weather became, both in temperature and dampness. I had wanted to stop in the city Algiers for a while, but instead, we had to roll straight through. However, while driving along, I did manage to see a few of those mysterious veiled women with tattooed faces, heels and hands. I always wondered if they were tattooed anywhere else on their bodies. I guess only Allah or their husbands would be able to satisfy that curiosity."

"My stress-level is now hitting new highs. With every other disaster that has now befallen us, Pokey's generator light is no longer flickering. Now it's bright red the entire time! I'm expecting the engine to fail at any moment. We managed to reach Setif on the 25th but as luck would have it, that day was another in an interminable succession of civic holidays. As a result, everything was closed. A nice cop we spoke with tried to locate an electrician for us, but no luck. We would just have to wait until the holiday was over. The cop even allowed us to sleep on the precinct's property, so that we would be safer. Once again, I had to revise my opinion of the police. Whereas, I referred to some as 'pigs', others were very generous and decent people. Just regular folks."

"The next day, we managed to find a repair shop and the mechanic rewound the generator's armature. While we waited outside of the shop, a kid threw a rock and hit me in the back. Before I could chase after him, a few of his companions, knocked him down and bawled him out. They didn't look like bad kids, but in North Africa, there is so little for young people to do, that they often get into mischief simply because they're bored. Once the engine compartment was closed and I settled the bill, we were back on the road. Unfortunately, we had only traveled a few kilometers when the generator light came on again."

"Fuming, I made a fishtailing U-turn, and drove straight back to the repair shop. The mechanic removed the generator once again and examined it closely. He then said that he couldn't fix it properly until he received certain replacement parts. Unfortunately, vehicle parts were expensive and hard to come by in Algeria. It would be at least a couple of days before he could acquire them. There was nothing for it, we were stuck. Leaving Pokey parked outside of the shop, we decided to go to the movies. Before leaving, we made a huge batch of popcorn and stuffed it into a plastic bag. Walking into the center of town, we managed to locate a theatre screening a Greek western dubbed in French. As we lined-up at the ticket booth, you'd think the aliens had just landed. Suddenly, there was a large group of people all around, ogling us in disbelief. I had been down this road before, so I knew what to expect. These people were harmless, they were just curious. I gave a couple kids some of our popcorn, and the next thing I knew, everyone else wanted some as well. By the time we seated ourselves inside the grubby, decrepit cinema, there was hardly any popcorn left for ourselves. The movie was unbelievably bad and the director should have been drummed out of the industry. The local audience snickered during all of the love scenes and gasped during the gunfights. The only entertaining moment during the entire flick, was when we observed a rat scurrying down the aisle and then run across the seat backs about four rows ahead of us."

"On February 28th, the garage owner returned to the shop with some parts and managed to rebuild the dynamo. While waiting, I did some more embroidering on my jeans, I read my book, wrote a few letters home, and also made a number of new fans while throwing around the Frisbee. I even drew a picture of a fictitious monster and presented it to the mechanic. He loved my drawing and taped it to the window of his vehicle. Finally, at 5:00 P.M., the mechanic announced that Pokey was ready to roll. However, when we tested her out, the red light returned. The mechanic now claimed that the regulator must be shot and that would cost an additional $25.00 dollars to repair."

"Well, let me tell you, I went ballistic! I completely flipped-out and was using every profanity that I could think of, plus a few more that I probably invented. I had already paid him $45.00 dollars for repairs to date, and I'd be damned if I was going to give him one penny more. Just when it was starting to get real ugly, one of the apprentice mechanics noticed a frayed wire at the side of the generator. For a third time, the generator was removed, the frayed wire soldered, and then reassembled. This time we were in the pink!"

"Pokey was finally repaired, and by 6:00 P.M., our German home on wheels, was chugging its way toward the Algerian/Tunisian border. After paying an additional 3-days insurance on the van, we later crossed that frontier without incident and barreled toward the port city of Tunis. On the 29th of February, I wished everyone a 'Happy Hippie Leap Year,' and we coasted into Tunis at 2:00 A.M. We pulled over to a secure location and crashed for the night. I was now running on adrenalin and determination, so the next morning, I was up early and already motoring forward while the others still slept."

McGuire now became reflective and deliberately offered Jesse information he felt was poignant. So far, he hadn't been overly profound or philosophical, but this seemed to be a very special moment for him. "You know Kid, sometimes when you're alone and least expect it, something touches you or you see something that remains with you for the rest of your life. This occurred to me that morning. According to the road signs, the next town approaching was Tabarka. It was a small coastal town in north-western Tunisia and nestled between the ocean and a lush green mountainous chain. As I reached the crest of the highway and began my descent, I was gob-smacked by the sight before me. Straight ahead and below me was the town containing a tight clustering of whitewashed buildings topped with orangey-brown terracotta roof tiles. I could faintly hear the rhythmic and resonating gonging of church bells, and in the foreground were several fishing skiffs on the water, with their crew members throwing out their nets overhead and in an arcing motion, into the sea. Variating on the depths of the bay, several different shades of blue and green entertained my eyes. To my left, the pale, yellow winter sun was just beginning to crown the snow-capped mountain peak. And to my right, to complete this magnificent image, was the mist embraced remnants of a Genoese castle dominating an offshore rocky outcropping that complimented the entire ancient community. Upon after-thought, I wished that I could have shared the moment with my sleeping companions. However, in some small way, I was also happy to have been the sole recipient of that idyllic view. It was something that I alone had witnessed and could treasure."

"Near 11:00 A.M. that same morning, we arrived in Tunis and immediately made our way toward the port. We managed to secure three Economy Class tickets and one Tourist Class ticket for our imminent voyage to Sicily. We still had a couple of days to kill before departure and this would have been a perfect time to do some last-minute shopping. Unfortunately, that thought was shelved, since Judy and I needed to save our money. After all, we were now supporting Rosco and Gee-Guy as well."

"It would be a major understatement to say that Judy was just a little 'pissed' over this unwelcomed development. In an attempt to soften the blow, I suggested that we drive to the Roman ruins at Carthage. First, I visited the post office to mail my letters and then we all headed to the public bath house for a much-needed soak. It had been nearly 2-weeks that I'd had a proper scrub. After our heavenly soaping, and a quick bite of street-food, we all piled into Pokey and drove northeast to the Carthaginian ruins. We spent a relatively relaxing day and night strolling along deserted beaches and marveling in the splendor of ancient Roman architecture. We returned to Tunis on the morning of March 2nd, only to be informed that the boat to Sicily was delayed by another day. We weren't prepared for the unanticipated delay, but neither were we surprised. I learned a long time ago not to expect anything to go correctly or on time, in North Africa. We spent that additional day just goofing around and arrived at the terminal extra early the following day. In fact, we ended up being first in line for the ferry. Once again, our departure was delayed several more times during the day. We wiled the time away, throwing the Frisbee and kicking around our soccer ball. Rosco was voted Mister Popcorn 1972 for the delicious batches of the snack he made for us throughout the day. We finally drove on to the ferry at 8:30 P.M."

"Before leaving the dock, we sold our soccer ball to another group of travelers for one American dollar. With that single green-back, we purchased a kilo of oranges, six loaves of bread, a tin of evaporated milk, a jar of apricot jam, a container of hot chocolate powder, a bag of cookies and a package of toothpicks from the line of vendors straddling the pier. Not bad eh?" Upon hearing the admirable extent of the grocery list, Jesse raised his eyebrows and expelled a low, extended whistle. McGuire was sufficiently satisfied that he had impressed his audience of one, and then continued.

"As we slept within Pokey's cozy confines aboard the rusty tub, we gradually bobbed across the Mediterranean Sea and reached Trapani, Sicily early the following morning."

"Before being permitted entry on to the island, we were expected to pay 3-months vehicle insurance. This was outrageous and I tried to explain to the Customs officials that we were merely traveling through Italy and would probably be off and away within 3-days. They were being stubborn as hell, but with my superior powers of persuasion, I finally managed to talk them down to 2-months insurance. This was still way more insurance than I needed, but at least I saved myself a few Lira. We then hit the highway and began playing a vehicular version of 'dodge ball', between us and all of the moronic Sicilian motorists. Without a doubt, I still thought that they had to be amongst the world's worst drivers."

"After a stressful drive over rugged mountains, winding through treacherous stretches of inadequate highway and plagued by an army of indescribably poor wheelmen, we finally arrived in Messina on March 5th and made straight toward the ferry. Shortly before the noon-hour we floated over the Strait of Messina and on to the mainland at Reggio-Calabria. I was dead tired but I could see that Rosco was sweating bullets of worry for his wife. I decided to keep driving. A couple blotter hits renewed my resolve, but by the time we arrived in Taranto, I had reached my limit. Although we were now quite close to Brindisi, I refused to go any further. We soon located a dumpy little pizza joint and ordered a few pies. They wanted 250 Lira per pizza which I thought was too much, but I was too tired to argue. We eventually received our mediocre fare and sat down to consume it. Receiving our bill, it now said 450 Lira per pizza. When I challenged the owner, he stated that there was an additional charge for us sitting down. If we had kept standing, the pies would have been the original quote of 250 Lira. I wanted to deck him, there and then, but like I said, I was dead tired and just wanted to catch some shut-eye. We regrettably paid the bill and left."

"The next day, we arrived in Brindisi and drove straight to the port. The boat wouldn't be sailing until 8:30 P.M., so we bought our tickets and decided to wait. At the ticket office, an employee said we couldn't board until we increased Pokey's insurance policy. He claimed that they sold us motorcycle insurance in Sicily, and that wouldn't cut it here. By now, I'd had enough of everyone trying to stick their hand in my pocket and I lost it on the guy. Hearing the commotion, another employee walked over and straightened everything out. No extra insurance required. I took a couple of deep breaths, smiled at our savior, gave the other guy my evil stare, and then walked back outside. I then bought 2-liters of wine from a dockside vendor and returned to the van."

"As we attempted to board the ferry that night, more or less on time, Pokey wouldn't fit through the entry door. She was too wide! It appears that we had both gained weight during our trip together. As a result, the crew had to hoist her on board with a boom and winch and guess what?" Before awaiting Jesse's reply, he continued. "You got it, Kid. Another 10,000 Lira grab out of my wallet. The way these guys were nickel and diming me, it was like 'death by a thousand cuts'. They were slowly bleeding me dry. Once Pokey was finally strapped down on the boat, we weighed anchor and sailed off into the night. I said a sarcastic goodbye to Italy and proceeded to get thoroughly hammered on the wine I had purchased."

"The 8-hour trip from Brindisi to Vlore, Albania over the Ionian Sea was relatively smooth and uneventful. It was still dark when we arrived but daylight was not far off. As we reached the Albania/Greece frontier, the Greek border guard we encountered was not impressed with Pokey's tattoos. Especially, the giant, smoldering Afghani joint painted on the side. As a result, we were directed into the secondary examination area and searched. I cashed my last $50.00 traveler's cheque at a nearby currency exchange booth while those filthy guards violated Pokey's private areas with their black, leather gloved hands. Naturally, nothing illegal was found, so, we were back on the road. It was now full speed to Athens and stopping only if necessary."

"Near Patras, approximately 215-kilometers west of Athens, the unimaginable occurred. You guessed it. Another flat tire. This time it was the left-rear. However, at least this time, I had a spare to exchange with the punctured one. Unfortunately, I was still without a functioning jack. About 30-minutes later, a Good Samaritan pulled over and loaned us his jack and waited while I performed the task. Before you could say 'Jack Robinson,' we were back in action and once again beetling toward Athens. I was now on a mission, and short of a Third World War, nothing was going to stop me from reaching my goal. Shortly after 5:00 P.M., we did indeed enter the outskirts of Athens and I gave Pokey a gentle pat on her dashboard for a job well done."

"I was 'pumped' to be back in Athens, and although I was tired, the first thing we did was to visit the 'Folk 17' nightclub for a few drinks and something to eat. The Retsina and Ouzo was afloatin' and it was near closing time when we finally staggered back out into the street. The next day, we went to the American Express office and Rosco received the money-gram from his mother. I received a package from my friend 'Seed', back home. It contained a brief letter and three audio cassettes."

"One cassette contained a compilation of 'oldie but goodie hit tunes,' while the other two were blank. I sent Seed a message, informing him that I was 'skinned' and asked him to wire me the money from my unemployment cheques."

"We then went out for a day of sightseeing in the 'Glorious City'. At one point during the day, we were at the Parthenon and Gee-Guy started goofing around. He shimmied up a ten or twelve-foot carved marble column and balanced himself on the top of it. I had just snapped a couple of pictures of him up there, when all of a sudden, I heard the shrill screech of whistles blowing. Suddenly, we were surrounded by a bunch of irate, uniformed, Greek cops, and they were frantically motioning for Gee-Guy to descend from his precarious perch."

"Once he climbed down, they grabbed on to us and were about to take us to their precinct office, when a plain clothes cop sidled up and intervened. After a brief, but heated exchange between the ranking uniformed cop and the plain clothes interloper, they removed their death grips and told us to scram. We thought the entire episode was quite comical, but rather then rub salt into their wounded pride, we decided to split. That night, we held a farewell celebration for Rosco and drove him to the main train station the following morning. For $73.00 dollars, he was taking a series of trains to France and then the ferry over to England. From London, he would fly back home to Vancouver, via Seattle. We all felt a pang of sadness as the train pulled away from the Athens station, and the look on his face would have made even a hardened criminal cry. The Afghani Bunch was now down to only three members."

"I later found out that Rosco's wife wasn't suffering a nervous break-down after all. That's just what his mother told him in order to get him home as quickly as possible. In actuality, she was having an affair with his best friend, and spending his money rather than wiring it to him at specified intervals during his trip. Rosco ended up leaving the wench, selling his barber shop on Main Street and using the proceeds to purchase a small bar in Barbados. I never heard from him again, but I'd like to think he's still down there sipping on a cool Pina Colada and watching the sun set into the Caribbean Sea."

"The weather in Athens was a complete reversal from the weather in North Africa. It was super cold at night, and one morning we awoke to find a dusting of snow on the ground. I didn't sign-up for that, and the sooner I got back on to a warm sunny beach, the better."

"A couple of days later, Seed still hadn't sent my money, so we all went and donated blood for 355 Drachmas a pint. That payout wasn't very much money, but it would keep us going for a while longer. Judy still had a bit of a nest egg, but we needed more. Reluctantly, she called home to her parents, and they wired her some money. A few freaks we met in the city, said that Mylopotas Beach on Ios Island was very nice and fairly quiet at the moment. Shortly after receiving that tip, we bumped into Jim and his crew. They had just arrived in Athens and were also game for another island adventure. Driving to the port of Piraeus as soon as possible, we all booked passage to Ios Island on the next available ship. We were going over 'on foot' and leaving our vehicles parked on the peninsula. After securing our rides and returning to the wharf, we stepped aboard the ship."

"Our anticipated 14-hour boat ride became 22-hours, when our captain received word that a helicopter had crashed into the sea, fairly close to our position. Unfortunately, our patrols and those of other craft that had also received the S.O.S., failed to locate the downed craft or any associated wreckage or passengers."

"Once on Ios, we first checked into the Youth Hostel and then later into a 50-cent per night hotel. We wanted to camp for free at some nearby caves, but were informed that the police arrested anyone residing outdoors. We dropped that former idea and remained inside the hotel. Anyway, it was nice having a private room with Judy that was only 100 yards from the beach. Our time on Mylopotas Beach was great, with lots of the same activities we experienced in Taghazout. However, I did have one shitty night at a local bar. After having a few drinks and something to eat, I was going to pay the bill. The old hag that ran the joint was overcharging me by 10 Drachmas and I refused to pay. Well, she was giving me the 3rd Degree, so I told her to call the police. That was my first mistake."

"The police arrived en masse, and without even hearing my side of the story, hauled us all away to their precinct station. Once there, I tried to explain, but the head honcho wouldn't listen. I thought he was going to have a coronary. He just kept screaming at me and telling me I needed to pay the bill or else, he would lock us up. Eventually, I could see that I was wasting my time, so I agreed to pay. He then threw another curve ball at me. Not only would I need to give that old bag 10 Drachmas she didn't deserve, he also expected me to apologize to her for causing all this trouble. Now, he was expecting too much from me and I was just about ready to say, 'Lock me up!'"

"But, then I realized my friends would also suffer that punishment, and that wasn't fair. I finally agreed to his terms. The cops took us back there and I ate a cold, nasty dish of crow, while the old crone just stood there smirking at me. It took all the control I could muster not to boot her right between her spindly, wrinkle-skinned legs. On a more positive note, during our time on Ios, one day I bumped into 'Irv', whom I had traveled with for a while during my Canary Islands/West Africa trip. He was on his way to Ethiopia once leaving Ios and we had a short but nice reunion. During our talk, Irv mentioned that out of the blue, he had recently bumped into Guitar Dave inside a small supermarket in Tel Aviv. Dave was living there at the time. I smiled, and remembered all of the good sing-alongs we had with Dave and his vast repertoire of tunes."

"He also mentioned that Ali, our Mauritanian buddy that we had met on the train leaving Port Etienne in West Africa, and who had traveled with us through to The Gambia, had been arrested attempting to smuggle drugs into Barcelona. For all Irv knew, he was now inside a Spanish prison. That was too bad, and I was sorry to receive that news. Ali was a good 'head' and his knowledge of Africa and a great deal of its culture, was invaluable to us."

"By April 5th, Mylopotas Beach was beginning to fill-up with a daily flood of arriving tourists. Our once tranquil Shangri-La was now beginning to resemble Grand Central Station. In addition, the restauranteurs and other business owners were raising their prices in an effort to maximize their profits from this steady influx of cash-flush visitors. That night, we all caught the last boat sailing, and returned to Athens after spending a breezy, open-air sleep on the deck. At least, we had the benefit of a couple bottles of Ouzo to pour down our throats in order to dispel the night chill."

"We left Athens a couple of days later and drove north to Thessaloniki near the Greek/Macedonian frontier. Leaving Pokey in a 75-cent per night parking garage, we then boarded a train bound for Istanbul. The tired, old locomotive that appeared to have been a throw-back from the Ottoman Empire, slowly but steadily pulled its line of rail cars over the rocky, barren terrain. Almost 24-hours later, we chugged into the Moorish citadel of Istanbul, the former jewel in the crown of Eastern Christianity. We all immediately headed toward the unfashionable but affordable Tourist Hotel and registered for the agreeable price of $1.00 dollar per night. Judy and I left Gee-Guy to do his own thing."

"I took Judy to the Bazaar and other places of interest during our first couple of days there. Her 21st birthday arrived on April 10th and I was hoping to make it a special event by showing her a good time. Unfortunately, I'd forgotten how ignorant some of Istanbul's citizens could be, but I was soon reminded. Once again, I began hearing all the mutterings and whistles from jerks who took exception to my appearance. But what really bothered me was the way men kept ogling Judy everywhere we went and undressing her with their eyes. I just wanted to lay into them. My previous good memories of the city were quickly evaporating and I just couldn't relax. The entire vibe I received from the place was completely negative and was really stressing me out. That night, I treated Judy to a deluxe Chinese dinner which we both thoroughly enjoyed. With not much to do afterward, we returned to the hotel and just played cards over a few drinks and a couple of pipes."

"Just to show you how small the world actually can be. One night, we met a couple of dudes from Vancouver, at Younie's Restaurant in the old city. Their names were Bob and Harry. We were having a great chat and discussing things we had in common back home. At one point, the other Bob started talking about a good friend of his named Gary Britten, who had recently died of severe burns in some sort of accident. When I heard the name, my ears pricked-up and I began asking him more questions regarding his friend. As he continued providing me more information, I shockingly realized that the Gary he was referring to, was actually my cousin, Gary Britten from my mother's side of the family. I hadn't seen him in years and could hardly believe Bob's story regarding him. Of course, I had no reason to question the other Bob's voracity, since he could not have possibly known beforehand that his good friend and me, were actually related. My mood immediately went from upbeat to depressed, and the remainder of my night, was a 'total downer'."

"After 5-days in Istanbul we all returned to Thessaloniki via train and retrieved Pokey from the parking garage. Leaving Greece, we drove in a northwesterly direction and entered Yugoslavia. The further we ventured into the country, the more random checks from police we encountered. On each occasion, all they wanted to see was our vaccination cards. As we neared the city of Pristina, what today is the capital of Kosovo, we came upon a police check-point. We were informed that the main road into the city was closed to all traffic due to a severe smallpox outbreak in the region. We were instructed to turn around and go back to the junctions of Routes 6 and 7, where we could then bypass Pristina via Peja. That was at least an hour in the opposite direction, so there was no way, I was heading all the way back there."

"I started arguing with the cops and our words were becoming very heated. Judy was grabbing at my arm and trying to calm me down. I eventually realized that I wasn't getting anywhere with the cops, so I figured I had two choices available to me. One was to continue arguing and probably end-up getting arrested, or, turn around and drive back to the junction. I didn't like either choice, but decided that the second was better than the first."

"I jammed Pokey in reverse and backed away from the roadblock. Continuously swearing under my breath, I tromped it, and headed back southbound. We hadn't been driving more than a few minutes, when I noticed a cop car coming up fast behind me. Wondering what I had done wrong now, I quickly pulled off the road and they stopped directly behind. One of the cops from the roadblock approached my side of the van and politely addressed me. It appeared that a secondary route around Pristine had just opened, and we were welcome to take it if we wished."

"Suddenly, we were all good buddies with the cops and smiles and waves were the order of the day. Once again, I pointed Pokey in the opposite direction and headed back northbound. We soon found the bypass road and left the main route. Unfortunately, the secondary road was all gravel and potholes, but at least we were going in the right direction. I kept Pokey's speed between 40 and 50 kilometers per hour, considering her delicate tires. It took a while, but we finally left the secondary road without incident and continued on to Dubrovnik, Croatia. A few kilometers outside of the city, I was pulled over once again by the cops. This time it was for speeding. According to the Slavic flatfoot, I was driving 77 kilometers per hour in a 50 km zone. He asked for my driver's license, my passport and 20 Dinars. I gave him my license but not the other two items. He just stood there staring at me until I finally also handed over my passport and the money. He went back to his car while we waited. I was getting so tired of people trying to rip me off in one way or the other. Whether it was cops, ticket agents, customs officials, shopkeepers, or anyone else that figured that I was an easy mark. Several minutes later, the cop was back at the van. He handed me my license, passport and surprisingly, my 20 Dinar note. He then told me to slow down and drive carefully as he walked away."

"Once again, I was confused by people's behavior. I thought for certain that the cop was going to strong-arm me for more money. Instead, he returned all of my belongings and didn't even ticket me, although I probably was speeding. We arrived near Dubrovnik late at night and had to sleep outside of the city walls."

"Vehicular traffic was not allowed within the city center. After a light breakfast and a cool but refreshing dip in the Adriatic Sea, we were back on the road, still holding our northerly heading. We hugged the gorgeous Adriatic coastline and passed through picturesque communities such as Split, Zadar and Rijeka. From the latter, we then cut back into the interior and through the tranquil and fertile countryside of Slovenia. As we had done so many other times throughout our trip, we stopped in a small village just outside of Ljubljana and stocked-up on edible supplies."

"Being a very small country, we soon drove through Slovenia and into Austria. On the 18th of April, we located a campground in the quaint town of

Alpbach, and stopped for the night. We awoke the next morning to a crisp outside temperature of 45-degrees Fahrenheit, and were really looking forward to a hot, leisurely shower before hitting the highway once again. We received a very rude surprise, when the only water available to us was cold."

"Once again, I was majorly peeved at having been lied to, but there was absolutely nothing I could do about it now. What was to have been a toasty douche under a torrent of steaming spray, turned out to be a soap-free, 2-minute 'in and out', under an impotent trickle of ice water. Naturally, the campground's owner was nowhere to be found. Once through Salzburg, it was just a quick jaunt to the Austria/German border. I was very surprised to see that certain portions of the mountain roads were free of snow and actually steaming in the cold air. I later learned that many mountainous roads in Austria, were thermally heated from below ground. What a great innovation in utilizing the region's geographic resources and a perfect way to eliminate the need for snow ploughs, all the while keeping the motoring public safer during winter."

"Arriving at the German frontier, the guards' attitudes were no-nonsense and all business. They examined Pokey's exterior with a fine-tooth comb, while examining our documents. When inspections were completed, I was informed that my vehicle insurance was insufficient, plus one of Pokey's tires required replacement. I knew from prior experience that it was no use arguing with those guys, so I looped back around and returned to Salzburg. Here, I purchased a better quality used tire to replace the inadequate one and I increased my insurance coverage for an additional 7-days. Back at the border, the prior improvements were accepted, but now they wanted to check us for drugs. I couldn't believe it, but didn't complain."

"Bells were ringing as we entered Munich and I took this as a friendly greeting to Pokey and the Afghani Bunch infiltrating its precincts. First thing, we booked a room in a cheap hotel and provided our soiled clothing to that establishment's laundress. We then immediately found the nearest Hofbräuhaus in order to wet our whistles. The smooth but potent Bavarian beer went down a little too easily, and being served in 1-liter steins, it wasn't too long before we were all afloatin' in an alcohol induced euphoria. We remained in Munich for a couple of days so that we could get away from the constant hours on the road and to partake of that city's hospitality. The food, the beer and the people were all great, but unfortunately, I acquired an addiction while there."

Upon hearing this last admission from McGuire, Jesse adopted a look of concern on his face. He then awaited further explanation. McGuire noticed the look of shock on Jesse's kisser, and chuckled. "No…no…no, Kid. It's not what you're thinking. I became addicted to German pastries and couldn't pass a bakery without going inside to buy some tasty morsel."

With that clarification, Jesse blew a noticeable sigh of relief and was ready for the story's next installment. Again, McGuire snickered, realizing that his rhetorical statement had made its desired effect on his enraptured audience. Satisfied, he then continued. "On the 22nd, I took Judy to see the Dachau Concentration Camp. As you know, I had been there before during my first trip, but I felt that it was worth seeing again. When we arrived, I had an idea of what to expect, but Judy was truly shocked. Seeing the photo collages in the small camp museum, the barracks, the gas chambers and the ovens that marked the final destination for many of the camps inmates, was almost more than she could handle. Outside, our guide requested that we not mark the walls with any writings. However, from the look of the abundant graffiti, in many different languages, already plastered to many of the camp's structures, it appeared that not everyone had previously respected this plea for restraint."

"We left Munich later that same day and motored through the remaining country in a northwesterly direction. Our final destination was to be our original starting point in Amsterdam, and as we cruised past the cities of Stuttgart, Heidelberg and Koln, our quest was rapidly nearing its conclusion. Before long, we were back in Holland and nearing Mokum, the nickname for Amsterdam. Approaching a large, empty garbage bin at the side of a road, we stopped and removed everything from Pokey's interior."

"Sweeping out the floor panels, tossing the garbage and giving the blankets and sleeping bags a good shake, we kept only what we needed and trucked off once again. Unfortunately, we accidentally left Rosco's brand new boots sitting at the side of the road when we drove off. Each of us thought the other had replaced them inside the van, but no one did. If I ever met Rosco again, I would have to make-up some sort of story regarding his lost kicks."

"Upon arriving in Amsterdam, we treated ourselves to a celebratory supper, a few drinks and a good smoke. Now primed for the night, we went to the movies and saw 'A Clockwork Orange'. It was a bit weird but well-acted, and although the overall film was okay, I liked the book better. The next day, I took Pokey to a carwash and gave her a thorough cleaning. Next stop was a gas station where she received a much-needed lube, oil and filter servicing. Although, I was going to miss her, she was ready to accept her next adventure with a new owner. My time with her had passed."

"It was cold sleeping in the van at night, but once again money was tight, and we still needed to get back to England, before returning home to Canada. At the moment, a hotel room was out of the question. Judy decided that she was going to call her parents again and ask them to send more cash. For the next 4-days, I stood outside of the American Express office with Pokey, shivering and holding up a 'For Sale' sign. I had a few interested enquiries, but no firm buyers. I was starting to become a little worried, since there were actually quite a few similar vehicles for sale in the city. To top things off, I was beginning to feel poorly, and it was becoming more difficult for me to get through each subsequent day."

"Finally, on May 5th, a couple of American freaks with long hair almost to their waistlines, began giving Pokey a real close examination. After a few minutes, they approached me and began asking a bunch of questions regarding my customized German fraulein. They asked me what I wanted for her, so I threw out a big number to see how they would react. 'Seven-hundred bucks,' was my response. Expecting to see them commence laughing, instead, they accepted. I tried to keep a straight face as we shook hands. Inside myself, I was performing cartwheels! I had only paid $300.00 dollars for Pokey and put thousands of kilometers on her with only a few added secondary expenses. I was now more than doubling my money. Such a deal!"

"On May 5, 1945, the Canadian Army with its American and British Allies, liberated Holland from Nazi Occupation. On the same day, 27 years later, this Canadian in Holland, once again re-enacted an American/German connection, with the liberation of his faithful Volkswagen van from his possession, by a couple of long-haired Yanks. Before they could change their minds, I quickly took them to an insurance agent, where I transferred the vehicle registration over to them. As they handed me the cash, I asked them about their plans. They answered that they were going to drive to Beirut, and purchase enough hashish to pack the rear of the van. Afterward, they would drive Pokey into Germany and sell the drugs, piecemeal, to American servicemen stationed at various military bases throughout the country. I considered their plan arrogantly bold and unrealistic. I couldn't see them getting a van load of dope, undetected, through the border crossings of several different countries and then into

Germany. However, that was none of my business, and Pokey was no longer my property. We went our separate ways, and I rushed back to Judy and Gee-Guy to give them the good news."

"I say, that I rushed back to Judy and Gee-Guy. However, in actuality, suddenly feeling drained, I dragged my ass back to them. With my new found wealth, we booked a room in a reasonably priced hotel, but I barely made it to the bed before I collapsed. I felt feverish and my joints were beginning to ache. The next morning, I started the day by puking out my guts and I noticed that my pee was discolored."

"Although, I hadn't done anything unusual, I felt exhausted and didn't leave the room that entire day. Judy brought me something to eat and drink but I either didn't feel like having it, or if I did, I couldn't hold it down. I just wanted to leave. I don't remember how they managed it, but Judy and Gee-Guy got me on to a ferry and back to England. Once in London, our friend Vicki acquired another room for us in the Hotel de France and I just vegetated there for the next couple of weeks. I was barely eating or drinking and Judy was really worried about me. She wanted me to see a doctor, but I refused. I felt that I just needed to rest and eventually, I would feel better."

"I spent most of the time sleeping, reading a little bit and playing to death a cassette tape of Neil Young's acclaimed 'Harvest' album. As May began drawing to a close, I still wasn't feeling any better and I finally admitted that I needed to get home as soon as possible. I nearly scared myself to death when I looked in the bathroom mirror and observed what I had become."

"My face was gaunt, I had dark circles under my eyes and my skin color had turned a pale yellow. I had also lost nearly fifty-five pounds. My El Gordo days of North Africa were now far behind me."

"With what little strength I still possessed, Judy managed to get me to Prestwick Airport and on a charter flight to Seattle. Looking in the shape I did, I don't know how or why I was able to get past all of the airport officials, ticket personnel, flight attendants and Customs Agents in either England, the USA or Canada, without being stopped and questioned. However, I managed it. As soon as we returned home, Judy took me to see my family physician, Dr. Britten."

"He was actually angry at my physical state and proceeded to give me an earful. I was diagnosed with hepatitis and given a specific course of treatment and medication in order to combat it. Without realizing it, I could have seriously damaged my liver and even contracted cirrhosis or liver cancer. And it all started from becoming run-down during my trip. Not eating or sleeping properly was probably what set that pendulum in motion."

"I did eventually kick the illness, with thankfully, no permanent damage to any of my organs, but it took months to do so. If you asked me 'Did I learn anything from my mistakes?' or 'Would I ever let myself go like that again?' I can't honestly say. I'm not deliberately on a mission to destroy my health, but I am what I am. I act in the manner of my choosing and I do what I feel like doing. I don't believe that I will ever change, but who knows for certain? I just live one day at a time and take things as they come."

This was now the third instance where McGuire's indulgence for pleasure and travel thrills, resulted in a negative impact on his health. It seemed that this was a common thread, and perhaps, it was the price he needed to pay for living that form of lifestyle. This thought concerned Jesse, but for the moment, he would keep that opinion to himself. The last thing he wanted to do was offend his host, plus he realized that it wasn't even his place to mention it. However, he had developed considerable affection for Bob during these last two weeks, and like any other good friend, this troubling detail bothered him. While McGuire's final, relevant sentence gradually trailed away, both men remained seated for several minutes without speaking. Before long, and as if on cue, they then rose simultaneously, bid one another goodnight and walked to their respective bedrooms.

Chapter 9
Rogues and Rastafarians

The next day unfolded almost identically to the one previous. After finishing breakfast, McGuire faithfully attended the greenhouse for his daily ritual while Jesse returned to his restoration of the yard and garden. The unpredictable island weather was once again asserting its prerogative by replacing a previously warm and glorious, sunny Summer morning with now a cool and gloomy, overcast early afternoon. By the time McGuire returned home several hours later, the first droplets of moisture were beginning to fall from the heavens. He quickly parked his truck, grabbed a pair of work gloves and a rake and then joined Jesse with his exertions. Over the past two days, the property's transformation was quite noticeable, to the satisfaction of both the homeowner and his diligent guest. Unfortunately, within the hour, the rain commenced in earnest, so the two men retreated to the protective environ of the home's interior. Judging by the strength of the current inundation and the solid, steel-gray canopy above, McGuire predicted that it would probably rain all night.

Once inside, both men went about their separate business. However, before the interior lights even required activation, they once again found themselves in their pre-assigned positions within the living room, all primed and prepared for their next bout of histrionic narration. McGuire was now handling his fourth journal. He scanned it briefly for reference purposes, and then commenced speaking.

"Like I was saying earlier, that bout of hepatitis really knocked me on my ass and it took quite a while to not only kick the illness, but to also gain back the weight I had lost. Unfortunately, after only a few weeks back home, Judy and I had fallen out. As a result, we decided to go our separate ways. I don't really know what caused the split, since I thought that I had opened Judy's eyes to a bigger world and increased her overall awareness of foreign peoples and their cultures. I guess I was wrong. I think what stung me the most, was the fact she left me for some unsavory character that abused her. Judy's parents were very upset with her, because they really liked me and felt that she definitely traded downward, with her next choice of beau."

"Before finally achieving my old 'fighting form', I missed the majority of the 1972 fishing season. However, I just managed to grab the tail end of that year's herring run, and was lucky enough to make some decent money. The better part of 1973 was fairly uneventful, with me taking things one day at a time, saving some money and planning my next adventure. I had been in contact with some of my previous traveling buddies, and after some debate, we all decided that the Caribbean would be this year's destination of choice. With my health back on par, I was able to work that year's entire fishing season, and by its end, I was more or less flush with cash. By November, when Vancouver's weather became increasingly more damp and cold, it was time to split."

"By month's end, I took to the skies once again and was soon stepping off the airplane and on to the tarmac of the Kingston, Jamaica International Airport. Leaving home, it had been an inclement 42-degress Fahrenheit. However, to my delight, that temperature literally doubled by the time I reached the island of rum and reggae music. Leaving the airport in a hired mini-bus, Rusho, Hager, Robbie, Gene and me, headed for the sleepy community of Negril, on the west-central tip of the island. This area was quieter than the more commercial regions of Kingston and Montego Bay. I'm always one for action, but the tourist scene, is not my thing. I prefer to be nearer the local population rather than a bunch of itinerant squares."

"Searching for suitable accommodation, we eventually managed to find just we needed at Ma Reynolds' bungalows for $10.00 dollars per week. Our suite was quite large and contained a room with five cots, a bathroom and a kitchenette. In addition, it was only two-hundred yards from seven miles of sparkling white sand and azure-blue ocean. From our location, it was also just a short stroll along the beach to a stretch of affluence from colonial times. Strategically placed in optimum locations all along the strand, were clusters of old, stately homes with high stone walls, wrought iron gates and lush, manicured grounds. The privileged few that had initially inhabited these estates, constructed them with fortunes derived from the sale of sugar cane, cotton, tobacco and rum. While the landed gentry amassed their fortunes, the indigenous population of the island were little more than slaves, barely eking out a living, while their European masters thrived."

"Our little corner of paradise was just what the doctor ordered. Days were spent sunning on the beach, body surfing, throwing a Frisbee or kicking a ball, or just plain laying around."

"On the beach or in town, we were occasionally approached by a dreadlocked Rastafarian looking to sell us some of the local ganja or hash oil. A few juvenile entrepreneurs flogging their handcrafted trinkets and tacky souvenirs, were also present. However, for the most part, we found relative peace within our sphere, with very few distractions or other people to disturb our tranquility. Even with my freckles and fair skin, I was acclimatizing myself to the radiant, Caribbean solar rays to the point where my outer hide was now beginning to develop a golden-brown hue. Every day brought me one step closer to the look I was eventually hoping to achieve. Unfortunately, the Christmas Holidays were rapidly approaching and this meant an influx of pale-skinned invaders would soon be arriving to defile our Eden. The last thing this Bronze God desired, was to be rubbing elbows with a pack of bandy-legged, floppy-fleshed outsiders, wearing their gaudy tunics, Bermuda shorts and knee socks."

"Ma Reynolds' thirty-year old daughter, Joy, was very helpful to us. She loaned us an electric, food blender and several other useful utensils for the kitchen. Plus, she took us to the nearby village of Savanna la Mar., or Savlamar, as it was referred to by the locals. It was situated directly on the coast and boasted an eighteenth century fort constructed for colonial defense against pirates in the Caribbean. It was in 'Sav' where we purchased the staples required for daily living. Overall, Jamaica suffered from numerous shortages, since the country needed to import almost everything it required. However, there was enough selection in the market to satisfy our needs. The local supermarket carried additional foodstuffs, but they were much more expensive there, compared to the marketplace. Plantains, breadfruit, papayas and mangos, were just a few of the items that I was now enjoying. I never even knew they existed prior to my arrival on the island. Merchants were practically giving away bags full of mangos for free, but a single egg cost the exorbitant price of 10-cents a pop. Go figure."

"We just needed to be a tad more discreet with our 'smoking habits' when Joy was nearby, since her husband was a sergeant in the local police force. Back at the bungalow, our market selections were transformed into tasty meals by Robbie, our designated, personal chef. He frequently cooked-up a pretty mean menu of conch soup, fried fish or chicken, savory rice, omelets and a variety of both cooked and fresh vegetables. Once again, I had to keep a watchful eye on my waistline, in order to avoid the return of El Gordo. Beyond our assortment of playground and water sports, the closest I came to work was cutting open my breakfast grapefruit. It was a tough existence, but I was willing to make the sacrifice."

"If and when we tired of our immediate surroundings, we would call upon Murphy, our local mini-bus driver, and for a few bucks, he would ferry us around to a few of Jamaica's popular sites. Ocho Rios, Dunn's River Falls and the Blue Mountains were just some of the hot-spots we visited. However, no matter where we went, I always considered our little beachfront oasis, the best place to be. One of our favorite pastimes was to travel along the island roads, stopping at the many rum shops along the way. A large glass of rum was generally cheaper than a bottle of Coca-Cola, and it nicely warmed the innards. After a few stops at these locales, me and the boys were comfortably afloatin'."

"There wasn't much of a nightlife in Negril, but we made do by frequenting the local bars and slumming in the Negril Yacht Club. In the latter establishment, we partook of their delicious Banana Daiquiris, in an attempt to slake our throats from the parching heat of the earlier day."

"One night after a serious round of bar-hopping and attempting to drain the island of its entire stock of Appleton Rum, we nearly made it back to our bungalow before Hager passed-out, face first, on to the grass. With him being a big sucker and us too drunk to lift his bulk, we just left him there for the entire night. Early the next morning as the rising sun gradually began evaporating the condensation from the lawn, Hager roused himself and stumbled into our suite. The front of his clothing that had contacted the grass overnight, was soaking wet and his cheeks glistened with moisture. Seeing him in that state, we couldn't stop laughing. As a result, we nicknamed him 'dew-face' for the remainder of the trip."

"There was one undeniable flaw in our vacation, and that was the lack of eligible women. They were outnumbered at least 2 to 1 by the men, and it was easier to get a date with a bush monkey in the forest, compared to a warm-blooded, human female. My bed hadn't felt that empty in a long time, and I wasn't happy about it."

"The vibe in Jamaica was extremely laid-back and no one was in a rush to accomplish anything. Sometimes, people's movements were so slow, it seemed like they were going in reverse. If you went to a restaurant, don't be surprised if you waited 30-minutes before anyone made the effort to first acknowledge, and then serve you. Shopkeepers aren't concerned whether you purchase something or not, and even most of the fishermen don't even bother going out in boats in order to catch fish. Instead, they merely hang mesh traps on ropes, from the wharves, and only check them every second or third day to see if they had trapped anything."

"It was difficult understanding the Jamaican/English diction unless they spoke very slowly and only if you were sober. Although the dope was afloatin' all over the island, it was still illegal, and the authorities didn't deal kindly with transgressors. In fact, while we were there, we heard about a Canadian kid from northern B.C., who was fined $150.00 dollars and immediately expelled from the country, just for smoking a joint. Needless to say, we were always very careful when toking, and constantly looking over our shoulders to see who was around."

"Postal delivery to Jamaica was nearly as slow as its populace. Hager and Robbie finally received some mail from home, but it took twenty days to arrive. I had already sent a postcard and a couple of letters to Gail, my new girlfriend, but I still hadn't received a reply. Since Gail was pregnant, I was a bit worried about her and hoped that everything was alright."

Hearing this last sentence, Jesse did a 'double-take' and asked the question requiring a response. "Bob, I thought you told me you didn't have any children?" McGuire just laughed, having once again hoodwinked the youth. "I didn't lie to you, Kid," was his reply. "Gail was actually pregnant, but I wasn't the father. She was already in that condition when I started dating her." Jesse bumped his forehead with the heel of his right hand and rolled his eyes. Satisfied with this reaction from his deliberate tease, McGuire continued.

"Like I said before, we spent most of the time on our little stretch of beach and more or less kept to ourselves. However, we weren't snobbish or introverted, so anyone that desired our company, received it. For about 10-days before Christmas, we had an American couple in the next bungalow, as neighbors. Dean and Audrey were from Manhattan and on their first major vacation together. They were close to our age, but with no pun intended, 'green as grass'. Before long, we had them out of their protective shells and enjoying the good life. In fact, Dean took to weed like a bear to honey, so by the end of their stay, Dean had won the 'Zig-Zag Rolling Papers Invitational Numbers Award', for the most dope smoked by a novice in the shortest amount of time. During their entire stay, the couple hardly ever left their bungalow. So, I remarked as they were leaving the island, that now they were even whiter than before they arrived. They got a big laugh over my comment, but actually, I wasn't kidding."

"As we neared Christmas, the weather was becoming a lot less sunny and there were even times during the day when we had to cancel our Frisbee tournaments due to the wind and rain. To top it all off, the invasion of the living-dead, mainland tourists had also begun. Our little piece of paradise was now becoming over-run with 'squares', and the town's merchants and shopkeepers had all raised the prices of their goods and services in an attempt to take advantage of all the new money now gracing their establishments. This was our cue that it was time to be moving on."

"Not long after arriving in Jamaica, we met an ex-patriated Scotsman named Guy Harvey. His family had settled in Jamaica years before and they owned a large house and property further up the island, in an area called Belmont. Guy was only about eighteen or nineteen years old, but he gave us an open invitation to visit him whenever we wished. We now thought that this might be right the time. A few days before Christmas, Hager met a cute little Jamaican girl named Charmaine. She was only about eighteen years old and noticeably naive and inexperienced. She invited us all to her church for a service and to sing some Christmas carols. I hadn't seen the inside of a church for sixteen years, and I wasn't really in the mood to break that streak. However, Hager managed to convince us that it might be fun. In any event, the Negril Yacht Club was now so full of newly arrived tourists, that you couldn't even get a seat in the bar after 6:00 P.M."

"We attended the church service and I was amazed by the congregation's enthusiasm. Everyone was jumping up and down during the worship, waving their arms in the air and regularly shouting 'Amen' and 'Hallelujah'. The minister was blasting out his sermon and sweating like an overheated plough horse. At the end of the program, we all began singing Christmas carols. Some of the parishioner's voices were flat, but overall, I think we did justice to the tunes. Once the service was officially declared over, and after shaking about one-hundred black hands, we were out of there and looking for the first watering hole with some empty chairs. We eventually found a tiny rum shop and quenched our thirst. The next day, we gave Ma Reynolds our notice, rounded-up Murphy, our trusty mini-cab driver, and made our way to Belmont."

"Once there, we thought we had been given the wrong address. The house was massive, with a large surrounding property and overlooking the beach. It was even named. It was called Horizon. You know that you've made the 'big time', when your house has a name. We entered the estate and were cheerfully greeted by Guy."

"He called a couple of the servants to help us with our bags and showed us to our accommodation. I couldn't believe it, we were provided three bedrooms, a kitchen with an oven, a dining room, a bathroom and access to a large verandah with a view of the ocean. We quickly made ourselves at home and then began exploring the area. Down at the water was a long, wooden plank wharf to which was tied a 24-foot Bayliner Cabin Cruiser with a 250 horse-power Mercury outboard motor. There were also several submerged lobster traps tied by rope to the wharf's support pylons."

"The next day, a couple of the guys left the house on their own and managed to wangle a ride in the Bayliner, heading out into the 'chuck', for a bit of fishing. When they eventually returned, it was with a 28-pound Kingfish, a 48-pound Wahoo, and a few 8 or 9-pound Barracuda. I was seriously jealous and the following day, Robbie and I made inquiries as to how we could try our hand at fishing as well. Unfortunately, the Bayliner was spoken for, but we managed to snag a ride with a local fisherman named Herman, who owned a 28' foot skiff, hewn out of a single Poplar tree. The boat looked a little primitive, but it sat well in the water and was powered by a 40 horse-power Johnson outboard motor. Before long, we were out in the deep water and cast our lines. I was using a Marlin rod with 80-pound test, wire line. The rod was secured by a safety harness around my torso, in the event of a big strike. Herman had baited my hook with 'stickfish', and within about 10-minutes, I realized my first bite. My line was out about 200-yards and Herman was giving me instructions on how to land the fish. Whatever I had on the other end of the line, it was mighty angry, and I was sweating like a pig trying to reel it in."

"It felt like I was battling with that fish for over an hour. However, in actuality, it had only been a few minutes before I could finally see the flash of silver just below the surface, being pulled ever closer to our boat. I finally managed to haul my catch onboard, and I took my first glimpse of a Barracuda in the wild. The fish weighed only about 8 or 9-pounds, but from the fight I just endured, it felt as though it had weighed ten times that amount. I was told those suckers were tenacious, but just how tenacious, I didn't know until then. I managed to catch a couple more a bit later on, but nothing of the sort of fish I was hoping to snag. Herman said that the recent bad weather had 'muddied' the water and as a result, the larger and deeper swimming fish couldn't see the lures or the bait. Unfortunately, Robbie didn't have any luck at all. Not even a nibble."

"Overall, I had a great time on my first tropical fishing excursion and on our way back to shore, we even saw a herd of Bottlenose Dolphins rip through a school of fish. It was something to witness, and I could imagine the panic within the school as the predators gorged themselves. At one point after feeding, the dolphins briefly checked us out. However, deciding we weren't very interesting, they then just swam away."

"That night, we feasted on my catch. The barracuda meat was firm and full-flavored. It was a bit like tuna but less oily and slightly sweet. It tasted a bit fishy but not as intense as anchovies. Paired with the other side dishes that were prepared for us, we thoroughly enjoyed our meal. We spent Christmas Day at Paradise Park, near the Horizon estate and we loaded up on numerous folk-art wooden carvings and locally made semi-precious jewelry. Overall, between a couple of us, I think we spent over $70.00 dollars. That night, I learned that Guy Harvey was leaving the next day for Scotland, where he was undertaking additional studies at the University of Glasgow. That was unfortunate, since I had really bonded with him and would definitely miss his companionship. Although, he was younger than me, he taught me a lot about Jamaica, its people and the region's fishing culture."

"That night, we enjoyed a traditional Christmas dinner and all the trimmings with Guy and his family. The food and the company was great, and it dulled a bit of my melancholy of spending another holiday season away from my own family in Vancouver. After several toasts with glasses of local rum and after consuming numerous bottles of Red Stripe beer, Guy asked if we wished to accompany him to a nearby church for the Christmas service. Once again, it wasn't really my 'thing', but for Guy, I was willing to make the sacrifice. After the service, Guy began circulating through the church and saying goodbye to its parishioners. Many of them were crying and you could detect the close relationship that had developed over the generations between them and the Harvey family."

"Guy departed for the Kingston Airport the following morning, and my gang also started making plans to leave. We decided that we would quit Jamaica on the 31st and make our way toward Trinidad. The night before, Guy's dad, Philip Harvey threw us a bit of a bash, complete with a delicious lobster dinner. We partied until midnight even though we needed to get up at 3:30 A.M. We had already arranged for a mini-cab to pick us up at that time, for our ride to the airport. Afterward, Mr. Harvey charged us only $10.00 dollars each for the entire time we were there. What a great man! Compare that price to our American neighbors who were paying $125.00 dollars per week, and their place wasn't near as nice as Horizon."

"Dawn had not yet broken when we received our early morning wake-up call. I could barely open my eyelids, and when I finally did, it felt like someone had thrown acid into my aching peepers. They burned like hell. After draining my bladder and splashing some cold water on my face, I grabbed my bags and joined the other guys outside to await our ride. In typical Jamaican fashion, our mini-bus driver was nearly 30-minutes late and now we needed to 'fly' to the Grace Kennedy Shipping Lines in Kingston, in order to arrange our voyage to Trinidad. As a cruel joke, our driver was nicknamed 'Deep Throat', since he couldn't speak above a whisper. At least he was a competent driver and he got us into Kingston in record time."

"Considering that Kingston was the capital city of Jamaica, it sure was a dump! Everywhere was piles of garbage, abandoned vehicles, as well as vacant and dilapidated buildings. I'm not squeamish by any means, but I don't think that I would want to walk Kingston's streets alone after sunset. As we arrived, the black of night had disappeared and the gray, overcast sky of dawn was beginning to show. We didn't have to wait long before the ticket office opened, and we soon booked passage on the earliest departing ship. We still had a couple of hours to kill before departure, so we made our way over to the Coronation Marketplace, in a part of town that was even sketchier than the docks. The vendors and merchants were still organizing their stalls when we arrived, but we managed to purchase a few of the staples we would need for the trip. I had learned from experience that most of the budget passenger carriers lacked fresh items such as fruits and vegetables, and what they did have, was usually expensive. By stocking up on these things from the market, we would forego any disappointment later on, and probably save a few bucks in the bargain."

"Soon back at the pier, we observed the sorry excuse for a ship that would become our home for the next several days. It was named the 'Maple'. We were traveling Deck Class, which literally meant that, at night, we would be sleeping on the open deck. Hopefully, we would luck out and it wouldn't rain. The scow's gang-plank had been lowered and several passengers were beginning to board. Once onboard, but still stuck in a very slow-moving queue, to my dismay, I soon noticed the cause of the hold up. Ahead of us were a couple of Customs Agents that were searching the containers and bags of every passenger, as well as performing a general pat-down. As was common, I often carried a small quantity of hash or weed inside my pocket, so I was always wary to avoid the law, or any other persons of authority. In this instance, my pockets were empty."

"However, inside my backpack, I was carrying a large box containing a chocolate cake laced with Lambs Bread, or in layman's terms, the remainder of our Jamaican pot, that we were unable to smoke before departure. In fact, just a few passengers ahead of me were a couple of new-found friends with an identical cake. We had met them on the beach during our stay, and it was the congenial lady of the pair, now standing before the Customs Agents, that had suggested the idea of the culinary delight. I admired her fortitude and poise as she held out the box for inspection. As for me, I was beginning to sweat, and visions of being locked into one of the many outdoor jail cells we had seen throughout the country, was flooding into my brain. I knew that I wasn't going to bypass those bloodhounds, so I discreetly removed the cake from my backpack and waited. Very nonchalantly, I then left the queue, walked around the roped-off loading area, and then toward the deck railing. My nerves were screaming as I expected to be challenged by the Agents at any moment, and told to return to the queue. To my surprise and relief, I reached the railing unmolested and reluctantly dropped the box over the side into the murky brine. I then calmly returned to the queue just in time to observe our friends pass through the checkpoint, still in possession of their cake. After witnessing their success, I wondered if I had overreacted in the disposal of my cannabis gateau.

But, seeing the relieved expressions on my buddies' faces, confirmed that I had acted wisely. Leaving Jamaica was bittersweet. On one hand, I was excited to experience whatever lay before us, but on the other hand, I would miss all of the marvelous friends we had made there. I wondered if I would ever return, and then sadly thought, probably not."

"Before long, we were underway. While on deck during the first couple of hours, the ship's captain was passing by. I stopped him and we began chatting. He seemed like a decent sort, and as he left, he offered us a free 24-pack of beer in order to properly celebrate New Year's Eve. I thanked him for his generosity and was pleased with myself for initially approaching him. We did indeed polish off that 24-pack between the five of us, but by 10:30 P.M, we were all out like a light. We were so 'bushed' that we didn't even stir when the ship's steam horn blew at midnight to announce the arrival of the New Year. I had completely missed the entry of year 1974. We had slept on the deck's wooden benches and Hager was none too pleased. I had reached the point through my many travels, where I could have slept on a bed of nails. However, not everyone could make that adjustment, and for the next several days, Hager was whining like a spoilt child."

"The next day, I snuck into the 3rd Class Lounge and grabbed a few seat cushions for us. This would soften our hard wooden benches a bit. On New Year's Day, we sailed past Haiti, but it was not overly visible because of the rotten weather. The sea was rolling and the wind was really beginning to blow. There would be cloud bursts every once in a while, and we would try to shelter in the 3rd Class Lounge. The weather continued to deteriorate and the majority of the ship's passengers suffered from sea sickness. Unfortunately, the toilet was so filthy that it was better to be on deck and just eliminate your waste over the side of the outer railing."

"As earlier anticipated, for two dollars and forty cents per plate, the food served by the galley wasn't fit for livestock. Unfortunately, we had soon run out of foodstuffs purchased at the market in Jamaica, so we had no choice but to swallow their swill. We tried to complain about the lack of quality and taste, but our protests fell on deaf ears. After three days at sea, we arrived at St. Kitts and we couldn't get off the ship fast enough. Everyone immediately headed to the nearest restaurant for a decent meal and then off to the open-air market in order to stock up on some fresh supplies. We were going to try and avoid the ship's indigestible offerings at all costs. I also took the opportunity to cash a few traveler's cheques, since the ship's exchange rate was similar to a loan-shark's, at 20%. Our lay-over in St. Kitts was far too short and soon we were back onboard."

"The following afternoon, we docked in Antigua and this time our lay-over was nearly 24-hours. We spent a great day in the town and ended up closing down one of the bars at 3:00 A.M., the following morning. In order to coax us out of his establishment, so that he could lock-up, the bartender agreed to drive us all back to the ship. We weren't sailing until later that afternoon, so as soon as we had steady legs below us, we were back in Antigua for a few more drinks. On this run it was me, Phil, Rusho, Robbie and the lovely Jeannie, a cute little thing we met on board the Maple. Once comfortably seated inside an accommodating whiskey bar and sipping on spirits and Red Stripe beer, I noticed a display case in the corner of the room purveying a variety of different liquors. Prominently positioned within the case was a huge bottle of Dewars White Label Scotch Whiskey, snuggly set atop an ornate brass carriage supported by two large metal wheels and resembling a colonial-era howitzer. Intrigued by the visage of this 'golden gun', I went over to have a closer look. Now, Scotch whiskey isn't normally my first choice of spirits, but when I saw the price-tag of only $10.00, I knew that I couldn't pass it up. For that cost, I could definitely acquire the taste. In the blink of an eye, the money was out of my pocket and I was now the proud owner of 160 ounces of liquid bliss."

"The Texas Mickey was quickly hefted atop our table that visibly sagged under its weight. However, before we could sample the contents, it was time to head back to the ship. Once safely back on board, we grabbed a ghetto blaster and some drinking glasses and promptly made our way to a deck-side table. It wasn't long before our libations were noticed by some of the crew, and they enviously eyed our socializing. Glancing at the formidable bottle on the table and then at Jeannie, I couldn't tell if they were thirsty, horny, or both. Being the genial person that I am, I beckoned the two deckhands to join us, and asked if they were interested in taking a drink. 'Yah Mon', was their immediate response, and we all then settled in to wetting our whistles. Eventually, the ship's steam horn blew announcing the release of the lines securing the vessel to the wharf. Our two impromptu guests recognized that sound as their cue to return to work. One of the pair did not hold his liquor very well, and was clearly three-sheets to the wind. We all laughed as his partner clumsily helped him rise, and then once supported on rubbery legs, half carried him toward the crew's quarters."

"That night at sea, we once again had difficulty getting comfortable on our makeshift beds, but eventually fell into a fitful sleep. Sometime along the way, we were awoken to the distinctive sound of the anchor chain being released. As we shook the cobwebs from our heads, we realized that this was not a scheduled stop, so we decided to investigate. As we went starboard, there was a lifeboat being winched into the drink. Inside the dinghy was a steersman plus the drunken mariner that we had earlier entertained with our Texas Mickey. Well, let me tell you, he looked none too pleased. Once in the water, the lifeboat steered toward what appeared to be the uninhabited portion of a small island not far from our position. Once on shore, the crew member was deposited on the sand, while the lifeboat returned to the ship. We continued scratching our heads as we watched the lifeboat returned to its mooring and secured. The anchor then rose from the depths, and once again the ship was on its way. We stood in disbelief as the marooned sailor remained on the shoreline, forlornly watching us slowly sail away. Far off in the distance on the other side of the island we could see a faint twinkling of lights, suggesting that it was indeed inhabited. However, they appeared to be miles away. It would be a long walk into town. We later learned that the two deckhands who drank whiskey with us that afternoon, were each scheduled to perform a three hour 'wheel watch', later that night. The one sailor who held his liquor well, successfully completed his watch while his buddy, waiting 'on deck' and batting next, attempted to sleep off his drunken stupor. At the end of the first watch, the now knackered crewman went down to rouse his partner. Giving the slumbering swabbie a good shake, he informed him that it was his turn to take watch. The booze hound, mumbled his acknowledgement and looked as though he was ready to rise."

"Just to be certain, the original crewman dragged his pal out of the sack and physically escorted him to the wheelhouse. Thinking that he had more than now fully completed his duty, the exhausted sailor returned to his own cabin for some much-needed rest. Unfortunately, his buddy was still feeling the effects of the whiskey, and enjoyed the comfort of his bunk a little too much. As a result, now finding himself alone, he simply stumbled out of the wheelhouse, returned to his cabin, and crawled back into bed. With the wheelhouse now abandoned, and no one piloting our metal monster, it was three hours before the dereliction of duty was discovered by the captain, as he arrived for his shift. By then, the ship was several miles off course, and it was a minor miracle that it had not run-aground on any of the numerous reefs or small islands that populated that stretch of water. Aside of outright mutiny, there was no more serious infraction a sailor could commit, and being exiled to that island was considered a lenient punishment. In days of old, such an offender would have been made to walk the plank."

"On the 5th of January at approximately 10:00 A.M., after a soggy night's sleep on deck due to rain, we arrived in the Dominican Republic. In lieu of shuttles from the shore, our crew launched a few of the ship's lifeboats and they ferried us to land. There was little to no gas on the island, so we couldn't rent a car."

"A taxi driver approached and offered to show us around the region for two bucks a pop. We couldn't argue with that price, so we all jumped in to the cab for an impromptu tour. We eventually ended up at the Charcos de Damajagua hot springs in Puerta Plata and decided to remain for a few hours. It was a bit of a hike in order to arrive at the mineral pools, but it was well worth the walk. I couldn't get over the dark green, densely treed hills or the lush and deeply trenched spreading valleys. Everywhere you looked there were colorfully feathered exotic birds, flowers, vines, mosses and wild fruits. The area seemed completely untouched by man and it was almost like going back to prehistoric times. The hot springs looked like they were on a foreign planet. Everywhere, boiling water was bubbling from cracks and crevices and even on the surface of the crystal clear, bottomless pools. Steam was rising from the ground at various locations and mineral deposits clung in clusters on the surrounding rocks. On top of that, the air was thick with the smell of vegetation and Sulphur. It was quite the place. Although our driver only initially agreed to take us around for an hour, he remained at the hot springs, and joined us for a couple cans of beer that we pulled out of our back packs. With our feet soaking in the hot mineral water, we all discussed world events, and how it would be a much better place to live in, if only we were in charge."

"Our driver stated that 15% of Dominican's population dominated the remaining 85%. They were the elite, and they decided who would lead the government, who could own property or businesses and who could even attend school."

"Although the Dominicans were oppressed, they were unbelievably friendly toward visitors. They were forever waving and smiling and I wondered whether their government was forcing them to behave in this manner in order to promote a positive impression of the country. After all, they were neighbors to Haiti, and that impoverished country had a notorious reputation for populace oppression and government corruption. After spending a relaxing day at the hot springs, we returned to the city for a delicious steak and baked potato dinner at the Alta Hotel. We were certain that this far surpassed what was going to be served that night aboard the Maple, and we took full advantage of it. We also ensured to drink as much as we could to help dispel the shock of being aboard that rusty old tub again."

"It was late when we finally got back to the ship and we decided to head straight for our favorite benches. I went to the alcove where we kept the cushions hidden, but they were gone."

"I rushed back to the 3rd Class Lounge just as it was closing, and saw all the seats were padded once again. I looked over at the bartender and he returned a greasy sneer. I immediately realized he was the bastard that swiped the cushions from me that I had originally swiped from him. I wanted to go over and smear that 'shit-eating grin' off his face, but the manager started nudging me out of the canteen. Back on deck, the boys and me hunkered down to a rough night's sleep on our unprotected benches. To top it off, it started raining a couple hours later. That's really what I call a wet dream."

"At 7:00 A.M. that morning, we arrived in St. Lucia. Still damp from last night's unexpected shower, the sun was now shining, and it was already beginning to dry my clothes and warm my stiffened joints. Luckily, we managed to grab one of the first launches ashore. The immigration check was light, and once through, I headed straight to a bank in order to cash another traveler's cheque. We began window-shopping as we strolled through the town. Passing by a camera shop, I went inside to see what they offered and what the prices were like. Robbie ran across the street to a pay phone. He was calling home to our buddy, Mags, to see if our unemployment cheques had already arrived in the mail."

"Robbie returned to us a few minutes later, looking like he had just seen a ghost. The color of his face was nearly gray and his expression was one of total disbelief. When I asked him what was wrong, it took him a while to answer. When he was finally able to speak, he confirmed that he had indeed been speaking with Mags, and that Mags informed him that my good friend, Jim Lowrey had just been murdered in South America. He was with another associate named Beaupre, and they had both been shot. Beaupre was wounded in the leg and would survive, but Jim had been killed."

"I could see from Robbie's expression and hear from his voice-tone, that he wasn't joking. At that moment, I didn't exactly know when or why, but I had just lost one of my best friends, and I couldn't wrap my head around the news. Man, the guy was in his 20s. People aren't supposed to die that young. And, especially in the manner that he did. I had just spoken with him a short time ago and we were actually planning to meet-up somewhere in South America, now that I was traveling close by. Within seconds, time and space seemed to stand still, and I guess I went into a temporary state of shock. When I finally snapped back to the present, I immediately ran out of the camera shop and across the street to the pay phone. I realized that by now, my parents had already heard the news about Lowrey, and I wanted to let them know that I was alright."

"I got my mom on the line, and she confirmed the news regarding Lowrey. By now, everyone was talking about it. She didn't have any additional details for me, but agreed that she would send flowers to his family for the funeral. We talked for a while longer and before terminating our call, I promised her, I wouldn't be traveling to South America. I had no problem keeping that promise, since now, I had absolutely no desire to go there. It goes without saying that my day was now a total bust and I didn't know if I would ever completely get over the loss. However, in some small way, I wanted to honor the memory of my friend. Do you know what the Irish do at a 'wake', Kid? No? Well, they get plastered, and that's what I was aiming to do."

"The group of us went to the nearest bar and 'launched' into it. The drinks were afloatin' and we were pounding them back like there was no tomorrow. For Lowrey, there would be no tomorrow, so we wanted to do him proud. While we toasted our departed friend and shared 'war stories' about our times with him, this big, black Rastaman in the corner, kept eyeballing us. Eventually, we struck up a conversation with him and he offered to sell us some Jamaican ganja. Wishing to make the celebration of life complete, we accepted his offer."

"Between the booze and the grass, by afternoon's end, we were well beyond the realm of mere mortals. We were now on Olympus with the other Greek Gods, looking down, droopy-eyed, on the world below."

"I don't remember how we made it back to the ship that night, but somehow, we did. At the time, I was so heavily sedated by the day's intoxicants, that I didn't even notice the discomfort of my wooden bed. It wasn't a restful sleep, but rather a state of grief and substance induced unconsciousness. The next day we landed in St. Vincent & the Grenadines. I wasn't in the mood for playing tourist that day, but I did want to get off the ship for a while. I didn't do much in town, but I purchased a nice rattan carpet for my bedroom floor back home. The shopkeeper was even nice enough to wrap it up for me, so that I could post it. I didn't want to be lugging that big thing around. Postage cost me a bit more than I wanted to spend, but I thought it was worth it."

"That night, I learned a new way of achieving a quick buzz on beer. Some guy in a local bar was doing it, and I asked him to show me. He took a can of room temperature beer and poked two small holes on either side just above the liquid within. He then sucked all the air out of the can from the holes, and pulled open the tab."

"Most of the carbonation had now been removed along with the air, and he was able to chug back the beer, essentially, in one large gulp straight into his stomach. Do that a couple of times, and you're well on your way to 'happyland'. I wasn't going to attempt the 'power-drinking' trick that night, but I memorized the procedure and tucked it away in my alcohol addled brain."

"Since they were all so close together, we now reached various island countries in daily increments. Our next stop was Barbados. As I trotted down the gangplank, I noticed that the Barbados Customs Officials were really giving debarking passengers a thorough search. Once again, I needed to think fast. Reaching into the right-front pocket of my 'cut-offs', I removed a single joint, or Montreal Toothpick as they're referred to back East. Popping it into my mouth, I chewed the bitter herb, including the rolling paper, until it was crushed enough to swallow. Naturally, I breezed through the check-point and we made or way into Bridgetown, on foot. Later, we heard that one of the ship's crew members was caught carrying a piece of hash. He was promptly arrested and carted off to jail."

"Once in town, we attempted to rent scooters from a car dealership, but our International Driver's Licenses were insufficient. We then rented a Ford Cortina instead. By now, the joint I ingested earlier was beginning to work its magic and I was heading for the stratosphere. It was decided that I should drive, and I boarded the left side of the car. I was amazed to discover that someone had stolen the steering wheel, only to be reminded that being in a former British Colony, the Cortina was a right-hand drive. Pretending that I had known this all along, I chuckled, and slid to my right. This was going to be an interesting drive. I had never driven on the left-hand side of the road before, and to increase that handicap, I was now higher than a kite."

"We got off to a pretty good start, and according to the boys, I only knocked over a couple of trash bins along the side of the road. I didn't recall hitting anything or anybody, but I took their word for it. With Hager navigating and Robbie ensuring that I remained on the road, we drove up to Bathsheba on the Atlantic side of the island. We were informed that the surf there was superb, so we wanted to check it out. Although most of the islands all looked similar, it was always amazing to see the disparity between the 'haves and have-nots'. One minute we were passing luxury mansions with expansive grounds, adjacent to cane fields or tobacco plantations, and next we were cruising past mini shanty towns, filled with inadequate housing, unpaved streets and scores of snot-nosed kids."

"This scenario seemed to replay itself in every land I've ever visited, where it had been previously colonized by Britain, France, Belgium, the Netherlands, or any other developed country for that matter. The occupiers always took the wheat and left the chaff for the local inhabitants. We enjoyed our afternoon at Bathsheba, but I wasn't overly impressed with the spot. I had seen bigger waves in other places I'd previously been to. As we headed back into town, we passed a Kentucky Fried Chicken stand. With everyone in the car suddenly chewing off my ear, I made a somewhat shaky, 3-point U-turn and went back to the restaurant. I've got to admit that it was a good decision, since the chicken and potato salad there, had been the best I'd eaten during our entire trip. We arrived back at the port just minutes before departure time. Leaving the rental car at the dock, we skipped aboard our floating garbage heap and prepared ourselves for another fun-filled night of sleeping under the stars."

"On the morning of January 9th, we docked at the island of Grenada. From the deck of the ship, the island appeared to be lusher with vegetation than our previous stops, and I was looking forward to debark. Once on dry land, we traveled to the capital city of St. George's, only to be dropped into the middle of a large, but peaceful demonstration. On February 7th, the island was preparing for its jump to independence from previous British rule. Irate citizens were venting anger toward their, soon to be, former masters. Not wanting to become falsely perceived as being affiliated with the occupiers, we made a hasty exit from the square and grabbed some breakfast well away from the political action. I ate a very tasty fried egg sandwich and drank a cold glass of the best, fresh-squeezed grapefruit juice, I'd ever tasted."

"After breaking our fast, we grabbed a transit bus to nearby Brandon Beach for a spot of swimming and a toss of the Frisbee. I've got to admit, that Brandon Beach was probably the nicest strand I'd seen anywhere. The ocean was fantastically blue and clear as glass. The sand was golden brown and spotless. I've never seen a cleaner shoreline. By the time we returned to St. George's, the demonstration had ceased and now there were just a few cars and pedestrians occupying the square. With time to spare and not wanting to choke down another S.S Maple supper, we stopped at a restaurant-cum-bar, and threw back a few drinks. Here, I tried out the beer can trick that I had learned in Barbados, and it worked well. After power-drinking only three cans of beer, it felt like I'd had doubled that amount. While feeling no pain, and also a bit adventurous, I ordered turtle steak with fried onions for dinner."

"Afterward, I was sure glad I did. It was absolutely delicious. I definitely wanted more of that, and hoped that I'd be able to find it in Trinidad as well. We were back aboard the Maple for its 10:00 P.M. departure, and completely pleased with our brief but satisfying visit to Grenada. I made a mental note, that if possible, I would return here someday. Nights aboard ship slipped by quickly, especially since we were always going to bed so late. Sleeping on deck, open to the elements, and on beds little better than wooden planks, meant that any kind of meaningful rest was simply out of the question. After barely having closed our eyes, it was now 6:30 A.M. on the 10th of January, and the welcoming harbor of Trinidad was coming into view."

"Trinidad was our jumping-off point, so before we docked, me and the boys started shaking hands with some of the crew members and traded addresses with a few more. Shortly after 8 o'clock, the ship's gangplank was lowered, and we were off to the races."

"We were subjected to a light search at Immigration and then into the Pan Am Airlines office to make enquiries for our flights to Central America. While Rusho, Robbie, Gene and Hager were arranging our tickets, I quickly popped into the Money Exchange Bureau to cash a traveler's cheque into Trinidadian dollars. Once that was completed, we all piled into a taxi and drove into town, arriving at the Pelican Inn. Rooms were $8.00 dollars per night, with breakfast. For that price, it was actually a good deal. The rear of the hotel faced the Governor's Mansion and its perfectly manicured gardens. I don't need to tell you that this was quite a nice change from our previous accommodations aboard the Maple."

"First stop after dropping our bags, was the hotel's bar. As we entered, all heads turned in our direction. Most of the patrons already inside were sipping on their designer drinks, and the looks we received were far from welcoming. Had we been outside, their upturned noses would have blocked out the sun. What a bunch of snobs! As far as I was concerned, they could all kiss my ass. After a couple of drinks and a game of darts, it was out to the patio and a well-placed cannonball straight into the pool."

"Later that night, we found ourselves back in the bar, and joined by Ron, an old buddy of mine that lived in Trinidad with his parents. By mid-evening, we'd already had a snoot-full, and as always, Hager, the consummate entertainer, commenced his floor show."

"A few tables away from ours, sat a middle-aged couple and their debutant, teen-aged daughter. She was a cute little thing, wearing a pale blue silk midi-dress and obviously on her first big night out on the town. As the house band began playing their version of a light, rock-n-roll tune, Hager staggered to his feet and careened over to the family's table. First expelling a poorly disguised belch, and then bowing slightly, he asked the young lady to dance. Unfamiliar with this form of social protocol, the vestal virgin first looked to her parents for approval. Obviously taken aback by the unexpected advances of this unkempt, tottering hulk, the girl's parents merely stared back at her without speaking. Accepting this silence as their form of agreement, the maiden joined Hager on the dance floor, and the fun began. Hager's dancing style was suspect at the best of times. However, after a few drinks, it bordered on the absurd. His first few moves were somewhat recognizable, but as he grew more confident and wild, he began to resemble a spastic ostrich attempting to fly. His arms were flailing about while he stomped his feet, and for good measure, he also threw in the occasional salacious hip-thrust in his partner's direction."

"At one point the girl simply stopped attempting to follow his lead, and kept glancing over her shoulder toward her parents. The look on her face resembled that of someone desperately begging to be rescued. Hager was now in his own little world and had completely forgotten about his pretty dance partner. He must have fallen on his butt at least a half-dozen times and at one point, slid into the band's drum kit, and then bounced off a wall. I just couldn't hold back any longer, and tears of laughter were streaming down my face. It was the same reaction for Ron and the other guys as well. Mercifully, the song ended and Hager clumsily escorted the young lady back to her table. To our astonishment, he introduced himself to the girl's parents, and with another slight bow, thanked the girl for her company. He then proudly staggered back to our table, as we continued roaring hilariously and practically falling out of our chairs."

"Over the next several days, Ron drove us around Trinidad in his car, showing us some of the island's highlights, its beaches and popular drinking-holes. The island also had a strong carnival culture, and one day, Ron took us to a large warehouse where they made floats and costumes for the yearly festival. That was actually quite an interesting tour, until one of the guys bumped into a table supporting a large, papier-mâché clown's head. The head went tumbling to the floor with a crash and we all cringed, thinking it would be damaged. After quickly replacing it on to the table and making certain that no one had noticed this potential disaster, we performed a quick examination of the oversized cranium and deemed it to be untouched. That was a relief."

"Surprisingly, during our days, we spent considerable time in the bar of the Pelican Inn, and I was becoming somewhat of a local celebrity at the dart board. I hadn't really played much back home, but I found that I was a bit of a natural at the game. There would always be a buzz of excitement when I entered the lounge and people were regularly buying me drinks, just so they could watch me play. I was also becoming addicted to the Pelican's French-fries and Stavanger soup. This was a fish soup originating in Stavanger, Norway, but given a special twist by the Trinidadians. I just couldn't get enough. I was quickly becoming attached to Trinidad, with its friendly people, great beaches and awesome cuisine. If I had to choose between Trinidad or

Jamaica in which place to revisit, I would have to choose Trinidad, hands down."

"One day, we rented a VW Karmen Gia and took a jaunt out to Roti Country. This was an area outside of the city where local families spent their weekends and which abounded in all manner of food and drinking stalls. Roti is a flat bread usually associated with the South Asian culture, but the Trini's adopted it as well. Just as they had done with the Stavanger soup, they also made these tasty, hand-held delicacies a thing of their own. The rotis were always spicy and usually filled with either shrimp or beef. They were similar to a Lay's potato chip. You couldn't eat just one. I was becoming quite expert at taking a bite of the savory handful, swallowing a raw oyster marinated in tomato sauce afterward, and then following both with a generous gulp of cold beer. While this was going on, I also heard the sounds of musicians tuning-up."

"Soon finding the source of this musical cacophony near a copse of trees, we found ourselves amongst the Catelli Trinidad All-Stars Steel Band. There must have been around one-hundred members with steel drums, traditional drums, plus electric bass and rhythm guitars. The steel drum is a commonly, recognizable instrument all through the West Indies and originated in Trinidad during the 1930s. They were originally constructed from cast-off 50-gallon oil drums and then cut-down to desired sizes. They are now a bit more sophisticated but still true to the traditional design. The bottom of the oil drum is first pounded into a bowl, then shaped and tuned with hammers to form distinctive resonating surfaces. Once the desired tone of the pan is achieved, they are then nickel or chrome plated for an attractive appearance. Smaller pans can be worn around the neck from leather straps, while the larger, bass pans are seated on the floor. Once the group were tuned and commenced practicing, they sounded great. When first heard, the sheer size of the band was very loud, but it was a pleasant sound, and you soon forgot about the volume level. Between the food, drink and music, we all had a great afternoon."

"After eleven days of sun and fun in Trinidad, it was time to pack-up our gear and search out new horizons. The day before departure, I attended a local medical clinic and received a Polio booster shot. My vaccination card was now up to date and I was ready for my Latin American adventure. I also attended the Mexican Embassy to complete my Visa application. Feeling a bit mischievous at the time, in the area requesting the applicant's 'Sex', I wrote down, 'Yes, twice a week!' On the morning of January 21st, I settled my account at the Pelican Inn, and headed toward the Mexican Embassy in order to retrieve my Visa. I did acquire the document, but not before I received an earful from the embassy official."

"He was not amused by my attempt at jocularity on my application form, and further stated that he had half a mind to decline my application. I silently agreed with him, that he did indeed have half a mind, but kept that assertion to myself. With an insincere apology, I grabbed the Visa from his hand and skipped out of the office. It was then off to the airport with the rest of the gang."

"It was slightly more than a one-hour flight from Trinidad to Curacao. Once arriving, I decided that it was a rather forgettable place, so we didn't hang around for very long. The same went for Aruba and the Dutch Antilles. When we arrived in Panama City, Panama, I thought we had been shoved into a sauna bath. I had never felt that level of humidity before. Every pore on my body was leaking sweat and no matter how much liquid I'd pour down my throat, I was still thirsty. There was a lot of Americans living in Panama, but they were completely segregated from the local population. There seemed to be no contact whatsoever between the two groups. The Americans lived in their own gated communities, they attended their own schools and churches, they shopped at their own grocery or clothing stores, and even had their own gas stations."

"Obviously, the most important and exciting attraction in Panama City is the Panama Canal. The original Panama Canal was completed in 1914, with the other two main channels being constructed years afterward. One of the three locks comprising the canal was the Miraflores Locks consisting of twelve separate sections where vessels are raised and lowered to various heights. This action essentially allowed them to cross the spit of land, via the canal, separating North and South America. This was where I went to visit on a blisteringly hot and humid day. Although the weather outside was brutal, I was glad I made the effort to go. Watching the ships pass through the various stages of the locks was amazing, and a real marvel of human technology. I snapped a lot of great pictures of a Swedish freighter being raised and lowered through these locks."

"Our stay in Panama was brief, as was our next stop in San Jose, Costa Rica. Most of the average citizens were living there in poverty and the marketplace was filthy, offering only a few meager selections of basic foodstuffs and other essentials. We couldn't get out of there fast enough. So far, our travels to Central America had been uneventful, and we were hoping that the other countries still to come, would have more to offer."

"It was now January 28th, and our next stop was El Salvador, with its capital city of San Salvador. Arriving at the airport, we discovered that the widely publicized boxing rematch between Muhammad Ali and Joe Frazier was being broadcast on television that afternoon. Looking forward to view the extravaganza, we luckily cleared Customs relatively soon and urged our taxi driver to get us to the Canada Hotel, pronto. We managed to check-in minutes before the start of the fight and then all congregated in my room, to watch it go down. After all the hype and build-up to the match, I was disappointed with the result. Ali eventually won by 'unanimous decision', but he looked sloppy and out of shape. He was a far cry from the youthful Cassius Clay we all watched crush Sonny Liston ten years prior, to win the Heavyweight Championship."

"The next night, we were out on the town and of course behaving ourselves. After closing down several licensed establishments, we found ourselves stumbling down the Avenida Roosevelt toward our hotel. Every street corner seemed to have a Military Police presence with heavily armed officers and machine-guns mounted on their Jeeps. Approaching an English-speaking local, we were informed that this week was the National Elections and things always got a little tense between the opposing parties during that time. Seeing the different political flags draped on buildings and lamp posts, along with posters of sinister looking characters in search of votes, all made sense to me now."

"We had picked-up a couple of stray American backpackers during our night of partying, and one of them now had the bright idea that he needed a souvenir of El Salvador to take home with him. His item of choice just happened to be one of the El Salvadorian national flags attached to a nearby building. Enlisting Hager as a step ladder, the goof climbed on top of Hager's shoulders, and started reaching toward the flag. All of a sudden, the concussive blast of a gunshot rang out directly behind us. We immediately hit the deck, quicker than Sonny Liston after receiving the right-cross from Cassius Clay during their epic fight in 1964. Looking for the source of the shot, I soon observed a uniformed Military Policeman walking toward us and pointing his rifle in our direction. I didn't like the look on his face and I wasn't certain what to expect next."

"As he purposely strode forward, he was shouting in Spanish and directing his anger toward the idiot who had been standing on Hager's shoulders. No harm or damage had been committed, so, I wondered what he was going to do. Thankfully, he just kept dressing-down the would-be thief, while fanning his weapon from left to right. I took this action to mean 'get the hell out of here,' so I started quickly walking away."

"The other guys followed my lead and soon we had distanced ourselves from the irate cop. The dude that attempted to steal the flag started laughing and that set me off. I wasn't going to end-up in some Central American prison because of the actions of some irresponsible asshole, and I told him as much. I then not too kindly suggested that he should bugger-off sooner than later."

"Needing a little something to settle our nerves after that incident, we eventually located a Disco whose doors were open, and with lights still on inside. As we entered, there was one couple on the dancefloor and another off in a corner, smooching. The Dee Jay informed us that he was just about to shut things down, but once we bought him a drink, he decided to spin a few more platters. He actually possessed a pretty decent selection of music, and after a few more free drinks, he kept his turntable employed. We ended the night grooving to Traffic's, 'Dear Mister Fantasy', and hit the cobblestones once again. The unpleasant incident with the military copper had now been forgotten, and we continued toward our hotel, feeling much better than we had previously."

"The next day, we checked-out of the hotel, ready to kiss El Salvador goodbye. Heading to the main bus depot downtown, we purchased our one-way tickets for Guatemala City, Guatemala, leaving within the hour. Surprisingly, the bus was near-new and very comfortable. The air conditioning worked well and the driver was piping in some good music through the vehicle's P.A. system. Arriving at the El Salvadorian/Guatemalan border a couple of hours later, we were detained by a lengthy and amateurish Customs examination. There were still several vehicles ahead of our bus, and the Guatemalan guards were in no great hurry to allow anyone entry into their country. While we impatiently waited for our turn to undergo their clumsy and unprofessional search, three pimple-faced teens, wearing peasant fatigues, arrived on scene and entered the Customs guard shack. It was two guys and a girl, with the oldest boy being perhaps fifteen years old."

"A few minutes later, these same three punks exited the shack, all dressed in shabby blue, ill-fitting uniforms and shouldering semi-automatic rifles. The weapon belonging to the youngest one, was nearly as long as he was. I couldn't believe it."

"Was this the best that Guatemala could do for an important job such as border guards? Three uneducated, teenaged peasants? If that was the case, I wondered what kind of people filled other important roles within the country. We eventually got through the check-point unscathed, and by 3:00 P.M., made it to Guatemala City without further incident."

"As we took turns registering into our reasonably priced pension, which included three full meals, I suddenly remembered that I had left our Frisbee onboard the bus. The depot was only a couple of blocks away from the pension, so I beetled back down there as fast as my feet could carry me. Unfortunately, our former bus driver had already picked-up passengers for his return trip to El Salvador, and left. I was kicking myself for that careless lapse of memory. I was going to miss our old friend and the hours of fun we had with him."

"Once again, this was another very poor country within the Central American region, with very little to offer guys like us. As a result, we weren't going to remain here long. One thing that surprised me was the attitudes of the people. They were all extremely friendly toward us, and although very few spoke English, they made us feel welcome. Most of the populace physically resembled indigenous tribes of the Mayan, Inca and Andes communities. In addition, the majority were also relegated to menial work within the urban center, or farming and animal husbandry, outside of the city. It was mostly women we saw on the streets, and they appeared to be in charge of things. They ran the stalls in the marketplace, handled all of the money in the shops and seemed to be the head of their households. Most were dressed in very colorful and elaborately stitched, handmade clothing, that appeared to be very clean, despite their surroundings. Almost all, also had a baby strapped to their back. Some even had two, one on their back and one strapped to their chest. It was obvious that sex education and birth control wasn't a high priority in their culture."

"The Guatemalan market seemed fuller than that of their neighboring countries, and we took advantage of that variety. We actually purchased quite an amount of their handicrafts at very reasonable prices. Whenever we bartered with the women, they would always giggle and whisper, 'Mira…Mira.' We later learned that 'Mira,' means 'to look' in Spanish, but we couldn't figure out if they wanted us to observe their wares, or them personally. Either way, we got what we wanted, and they made a few bucks. Two things I wasn't receiving on our Central American jaunt, was beddable women or dope. Both were in very short supply and I was becoming annoyed with this drought."

"The next day, we hopped on a tour bus and ventured outside of the city, in order to visit the ancient Mayan and colonial Spanish ruins at Antigua. Our guide was well versed with both cultures and turned our visit into an educational experience. It was one of the better tours I'd taken during our entire trip."

"The following day found us at 35,000 feet on an Aero-Mexico 727 jet, flying to Mexico City. As we began our descent through the clouds, our vision was soon obscured a second time. Only on this occasion, it was caused by the thick layer of smog that blanketed the sprawling metropolis. With a population of nearly 8-million citizens at that time, Mexico City eclipsed that of New York City, making it the most densely inhabited city in all of North America. I didn't like that. I valued my privacy and needed some elbow room. Millions of people constantly looking over my shoulder and sticking their noses into my business, wasn't going to cut it for me."

"Upon our arrival and once we were settled, it was just as I suspected. The city was massive and super crowded. There were quite a few interesting sites throughout the town, but it was a real drag trying to get to them. We saw what we could, but overall, I wasn't impressed with the place. On our second day there, we found ourselves near the Liga Soccer Stadium on Avenida Fernando Iglasias y Calderon. The arena was beginning to fill with fans and we learned that two, regional, rival soccer teams were playing that night. Always 'up' for a good soccer match, I suggested that we take a look. Paying the astronomical entry fee of 40-cents for bleacher seats, we made our way inside."

"The crowd was restless and extremely rowdy. As we made our way along the lower bleachers, we were pelted from above by apple cores, food wrappers, vegetables and even an empty beer bottle. Not daring to look upward, we just continued on our set path and eventually found an empty bench. As the game commenced, the crowd became even more frenzied, and additional debris was not only aimed in our direction but everywhere else as well. Garbage, fireworks and even flaming seat cushions were being hurled, with some of it landing on the playing field. Observing this onslaught, we realized that the crowd wasn't just targeting us, they were targeting everyone and everything. It turned out to be a pretty good game and we attempted to enjoy it as best we could despite the missiles constantly flying through the air. I bought a couple cans of beer to drink during the game. There was a young girl, maybe six or seven years old, watching me like a hawk. She stood there patiently for nearly half an hour, and as soon as I dropped the empty cans on to the ground, she ran up, grabbed them, and ran away again. I realized that in Mexico, every penny counted."

"At one point, I got hit in the head with a limp banana peel, while Rusho was struck on his arm by a tomato, ruining the brand-new peasant shirt that he had just purchased in Guatemala. We left the game shortly before it ended in order to prevent being trampled by the departing crowd, or maybe even being murdered by them. What a place! I thought sports was supposed to be fun."

"Quickly fed up with Mexico City, the next day, we took a night flight to Puerto Vallarta for $40.00 dollars, with a smaller regional airline. I was looking forward to finally getting in some more time sunbathing and frolicking on the beach. Unfortunately, I was sorely mistaken. Puerto Vallarta was actually quite a depressing place and definitely didn't live up to the hype its promoters were attempting to sell. It was over-priced, the water was cold and unclean, the beach was strewn with rocks and it was overcast during most of our stay. The locals weren't very friendly and the place was full of American college students, high on Quaaludes and stumbling around the town like a bunch of zombies. Once again, we were totally bummed-out by this turn of events and on February 5th, we pulled the pin on that part of the world by flying back home to Vancouver."

"During the flight home, while sipping on a beer and munching on a bag of peanuts, I ruminated on the entirety of my trip. The majority of the Caribbean portion had been enjoyable. However, most of Central America and Mexico, had been a bust. I decided that I was definitely not returning there. Finally, was the reminder that I'd never again be seeing my old buddy, Jim Lowrey. I didn't know it at the time, but his murder affected me deeper than I had realized. Although my burning need for future travel still persisted, the full flame of that desire actually sputtered a bit, and its heat cooled somewhat, with his passing."

"Our flight home was unremarkable and I attempted to sleep most of the way. Unfortunately, rest seemed to be the furthest thing my body craved and I remained awake the entire time. I realized that I was extremely tired, but strangely, that overwhelming fatigue inexplicably prolonged my inability to sleep. Upon landing in Vancouver and clearing Customs, I said goodbye to my traveling companions and suggested that we get together again soon. I then took a cab to my girlfriend, Gail's house. I hadn't received any mail from her during my entire trip."

"I had barely completed my second rap on her front door, when it suddenly flung open. There stood Gail in all her maternal glory and beaming from ear to ear. I hadn't even greeted her yet, when she suddenly pulled me into her living room and began ripping at my clothes. Before she actually damaged something, I dropped my bags and attempted to assist with her struggles. We didn't even make it into the bedroom, but performed the act right there on the shag carpet. Once spent, I thought to myself, *Welcome home, Bob.* Obviously, I had heard the term 'rape' used before, but this was the first time that I had personally been the victim of such an act. At the time, I didn't know what possessed her, but later that night, the purpose became clear."

"By now, the full weight of my weariness bore down on me and I immediately needed to find a bed. It seemed that I had just drifted off to dreamland, when Gail burst into the bedroom and woke me. Blearily looking at the alarm clock on the nightstand, it was nearing 2:00 A.M. Gail's 'water' had broken and she was going into labor. Her first child had been born at home and she was expecting the same for her second. To top that off, she wanted me to deliver the baby. I told her there was no way that I was going to be able to do that, and suggested we go to a hospital. Gail had already telephoned a friend who assisted birthing her first child, and that person was now on her way over. Gail assured me that the two of them would talk me through it. Afterward, her family doctor would pay her a visit to remove the placenta, cut the umbilical cord and ensure that the baby was healthy. I thought this chick must be nuts, but for some reason, I went along with her."

"Well, it was intense, but a few hours later the baby was born and upon leaving, the doctor pronounced both baby and mother perfectly fit. Now, I could add 'midwife', to my resume. When Gail and her friend suggested that I eat the placenta, that was my cue to split. Truly, both of those 'broads' were nuts! As I left the house, Gail requested that I name her new baby boy. Dog tired, I just couldn't think straight, so I told her that I'd drop by the next day and we'd talk about it. Well, the next day, I did return to Gail's house, to find her female friend still there, and both cooing over the baby. Gail announced that during the night she had received an epiphany regarding her infant son's name. He would be called, GALAXY TRIPPER MORRISON. *The poor kid*, I thought to myself. With a mother like that, he was doomed. That was it! I was out of there, never to return. I don't mind a bit of kinkiness or a touch of the bizarre once in a while, but Gail was downright mental. In my current cycle of life, I didn't need any of that!"

Chapter 10
Bobby Thai's One On

McGuire loudly slapped shut his fourth journal and drew a deep breath. He had been talking nearly non-stop for hours, and it was now well into the early morning hours of the next day. He seemed satisfied to have discharged the entirety of that chronicle in a single protracted session. The fact that neither of them partook of supper that night didn't seem to be an issue. The numerous glasses of wine drained by each throughout the entirety of the evening, probably helped to belie any hunger pangs. Now, it was time to finally retire to their respective bed chambers. But not before McGuire prepared them both a substantial breakfast. From past experience, he knew all too well how difficult it was to sleep with an empty stomach. Before heading to his bedroom afterward, he removed two sirloin steaks from within the refrigerator's freezer compartment for that evening's supper. They would slowly thaw in the kitchen sink while the pair slept.

By the time they rose later that day, it was already well past the noon-hour. At least it had stopped raining. After they refreshed themselves, puttered around a bit and polished off the steaks with potatoes and fried eggs, they decided to forgo any further yard work. Delving into the past and revealing the contents of McGuire's travel journals was now, and had actually always been, their main focus. They both knew this to be true, so, there was no sense pretending otherwise, or, spending time on less motivational concerns. Besides, Jesse had actually already completed far more work than had been expected of him and his congenial host was fully satisfied with the results. There was just a bit more to clear-out behind the house and then later trucking the entire debris away to the landfill. McGuire could easily perform those tasks at some later date. Bob then slipped out of the house for a short interval in order to check on 'his girls' at the greenhouse.

He returned shortly afterward and found Jesse sampling the sounds from a few of his classic record albums. Sensing the time was right, he then entered his bedroom, and moments later, reappeared with his fifth and final journal. Jesse was just exiting the kitchen with a jug of wine and two empty glasses. Once the glasses had been filled, both men then assumed their rightful places in preparation for their now, nightly ritual. McGuire opened the black compendium with its red spine and corresponding red corner protectors, and then commenced speaking.

"So, Kid. As you already know, on my third excursion, my girlfriend Judy accompanied me on my adventure. And as before, I then hooked-up with other travelers along the way. This journey was a bit of a repeat to that one, since my sweetheart of this period, would also be traveling with me. Usually, I like to spread myself around when it comes to the ladies. It wouldn't be fair to limit myself to just one person and deny the rest of the female population the heavenly experience of romantic bliss with a Love God. However, in this instance that's exactly what I did. I limited myself to just one."

Here we go again, Jesse thought to himself. If grandiose confidence and self-exultance could be equated financially, then McGuire would have been a multi-billionaire.

"As I said. I'm usually quite happy as a lone wolf on the prowl, but for nearly five years, I remained loyal to a single woman. Her name was Carol and her family lived just up the street from us on East 44th Avenue. I met Carol shortly after Gail and I had called it quits. There was quite an age separation between us, but my motto has always been, 'if they're old enough to bleed, they're old enough to breed.' Carol was a beautiful girl with a great personality and an even greater physique. It's kinda' weird when you think that for years, she had lived just up the street and I never gave her a second look. Previously, she had just been another young girl, in the neighborhood. Now, suddenly, she had become a young woman and she really knocked my socks off."

"I truly became acquainted with Carol in mid-1974, while I was visiting Larry and Brian, two close friends of my two younger brothers, Ken and John. They had also grown-up on East 44th Avenue with the rest of us jokers. However, they had recently moved out of their parents' homes and now lived in a large, older rental house, kitty-corner to the Canada Dry bottling plant, at Moss Street and Kingsway Avenue, in Vancouver. I dropped into 'The Pit', as it was aptly named, just to say hello and to see what was 'shaking'."

"While I was there, in walked Carol to see her brother, Brian, and also bearing some treats for the boys. At first, I hadn't noticed her entering, but when I finally did, I nearly choked on my mouthful of beer. What a 'dish!' I had now completely forgotten about Larry and Brian, and focused all of my attention on Carol. We had a great chat and I discovered that we also shared some common interests. I was smitten! Before leaving, I was able to get her phone number and I vowed to call her up as soon as possible."

"A couple of days later, I did exactly that, and asked her out on a date. To my good fortune, she accepted, and the rest as they say, was history."

"Eventually, Carol and I shared a house on Arbutus Street, in the Kerrisdale district of Vancouver, and essentially lived there as man and wife. Although we definitely loved each other, I don't think either of us actually considered getting married in order to make our union official. Over time, my passion for traveling had never waned, and I was still constantly thinking about faraway places and exotic locales. I continued working as a commercial fisherman, when not on the road, and I made some pretty decent coin. Much of it, I was able to save. By the Fall of 1978, I was once again getting that restless feeling, and that meant only one thing. I needed to get 'out there'. I couldn't very well leave Carol alone at home, so I invited her along. She was more than happy to join me. She was well aware how much my trekking meant to me, and she worked hard at keeping me satisfied."

"Since, I had already been to Europe and North Africa a couple of times, I wanted to experience something different. After discussing alternatives, Carol and I decided on the Orient. So, now that our destination was decided, it was time to establish an itinerary. One thing I already knew about this next excursion, it was going to cost me a lot more than my previous ones. As a matter of fact, I was probably going to need to take more money than all of my previous trips combined. Inflation and the cost of living had risen considerably since my last jaunt, and women as classy as Carol, weren't used to living 'rough'. Carol wasn't careless with money, but she wasn't a skinflint like me either. It would take all of my conniving and powers of persuasion to keep us on a reasonable financial keel."

"We left Vancouver at 10:30 A.M. on October 14th and soon reached our cruising altitude of 35,000 feet. I was hoping to achieve a good booze buzz during the flight but service was poor. So much for flying with Canada's national airline. The previous negative rumors I had heard concerning Air

Canada, all proved to be true. Even the inflight movie, 'Grease', with Johnny Revolta and Olivia Newton-Hoop, sucked big time."

"Five and a half hours later, we mercifully landed in Honolulu, Hawaii and were eventually deposited on to the steaming airfield tarmac. Looking up, the normally reliant and clear sapphire-blue skies were replaced by a thin layer of opaque cloud. Thankfully, this atmospheric veil didn't diminish the warmth of the sun, and the 83-degree Fahrenheit climate, was quickly beginning to loosen our former air-flight induced, stiffened joints."

"After retrieving our bags and clearing U.S. Customs, we grabbed a cab into Waikiki and commenced looking for a place to sleep. I was already deviating from my regular travel routine by taking a taxi into town. Normally, I would have hitch-hiked or coaxed some truck driver into giving me a lift. That would have been fine for me, but I couldn't expect that of Carol."

"Most of the hotels along the strip were affiliated with luxury, high-end chains. However, if you looked hard enough, there were still a few smaller, more reasonably priced locales. We eventually spotted a little 'hole in the wall' joint called the Traveler's Inn. It didn't look too bad from the outside and it was only a block away from the beach. We registered at the reception desk and went to our room. Once inside the room, the reality of the place hit home. What a dump! If we had been staying longer than just one night, I would have demanded a refund and gotten the hell out of there. As it was, we both felt that we could stomach it for a few hours."

"After quickly freshening-up, we hit the bricks and headed for a stroll along the beach promenade. It was comical to watch all of the alabaster, skin-toned 'Haoles' strutting around in their Ray-Ban sunglasses with newly purchased tropical patterned, short-sleeved shirts and polyester slacks. They looked about as natural there as nun in a whorehouse. Our meandering eventually brought us to the Kodak Show Bowl where a Polynesian Extravaganza was about to begin. After paying our $25.00 entry fee, we were treated to a visual feast of hula girls, fire dancers and ukulele players. It was a great show, but only lasted an hour. We grabbed something to eat en route back to our flea bag and then made an early night of it."

"I was up at 8-o'clock the next morning and went straight to the beach, where in the shadow of Diamondhead Mountain, I took a swim in the Pacific Ocean. After that, it was a 99-cent breakfast at one of the many cafes in the district. Carol stayed in bed for some extra beauty sleep. Within a few hours, we were back at the airport and boarding the Pan Am 747 bound for the Orient. The flight to Tokyo, Japan was lengthy but at least the service was good. The food and drink was afloatin' the entire time and I actually got a bit of a buzz from the booze. Although, I never mentioned it to Carol, the stewardesses were pretty 'hot' as well."

"There was only a one-and-a-half-hour layover in Tokyo, so we had to remain on the plane. That was a bummer. I wanted to get off the jet and at least be able to tell everyone that I stood on Japanese soil. The final leg of the flight from Tokyo to Hong Kong, China was a bit hair-raising. We caught the tail end of Tropical Storm Nina and it was tossing around the 747 as if it were a toy. By the time we landed at Hong Kong Airport, we had been rooted to our seats on the plane for nearly 13-hours. Carol was nauseous from the turbulence we had just experienced and wasn't in any mood to wait in a taxi queue for another 45-minutes in order to reach the city center. As a result, we remained near the airport and checked-in to the Grand Hotel for $215.00 per night. The room was nice, with a marble bathroom and a Touch-Mate television set, but big deal. A room is a room. When I considered it, the cost of that room for a single night, was more than I paid for my entire accommodation during my last trip. I guess that's the effect of inflation over the years and women with expensive tastes."

"I did a little research the next day and located a reasonably priced rooming house in my Student Guide Book, called the Chungking House. It was pretty close to the action and only $12.00 per night for a double bed. As a result of the tropical storm, the rain was still coming down nearly non-stop, with only the occasional break for a couple of hours. Everywhere on the streets there were pumps sucking out water from flooded shops and apartment building cellars. We tried our best to ignore the weather and actually got to see quite a few of the important sights around the city. The crush of humanity here was unreal. On one occasion, we took the Star Ferry from Kowloon to Hong Kong Island. There were hundreds of vessels in the harbor, including ferries, commercial freighters, tug boats, and Chinese junks, just to name a few. All were sailing in various different directions, and amazingly, avoiding one another. I believed it might be a different story, if it were ever to become foggy in the bay."

"The British presence in the city was a big plus for us, since many of the street markers and other signage was in English as well as Chinese. In addition, many younger people, business owners, and professionals spoke English, as well as their native tongue. Having been regularly exposed to Chinese food back home, we actually enjoyed most of the cuisine we had sampled so far. However, it was a bit of an eye-opener walking through the public markets. Everywhere you looked there were cages of snakes, birds, weird looking rodents and even small monkeys. Butchers were dissecting pigs and chickens out in the open air after having first ripped-out their organs and entrails and then hacking the body parts into sections. It was quite the scene and not for the faint of heart. I won't even go into what it smelled like."

Jesse vividly imagined the scene as McGuire described it in this last representation. Just the mere thought of it caused him to crinkle his nose in disgust.

"The holistic medicine shops sold everything from natural roots and herbs to shark fins, ground rhinoceros horn, bear claws and gazelle antlers. Whereas, many of the natural ingredients apparently have health benefits, others like rhinoceros' horn and even snake venom, are of no demonstrable value at all. So far, I haven't purchased anything here for myself, but Carol has managed to buy a few clothing items for a good price. Many of the shopkeepers seem to care little if they make a sale or not and words such as enthusiasm or customer-service don't seem to be a part of their vocabulary. However, the short top that Carol wore, baring her magnificent midriff, seemed to be a hit with most of them. Judging by the condition of certain streets, it appeared that some people just throw their trash from the windows of their homes to the road below. At night from our room, we often heard the sounds of breaking glass bottles and related noises. I guess that's what it was."

"It continued raining over the next several days but we tried visiting as many locations as possible. Some of the highlights were Macao, Stanley, Aberdeen and Ocean Park where we rode the Ocean Express funicular railway. This cable car traveled from Victoria Peak to Hong Kong Island waterfront. Although, only an 8-minute journey, the views were fantastic and it was well worth the price of the ticket. Rising nearly 500 feet at its apex, we were treated to spectacular views of both Deep Water and Causeway Bays, with its myriad of watercraft sailing about. Everything ranging from traditional junks and hydrofoils, all the way up to modern day luxury yachts. As soon as you leave the funicular, there's a huge aqua tank, filled with seals and penguins. What a great place to spend some time and we seriously considered going back for another ride before leaving town."

"On the morning of October 20th, we ate at our favorite Indian restaurant and then took the ferry back to Kowloon where we then attended the Thai Embassy in order to secure our Travel Visas for that country. Afterward, we headed to the Kowloon Railroad Station and boarded a smooth riding train bound for Sheung Shui, New Territories. This was close to the Mainland Chinese border and we had previously arranged to meet a local guide who would show us around the area. As soon as we stepped off of the train, Peter, our English-speaking guide was there to meet us. Carol and I entered the rear seat of his Volvo sedan like a couple of VIP'S and he proceeded driving north."

"En route to the frontier, we stopped at a duck farm, where seemingly millions of the canards were being bred for human consumption. Right next door was a worm farm, but Carol wanted nothing to do with viewing this somewhat creepy-crawly installation. All I knew was that worms were good for the garden."

"We continued northward and eventually reached the virtual 'no-mans-land' that separated Hong Kong territory from mainland, Red China. There was a vast swampy area dividing the two domains but the sight of concrete guard towers placed at strategic intervals and military patrol boats continuously cruising the river's length, reinforced the fact that the two region's separation was far more than just a tract of land. Although their peoples shared common heritage, their ideologies were like night and day. It was a bit intimidating to know that a mere mile in front me was one of the world's most oppressive regimes, and the term 'human rights', was as unfamiliar to its government and citizens as was rock-n-roll."

"Driving back toward Kowloon, there was still a good portion of the day remaining. When I remarked on this, Peter further offered his guiding services for a slight increase in fee. Carol and I accepted his offer and we went off the beaten path. This is what I was after. In essence, I was a tourist, however, I wasn't interested in seeing the usual shopping malls and tourist traps. I wanted to absorb the culture of the places I visited and go to areas mostly attended by the local inhabitants. I wasn't disappointed. Before arriving back in the city, we had visited an ancient walled village, the largest Buddhist Temple in the Hong Kong region and sailed aboard an authentic 'junk' in the fishing community of Tai Me San. We met local fishermen and also checked-out the huge, fish sorting market. I had never before witnessed an operation on that scale and I marveled at the many different species of fish, eels, rays and crustaceans. It was a real eye-opener for me. Peter even turned us on to Chinese beer. I was surprised to discover that it was some of the best tasting beer that I'd sampled anywhere in the world."

"It was near 8:00 P.M. when we arrived back in Kowloon and Peter had one more sight for us to see. It was the weekly 'night market'. Once again, everything was on a grand scale. There were vendors' stalls as far as the eye could see and selling everything from nuts to bolts. This place was a shopaholic's wet dream. We started browsing, but there was no way that we were going to be able to take it all in. Carol ended-up buying a few things and I purchased an Orion cassette deck player complete with a digital clock. By now our arms were full, so we said good-bye to Peter and returned to our hotel."

"Depositing our treasures inside our room, we then quickly returned to the market and commenced sampling as many different types of street food as we could. Of course, I couldn't forget to wash it all down with my new discovery, Chinese beer."

"By the end of the night, I must have been really pissed, since before leaving the market, I bought a side of fries from a nearby McDonald's Restaurant. I never eat that crap when I'm sober. It was shortly after midnight when we finally split the sheets and I turned-on the television as a non-pharmaceutical sedative. The 1960s Hollywood sitcom, 'Bewitched' was playing. But instead of its stars, Elizabeth Montgomery and Dick York speaking English, their voices were dubbed in Cantonese. What a hoot! I couldn't stop laughing. Not because the program was comical, but because it sounded so strange in an Oriental language. It had been a busy and very interesting day. Carol fell asleep nearly as soon as her head hit the pillow. Not being far behind, I was out like a light before the end of the T.V. show."

"With our bacon and egg breakfast the following morning inside the hotel's restaurant, we received an unexpected floor show in the form of a Chinese waiter and South Asian guest arguing over an order of toast. We never did discover what precipitated the debate, since between the waiter's heavily accented English and the Indian guest's incomprehensible diction, neither were able to get their points across. I quickly lost interest when it became obvious that blows would not be exchanged. I returned my attention back to that morning's Hong Kong Standard newspaper, and received my fill of the multitude reports of break and enters, robberies, car accidents and even a few murders from the previous evening."

"Over the next several days, we repeated activities that we'd already enjoyed and re-visited sites we found either interesting or entertaining. And, not to neglect the recurrent shopping-sprees and repetitive visits to some of our favorite eateries and street vendors. Where at first, I was a bit apprehensive at our initial introduction to the city, we were now truly beginning to love these 427 square miles of organized chaos. Something new for me, but nevertheless unforgettable, was our visit to the Wan Chai District of the city and the Tiger Balm Gardens. These were developed by multi-millionaire tycoon, Aw Boon Haw, the creator of Tiger Balm ointment as well as several other medicinal salves. The beauty of those 6-acres was incomparable, with an unbelievable array of exotic plants, tropical birds and water features. It was fantastic!"

"Since useable land in Hong Kong is so finite and so expensive, the average person does whatever they can to sidestep this issue, yet still be able to meet their daily needs. Rooftop gardens was one way that people could still grow a portion of their desired vegetables. Although ingenious in its concept, unforeseen events can still result. That's what happened to us one morning when Carol and I slipped out of bed, only to find ourselves standing in a virtual lake, that was once our room. We later discovered that full-time residents of the hotel had created their own rooftop garden on top of the building but did not first seek anyone's approval. Nor did they attempt to determine if the roof was even equipped to support such an undertaking. As a result, over time and daily watering, that confluence had accumulated and eventually began leaking through our ceiling and into our room. Once apprised of the situation, the Management graciously moved us to another room. We never did find out what, if anything, was done to discipline those would-be gardeners."

At this point, McGuire discreetly glanced into his journal, flipped a few pages, and after perusing it briefly, recommenced his dissertation. "The night of October 24th was to be our last in the Pearl of the Orient. We sent our soiled clothing to the hotel's laundry service and for the reasonable price of only 40 Hong Kong dollars, they were going to clean the whole kit and caboodle. Not bad! We then rushed down to the night market and visited our favorite eatery. After informing the proprietor that this was our final night, he pulled out all of the stops and created us a feast fit for royalty. It was absolutely out of this world. The next morning, we retrieved our laundry and it looked better than new. Cleaned to 'the nines' as well as ironed and folded. We couldn't get that back home for that price. We then said goodbye to the lodge's accommodating staff and hit the bricks."

"With a little time to kill before our flight to Bangkok, Thailand later that afternoon, we grabbed a hydrofoil from Hong Kong Island to Macao. The ride was as smooth as a baby's bottom, as we cruised at 32 knots on our 2700 HP inboard engines. Approximately 75 minutes later, we tied-up at Macao and grabbed a 1 horsepower, or should I say, 1 human power rickshaw. Tommy, our 'peddle pusher', was an older chap but still pretty spry. He took us around to all of the interesting spots and charged us only $6.00 USD for 2 hours. He was sweating like a dog by the time we finished. Before leaving the island, I decided to check-out one of its many casinos. Not to gamble, but merely out of curiosity. It was very lavish inside and I could see quite a few serious gamblers. But overall, I wasn't impressed."

"Shortly before the supper hour, we boarded the Air France 747 at the Hong Kong Airport, bound for Bangkok. It was only a two-and-a-half-hour flight, but I actually wished it had been longer. The service was great and it was one of the best flights that I'd had in a long time. As we boarded our cab for the city shortly after 8:00 P.M., it was still 32-degrees Celsius outside. We later registered at the Century Hotel with a reasonably priced room. Before paying, I made certain that it had air-conditioning. Anywhere in this country without air-conditioning was tantamount to a death sentence. After stowing our gear, we took a long walk down a few very crowded and smelly streets. Most of the shops were already closed, so we gradually just meandered back to the hotel."

"The morning of October 26th was our first full day in Bangkok. It broke clear and hot. After a decent breakfast in the hotel, we were ready to start our adventure. After walking for a couple of hours, we weren't really getting anywhere and hadn't seen anything interesting. We hailed a cab and headed in to Chinatown. Funny, we had just left China, and our very first stop in Thailand, was Chinatown. Go figure!"

"What an eye-opener being in that traffic. I couldn't believe that anyone there even possessed a driver's license. The cars were driving in every direction, and forget about road courtesy or right of way. The noise was deafening, and many of the cars were in rough shape. It appeared as though mufflers were non-existent here and the motorcyclists behaved as though they were on a drag strip. After finally arriving in Chinatown, we went down to the river and took a scenic boat ride. Everywhere we looked, flooding from Tropical Storm Nina was evident. Even the sewers had overflowed as a result of the storm and the entire place stunk like a backed-up toilet. The floating market was fascinating, including the vendor who handled all manner of nasty reptiles. King Cobras and other poisonous snakes, were all to be sold to those most discerning of diners."

"Fewer people spoke English here as compared to Hong Kong, but most shopkeepers were friendlier and went out of their way to barter in good faith. The frustrating thing was that just when we thought we got something at a good price, we'd go a little further on, only to find another shop selling the exact same article at a cheaper price yet. As we entered the Wat Po District, a female university student offered to escort us around. She took us to see the famous reclining Buddha that was struck by lightning thirty years prior and it melted-off the statue's gold leaf coating. She also explained how Thai Buddhism was a combination of Indian and Cambodian religions."

"She took us to an open-air market and it was even more bizarre than those of Hong Kong. Here they were selling everything from plastic shoes to deep-fried baby birds. There were stacks of gutted frogs coated with black flies, along with others that were still alive sitting in baskets and had their hind legs tied together with straw. There were numerous species of birds crammed into hundreds of wicker cages. There were dried mice, all manner of fish and even puppy dogs, ready to grace someone's dinner table. There were people placing bets on fighting cocks and even some wagering on Siamese fighting fish. It appeared that no matter what they sold, nothing went to waste. Tongues, eyes, feet, tails, everything could be cooked and eaten. The populace didn't seem to have any regard for animals. They were just either used for work, or to be consumed."

"The next day, we kept most our belongings at the Century Hotel and I placed the bulk of my traveler's cheques in its safe. We were going to be taking a little side trip to Chiang Mai, in the northern portion of the country. After purchasing 1st Class tickets, we boarded the train. Carol is now dealing with a painful looking sunburn from yesterday. The sun in these parts is deceiving and if you're not careful, you could be in serious trouble."

McGuire now paused a moment and addressed Jesse directly. "Hey, Kid. Did you hear me say that we purchased 1st Class train tickets? From my previous travels, did you ever know me to do that before?" In response, Jesse merely shook his head from side to side and shrugged his shoulders. McGuire continued. "I didn't think so. It's like I said before, traveling with a classy woman was a whole different ball game."

"Anyway, our berths were quite comfortable but the bar coach was woefully understocked. To top it off, I'm not feeling very well. I don't know if it's because of the heat and humidity or as a result of all the Scotch, Grand Marnier and Kahlua I drank the previous night. I've now got a serious case of the 'trots'. After ruminating on the totality of the matter, I decided that it was probably as a result of unfiltered water in the crushed ice, we dropped into our drinks, at the open market the day before. Either way, it was as uncomfortable as hell and I didn't stray too far away from the WC."

"When we eventually reached Chiang Mai, we were greeted by a representative of the President's Hotel and he chauffeured us there in the hotel's limousine. I felt like a king." Once again McGuire directed a question toward Jesse. "Kid, did I ever tell you that I was descended from royalty? No? I must have accidentally left that out." Jesse just snickered.

"Our room was quite reasonably priced, considering that it was well outfitted and afforded a nice view overlooking the impressive swimming pool and some beautiful landscape in the distance. Being so close to the Burmese border, I didn't expect to find such luxury. We were hoping to book a 2-day trip through the Golden Triangle passages, that would have included transport via, taxi, bus, boat and plane. Just as I was about to seal the deal with payment, the guide reneged on the airplane portion of the trip, so I just cancelled the entire venture. Disappointedly, we returned to the hotel and spent the remainder of the day sitting around the pool. Maybe it was for the best, since Carol's sunburn was still giving her pain and my intestinal disorder had still not completely disappeared."

"Considering it was called the President's Hotel, the food there was substandard and not worth the effort. Everything we ordered was tasteless and the entire staff barely spoke a word of English. What I had hoped was going to be an interesting side-trip was quickly becoming a bust. There wasn't even any decent smoke to be had. The couple of Thai sticks I tried were disappointing, but it seemed that every other cab driver was offering me heroin. No thanks."

"With not much happening in Chiang Mai, Carol and I decided to head back to Bangkok and purchased return bus tickets. Out on the street afterward, a cab driver offered to take us to the Mae Klang Waterfall in Doi Inthanon National Park, a little further to the north. He also volunteered to throw-in a trip near the Thai-Burma border to check-out one of the Karen hill-tribes. With nothing else going on at the moment, we accepted his invitation and paid him the 300 Baht fee. The 100-metre-tall Mae Klang waterfall was very impressive, as was the surrounding park. We walked a few of the trails for a bit and then jumped back into the taxi. The driver handed us each a can of beer and cranked-up the tunes. Before long, we left the main road and traveled along a poorly maintained dirt track. The bush was becoming denser the further we drove and the humidity was also more notable."

"Our butts were pretty sore by the time we finally arrived at the Karen village. The taxi was immediately besieged by a riot of half-naked snot-nosed kids and I thought they were going to remove the skin from the palms of Carol's hands, as they quickly snatched the coins she offered. The Karen were the indigenous peoples of Thailand and Burma and numbered approximately one-million souls between seven distinct groups. Each group possessed their own language, culture and customs and were spread throughout numerous villages within the region."

"Their level of poverty was as bad as I had seen anywhere else in the world. Beyond the confines of their own village, prospects of a better life were nearly non-existent. Over the years there had been continual conflicts between the Thai and Burmese governments with the more militant units within the Karen populace, as they struggled to keep their identity and autonomy. The Karen's persistent cultivation of opium poppies as a livelihood, also created problems for authorities."

"Overall, it was quite a depressing scene in the village. Aside from taking a few pictures and handing out cigarettes to the elders, we didn't feel like remaining there any longer. At one point, I gave Carol a brief kiss and the kids started cheering and screaming for more. When I kissed her again, they went nuts! We took this as our queue to split and started walking back toward the cab. As the mob of kids followed close behind, I reached into my pocket and pulled out a wad of small denomination bills. I tossed the bills into the air while Carol began chucking single cigarettes out toward the kids. The frenzy that ensued as a result, allowed us to get back into the taxi unmolested, and we drove off leaving a virtual dog-pile of kids all scrambling for the treasures we had left behind."

"The drive back toward Chiang Mai was very picturesque and a welcomed change from the squalor of the Karen village or the polluted, overcrowded bustle of Bangkok. The sight of sprawling rice fields, gently swaying palm trees and plenty of local flora and fauna, soothed our frazzled countenance. We arrived back in Chiang Mai just in time to check-out of the hotel and catch our 7:30 P.M. bus back to Bangkok. Unfortunately, the bus was running late, so we didn't actually depart until 8:45 P.M. The 1st Class coach was filled with a cadre of gentile, mostly Western European travelers. Two female attendants handed us the obligatory steaming towels, followed by coffee and snacks. Overall, it was a pleasant ride but ruined by the air-conditioning system that was blasting-out an icy flow the entire way to Bangkok. Now I know how those fish felt when we pulled them out of the ocean and packed them on ice in the purse seiner."

"We traveled on this six-wheeled refrigerator throughout the night, and under the circumstances, it was impossible to grab even a few minutes of sleep. It was nearly 7:00 A.M. when we pulled into the Bangkok, combination bus depot and train station. Slowly but gratefully, we eventually reached the open air in order to thaw our frozen extremities. After losing the icicles that had formed on the end of our noses, Carol and I hung around the train station until the tourist office opened. Once it did, we made inquiries about booking a flight to Singapore."

"We were in luck. There was a Singapore Airlines DC-10 departing later that afternoon at 4:40 P.M. That left us plenty of time to check-out of the Century Hotel and still do a bit of shopping before flight time. We immediately purchased tickets. Completing our chores and eventually reaching the airport, I moseyed over to the Currency Exchange Booth and substituted my remaining Thai Baht for Singapore Dollars. Retrieving my luggage from where I had left it, we then passed through Customs and made our way to the Departure Gates. We eventually boarded our flight and to my immense pleasure, the liquor service began as soon as we planted our backsides into our seats. I was stoked! That was, until we reached about 20,000 feet, and I suddenly remembered that I had left my Hong Kong purchased, Orion cassette deck player and brand-new collapsible umbrella leaning against the wall near the Currency Exchange Booth. I wanted to scream, but extreme control of will, prevented me from doing so. What a bummer! That realization just ruined my day and I then proceeded to drink as much alcohol as I could during our short flight, in order to dull the biting ache of my loss."

"We arrived in Singapore at approximately 8:00 P.M., just as I was beginning to achieve a pleasant buzz from the booze. I was still choked regarding my negligent loss at the Bangkok Airport, but I realized that now I needed to forget about it. As soon as we commenced our ride into the city, I immediately noticed a distinctive difference between Bangkok and Singapore. Where the former was a hectic, noisy and polluted madhouse, Singapore was clean, refined and calm. It was extremely modern and organized. It might not be nice to say, but the citizens here also appeared to be more civilized. Not knowing where to go, I instructed the taxi driver to bring us to the nearest reasonably priced accommodation with which he was familiar. A few minutes later, he stopped in front of cluster of row houses and explained that they were newly renovated guest houses. They were available by the night, the week or the month. Once meeting the custodian, we learned that a suite was $47.00 dollars per night. It was more than I wanted to spend, but it was getting late and at the moment, I didn't have an alternative."

"The next morning, we were up and at 'em nice and early. The weather was gorgeous, with a wonderfully warm temperature but also a constant, cool and gentle breeze to keep things tolerable. After searching for more reasonably priced living quarters, we eventually located, and then checked in to the Mitre Hotel. The cost of a room here was almost half that of our previous night's stay. You know me by now, a penny saved is a penny earned."

"As soon as we settled ourselves, we were out the door and heading downtown. There appears to be plenty of public transit to take people around, but unless you know where you're going, it's a bit daunting. For the moment, Carol and I are cabbing it, but this is dramatically cutting into my bottom line for daily expenses. Singapore is definitely a city on the move, with brand new office towers, high-rise apartments and shopping centers almost springing up overnight. Everything is modern and there's almost no heritage or older-style architecture to be found anywhere downtown. A few areas such as Chinatown, that still have a few older buildings, were quickly being swallowed-up by the city's insatiable development monster. Naturally, all of this progress comes with a price tag, making Singapore one the most expensive cities I've ever visited."

"All of the streets are super clean and everywhere you look, there are signs bearing slogans encouraging people not to litter. Or, there were other signs warning offenders that the penalty for littering could be as high as $500.00 dollars. You can't even chew gum in Singapore without potentially facing a hefty fine if you're caught. Entrepreneurs found importing or even selling chewing gum could face jail time. Certain therapeutic, dental or nicotine chewing gums may still be acquired, but only with a doctor's prescription. Thinking about how serious they considered the offence of chewing gum, I wondered what they would do if they caught me with a gram of hash or smoking a joint? I would probably be stood-up in front of a firing squad at sunrise."

"Actually, the crime-rate for a city-state of that size is quite low. Laws are strictly enforced and punishments for serious crimes are severe. Even jay-walking is seriously frowned upon and immediately dealt with. Also, the police there don't accept the crap from miscreants that their North American counterparts do. We spent most of our first full day just getting acquainted with the surroundings and did a bit of window shopping. Most of the clothing and merchandise prices there were out of my league. Singapore has the 3rd largest, and busiest, deep-water port in the world, so we slowly meandered our way down there to look around. It was amazing to see all of the international ships being serviced simultaneously, loaded and off-loaded. Giant cranes that looked like Meccano erector toys from a distance, were swinging from one side to another. First lifting and then depositing huge storage containers and other cargo, here, there and everywhere. It was quite impressive to watch how swiftly and meticulously they worked."

"Ships that couldn't find a place to berth, were on and off loaded from a distance. While they dropped anchor in the outer harbor, large wooden, unpowered barges would travel back and forth from dock to ship and vice versa carrying their goods. You could even occasionally tell the ethnicity of the barge's crew from markings it may possess. For instance, a large eyeball painted on either side of a barge's bow with two men handling the large rudder tiller at the stern, meant that it was of Indian origin. I don't think Carol was too impressed with the whole maritime operation, but I was fascinated."

At this point in time, McGuire paused his monologue, thought for a moment and then resumed his discourse. "I'll be honest with you, Kid. Singapore was a nice place to visit but way too straight for me. I need action, excitement and maybe even a touch of danger. Singapore just didn't offer any of that. Most of our days were spent doing the 'tourist thing' while the nights were spent having a few drinks and going to the movies. Carol and I had originally planned to leave Singapore on November 8th and then fly to Denpasar, Bali. Unfortunately, the tourist agent in Bangkok messed-up and didn't schedule us to leave Singapore until the 11th. Our Thai Air flight tickets were now booked for that later date and we couldn't alter them. That meant a few unexpected days here at prices that I would rather not have to pay."

"Trying to make the best of our situation, Carol and I put together a brief itinerary for our remaining time in Singapore. Although this was impromptu, we actually enjoyed the additional days spent there. One of the highlights was the 'Big Splash Water Park' with its 100-foot-high and 150-foot-long mega-waterslide. It also boasted a huge circular pool with a 5-knot current to stimulate its swimmers. What a rush! Another great day was spent at the Singapore Tiger Balm Gardens. This was a much grander version of its Hong Kong cousin. The only downside was the price of extras after already having paid the entry fee. I thought that I would need to donate a pint of blood just to afford a couple cans of beer and a bag of peanuts. Later, we left the Gardens hungry, but at least I still had a few bucks left in my wallet."

"Our last full day in Singapore was chock full from morning until night. We started off visiting a crocodile farm on the outskirts of the city. The reptile show was interesting but we quickly bypassed the gift shop with its extravagant offerings of alligator hide clothing accessories such as belts, purses, wallets and shoes."

"Next, we headed to the Singapore Cable Car that linked Mount Faber from the city-side, with Keppel Harbor on Sentosa Island. The 15-minute ride at a maximum height of 300 feet, offered a fantastic, 360-degree view of the surrounding area. This was yet another funicular ride. I had taken others in several different countries, but I never grew tired of them. After the cable car ride, we rented bikes and toured parts of Sentosa Island. The place was packed with Japanese tourists and several of them asked to have their photo taken with Carol. Yah, she was pretty 'dishy'. By the time we got back to our hotel and cleaned-up, we were both pretty bagged. Carol said she had a surprise for me and suggested that we go to Jack's Place for supper on Killiney Road. Jack's specialized in Italian food and I thought that was a great idea. After eating nothing but Asian food these last couple of weeks, I needed a change. As we arrived, I had to admit that we made a handsome couple, and as we took our seats inside the restaurant, all eyes were upon us."

"After placing our orders with the immaculately dressed waiter and beginning to sip on our crystal glasses of red wine, Carol reached into her bag and handed me what she referred to as my early Christmas present. I was thinking maybe a bottle of men's cologne or an Old Spice gift pack. Instead, what I saw caused my eyes to bug-out. What she actually gifted me was a 5-gram Swiss gold bar and a 24K gold-link neck chain. I just about crapped. What a surprise! I've got to say, that gesture set the tone for the rest of the night, and I couldn't have been happier. Our steaks were cooked to perfection and the side-dishes of oyster soup, spaghetti Bolognese, baked potato, salad and fresh-baked Italian bread were out of this world. After dessert and a few more glasses of wine, we slowly made our way back to the hotel. The evening's 'piece de resistance', occurred after I turned out the bedroom light and slipped between the sheets with my gracious and generous travel companion."

"After paying our $700.00 bill at the Mitre Hotel the following day, we headed toward the airport. It seemed that the Gods on Mount Olympus were definitely arguing. Lightning bolts cracked regularly and the thunder was so loud, I thought my eardrums would burst. When the rain started falling, it was something like I had never seen before. You couldn't even see individual drops falling from the sky. It was literally like someone took a gigantic bucket of water and was simply pouring it onto the earth in one sustained flow. Within minutes, the entire place looked like it might be sinking into the sea, and the cabbie's windshield wipers, although turned on full blast, couldn't remove the deluge quickly or sufficiently enough for him to be able to see more than just a few feet ahead. I believed that he was simply driving on instinct and just hoping for the best."

"Wheels were up on our silver bird right on time and the ride was smooth. The booze flowed and the T.V. dinner that constituted our meal, was more or less edible. We touched down in Jakarta, Indonesia for 40-minutes to off-load some of our passengers. Once the plane was back in the air, we were down to about a quarter full, since no additional passengers boarded in Jakarta. A few drinks and a couple bags of peanuts later, we landed safely in Denpasar, Bali. We breezed through the Customs check-point and grabbed a cab for Kuta Beach. Before leaving Singapore, I had exchanged their currency for Rupees, which was now the legal tender in Bali. The ride to Kuta cost me 1000 Rupees.

I told the cabbie that I wasn't interested in buying his taxi, I just wanted a ride. My joke was lost on him, as it went right over his knobby, bald head."

"Kuta Beach was the mecca for tourists while in Bali and I always like being near the action. We checked into the Kuta Cottages for $15.00 USD per night plus a 21% service charge. I thought that was a bit pricey for a room without air-conditioning or hot water, but it would be a good base of operations. After sorting our gear, we hit the beach. The strand was impressive, with golden sand that seemed to stretch out forever. The water was nice and warm as well. We weren't there for more than a few minutes when we were already assailed by the usual pack of young male and female peddlers offering us everything from cheap trinkets to wooden carvings. I couldn't believe how quickly they sniffed us out. They were like sharks smelling fresh blood in the water from a mile away. They were persistent as hell, but so were we. After a while, they lost interest in us, and left."

"Where we stayed was not serviced by public transit, so we had to rely on taxis or our own shoe leather to get around. I could see that the cost of taxis would begin adding up, so I decided renting a motorcycle for the week, would be the way to go. There was a motorcycle shop in the small community of Kuta Town and the owner was willing to rent me a 1977 Suzuki 185CC GT. Unfortunately, my International Driver's License wasn't going to cut it. After speaking with another cab driver, he suggested that I go into Denpasar and apply for an Indonesian license. He told me that it would be a piece of cake, so long as I could pass the maneuverability test. He even offered to drive me into town. I thought, *what the hell. Let's do this thing.*"

"Before I knew it, I was delivered to a police station in Denpasar and started the application process. They had an old Honda 100CC bike there for testing purposes. I completed the form to the best of my ability."

"The driving assessor then took me to a paved parking lot at the rear where an obstacle course had been pre-set. He instructed me to start weaving through the course and he would return in a few minutes to test me. I was a little shaky at first, but soon I was gliding through the course without knocking over any of the cones or using my feet to balance myself around the turns. Luckily, the assessor was gone for more than a few minutes, which gave me extra time to practice. By the time he returned, I blew through the course like nobody's business. I received a 'two thumbs-up' and we headed back into the office to complete the process. The test itself and registration fee cost me 1500 rupees.

One week's insurance cost an additional 2750 Rupees, and the document photo was 1800 Rupees. Finally, the tip for the assessor's trouble was 800 Rupees. With the rental fee for the bike being 11,000 Rupees for six and a half days, I wondered if it would have been cheaper just to cab it after all."

"After returning to Kuta with my freshly minted Indonesian motorcycle license, I headed straight to the bike shop. As I was sealing the deal, I received another kicker. I had already paid 2750 Rupees for insurance but then I was told, it was valid only for me. If I chose to take on a passenger, the insurance would then be null and void. What a drag! I decided to keep my mouth shut and just take my chances. After jumping through all of those hoops, I wasn't going to be defeated now. I raced back to our cottage with the bike and offered to take Carol for a little spin. She was very hesitant at first because she had previously been in a motorcycle accident and was afraid that history would repeat itself. I assured her that everything would be fine and eventually managed to coax her aboard. In order to allow her to gain her confidence, I didn't take her too far, and I drove like a senior citizen."

"The next day we decided to take a ride out of town and pointed ourselves southbound. We were told that Tabanan was a good place to go and it was only about 35 kilometers away. After obtaining directions, we headed out. I initially chose a larger street bike, since I thought having the extra horsepower would be better in the city or on the hills. But most of the roads we drove on now were really rough, and a lesser powered dirt bike would have given us a smoother ride. With all of the ruts and bumps, we were now bouncing over, I thought the fillings would drop out of my teeth. En route to Tabanan, we drove past numerous small villages. I was beginning to think that they must have radar installations in each one, since whenever we approached one, a gang of kids would always run out to the road and attempted to wave us down. They were all so excited you would think they hadn't seen another human being in months. And when they saw my red beard, they went crazy."

"Once again, it was like being in an episode of National Geographic. It appeared as though as much as 75% of the country was devoted to cultivating rice, and the majority of country folk were subsistence farmers. There were large ditches on either side of the roads and everywhere women were seen washing clothes in those trenches, while right beside them would be young boys and girls washing oxen and other livestock. Older girls and grown women would carry large containers of water on their heads to take it back to the village, while the men sat around, essentially doing nothing. The women and children did the bulk of all the work. Carol would wave at people as we cruised past and they would always respond in kind. Although they were dirt poor and lacking almost everything, they still appeared to be happy with their lot in life. I always found it comical when one of the old grannies would smile and wave. Most would dye their teeth red, if they had any teeth at all, and it would always remind me of a kid who had just eaten a big, cherry flavored lollipop and it stained their entire teeth and tongue bright red."

"Back at Kuta Beach, we fell into a routine of easy living and doing only what we wanted. Sometimes we leisurely strolled the few streets in Kuta Town, peering into its few shops or grabbing a bite to eat in its several cafes. I especially liked the roasted turtle steaks. We would always spend part of the day swimming and sunning, but this cost us dearly. Both Carol and I are fair skinned and most of the time we were trying to sooth our sunburnt skin with a myriad of lotions and ointments. This didn't stop us from enjoying ourselves, but it did make for a few painful moments. We even managed to score the odd bit of decent smoking dope, but this was hit and miss, since a few unscrupulous dealers would often pass-off perfumed cow dung for hash or banana flowers and leaves as Thai sticks."

"The abundance of exotic birds was breathtaking and we wished that we could take some home with us. But on the downside, there were ants and mosquitoes everywhere, and we were constantly being bitten while on the beach or inside our room. One night in bed at around 1:00 A.M., I was bitten so hard on my side that I jumped up. Do you think I could find the culprit anywhere?"

"At one point on the beach, we met Trevor and Sam, a couple of Americans with a bag of magic mushrooms. For a small fee, they made us a delicious mushroom omelet that had Carol laughing like a mad woman for half the day. Myself, I felt nice and mellow. Being older and more experienced than she was, made all the difference. I guess they added a few too many mushrooms."

"I offered to buy the remainder of the 'shrooms' from Trevor, but he warned against it, if we were planning on taking them to Australia. He said that their Customs Agents were always on the look-out for drugs coming into the country, and there would be a good chance I'd be caught. I thought about it for a while, and then decided to take his advice."

"One day we visited Simon at his low-budget squat on the beach. He's another 'head' that we became acquainted with during our stay. Simon had a pet monkey named Susan, and she was enamored with Carol and her facial jewelry. Susan immediately jumped onto Carol and took one of her gold earrings into her mouth. We had a helluva time prying her mouth open, and when we finally did, the earring was bent. Thinking proactively, Carol then removed her diamond nose stud before Susan could glom on to it as well. Unfortunately, she then accidentally dropped it into the sand. We frantically searched the area but couldn't find the lost diamond. I offered a 5,000-rupee reward for anyone that could locate it. Even with that incentive, and a lot of people searching, the stud was never recovered. Carol tried to accept the loss like a trooper, but I felt like crap."

"After a few days at Kuta Beach, it started raining. Without trying to be a smartass, that really put a damper on things. As a result, we checked out of our cottage three days early, returned the motorcycle, and went into the Jensen Hotel at Denpasar, for the remainder of our stay. That same day, we attended the Bali Beach Hotel and purchased our airline tickets to Australia via Qantas Airways. One night after supper we were strolling down one of the main thoroughfares of Denpasar when we were approached by the usual street entrepreneurs, intent on selling us something. The boys were pitching theatre and ballet tickets along with rings, swords and wooden carvings, while the girls were flogging fruit drinks and batik fabrics. We had never succumbed to the entreaties of these street urchins before, but tonight, in a moment of weakness, we purchased theatre tickets for a Barong Dance performance as well as a Monkey Dance performance. In my opinion, the Monkey Dance show was a waste of time, but I rather enjoyed the Barong Dance. The costumes for the sixty or so cast, were very colorful and intricately made. The main character Barong, a lion, and the king of the spirits, was the arch enemy of Rangda, an evil witch who devours children and is intent on destroying humanity. There were many other animal characters in the dance all possessing supernatural powers and placed on earth to save civilization. The dance culminates when Barong and Rangda battle it out to see whether the world will be destroyed or saved. Essentially, it is the eternal fight of good against evil. Heavy stuff!"

"It was now the night of November 24th and the following day, we would be flying out to the land Down Under, our final destination. I received a tip that there was a Cock Fighting event being held in an old warehouse-type building just outside of Denpasar's main shopping district. Carol wasn't interested in coming along, so I cabbed it there myself. The place was packed and the level of excitement inside was like electricity unleashed. In the center of the large open space was a makeshift ring, with its floor covered with dirt and surrounded by wooden boxes in order to keep the birds localized. Before each fight, there's a grand production where the handlers display the cocks above their heads and then parade around the arena so that everyone can see what they're betting on. The handlers then bring their birds ringside, where they begin to antagonize one another and work themselves into a frenzy. An official examines them both to ensure that they are suitable opponents. Size and weight must be similar. Once it's decided that the birds are compatible, a singular curved, razor-sharp blade is chosen and then expertly tied in just the right spot, to one leg of each cock. The blades are meant to either lame or kill an opponent, so their placement must be perfect. After this is done, the birds are lofted on high one more time, so that last minute odds can be determined and final bets can be placed. By this time, the crowd is at a fever pitch and wads of bills are being tossed back and forth between the betters and the bookies."

"When everything is set to go, the official bangs a gong and the cocks are tossed into the ring. With their wings flapping furiously and their claws thrust toward each other, each bird attempts to fell the other. If one bird is wounded and retreats, the handler has an opportunity to get him back into the fight. If the wounded bird refuses to fight any further, then the other one is declared the winner. If one bird is killed, then the opposing handler is given the carcass. He then removes the prime feathers from the loser and cuts-off both feet. The foot with the blade still attached, is then handed back to the original owner, while the other foot is kept as a trophy by the winning handler. I'm not really into 'blood sports', but overall, it was a very interesting night."

"The following morning on the 25th, we checked out of our hotel and made our way to the airport. As it was, our flight was delayed by thirty-minutes, so as a result, the captain declared that all drinks would be free. What a treat! I was in heaven. It was a great flight and the canned-music choices were pretty decent as well. We had a sixty-minute stop-over in Perth, Australia but it was a bust, since the passenger lounge was dry. Can you believe it? No booze at an Australian airport. Not even one stinkin' can of Foster's Lager!"

"It was 6:30 A.M. when we finally touched-down in Sydney. I don't often take advice from people, but I was sure glad that I had listened to Trevor back on Kuta Beach regarding those magic mushrooms. The Sydney Customs Agents went through our luggage with a fine-tooth comb. Had we possessed any illegal contraband in any of them, our fabulous vacation would have immediately ground to a dramatic and unceremonious halt. Thanking my lucky stars once again, we blitzed out of the airport and into a cab directly to the Sydney Central Train Station. From there, we located a cheap hotel nearby and checked-in. Carol was not impressed. There was no wash basin in the room and there was only one centralized bathtub, for communal utilization, by the occupants of all six floors. Bondi Beach was only 7-kilometers from where we were, so we decided that would be a good way to spend the day. We got our bathing accessories together and then took the bulk of our luggage back to the train station. We were going to be leaving for Brisbane from that location the next morning, so why not leave most of the stuff there overnight. It would save us having to lug it there later. We purchased our 2nd Class tickets at the same time and I quickly placed a long-distance call to my brother Ken, to inform him when we would be arriving. He was going to pick us up at the Brisbane Station."

"After that, we jumped on a bus, en route to Bondi Beach. Since, I hadn't slept on the airplane during our entire flight, I was mighty tired. I was just starting to doze-off, when Carol jabbed me in the ribs. As I looked out of the window, we were approaching the beach. I was immediately disappointed. It seemed as if there were twenty-thousand people all crammed together on the sand with barely a few feet between them. When we eventually arrived and I stuck my toes in the water, it was cold. Definitely not like the wonderfully warm water on Kuta Beach. Between all of the surfers, boogie boarders, swimmers, kayakers and life guard vessels, I barely had enough room to move." At that moment I thought to myself, *What I wouldn't give to be back on the beach in Las Palmas.*

"The following morning, we caught the 7:45 train to Brisbane and arrived at the Roma Street Station, fourteen hours later. There to greet us, was middle brother Ken, with his movie star looks and shoulder length, wavy brown hair. After spending a few minutes getting re-acquainted and catching-up on a bit of gossip, we all piled into Ken's ride and headed for his flat. He was still single at the time, much to the chagrin of half the Australian female population.

I've always considered myself a bit of a 'lady's man', but Ken put me to shame. He literally had to beat them off of himself with a cricket bat, and his cast-offs were usually better than my first catches."

"Once settled at Ken's place, life slowed down for me and Carol. We had been running at a hectic pace the past few weeks and didn't realize how much it had taken out of us. We were totally cool with relaxing a bit. Ken had taken some time off from his job as a gym teacher at one of the local high schools, and during that week, took us around to most of the interesting sights in the city. We also had a chance to see him play hard ball with his teammates in the municipal baseball league. He still plays ball even today at 70-years-old, and faithfully roots for the L.A. Dodgers via satellite television."

"We had a great time in Australia and returned to Vancouver via Sydney and Honolulu in early December. I didn't see it coming, but not long afterward, Carol and I called it quits. I don't know if it was any single event that caused the split, or if it was just the fact that things had run their course. Whatever happened, it was a great five years we shared together and we parted without animosity or hard feelings toward one another. One thing I can honestly tell you Kid, every woman that I've shared my time with, was very special to me, and I'm certain if I was to call-up anyone of them, even today, we could still have an amicable conversation and a few laughs with one another."

"There you go, Kid. I've laid myself bare," concluded McGuire while inhaling deeply and draining the remnants of his wine glass.

On that note, Jesse realized that they had finally come to the end of their momentous, 3-week retro-fest of Bob McGuire's prior travel adventures. He felt somewhat awkward, but realized that he needed to say something. Lacking anything more profound or substantive, he merely stated, "You know Bob, those were some pretty interesting stories you shared with me. In fact, they were down-right excellent! I think that you should put them down in a single edition, so that they're not lost. Have you ever considered writing a book?"

McGuire considered Jesse's words for a moment and then responded. "I hadn't really thought about it until right now. However, that might not be a bad idea. I don't know if anyone would be interested in reading the book, but maybe I'll just do it for myself. Or maybe my brothers Ken and John would like to know what I'd been up to all those years. Did I ever tell you about my brothers?"

Jesse reminded McGuire of the few times that they were briefly mentioned during his reminisces, but they were still both quite young then. He hadn't really spoken much about them as adults.

"Okay, Kid," offered McGuire. "Let me tell you a bit about them and how they've made me so proud. I'm going to start with my brother, Mike. He was only one year older than me, but completely different in so many ways. Of the four boys, Mike was the 'Rock'. He was the one that my parents would have considered the closest to their mindset and traditional values. Mike was extremely dependable and super conservative. We both first attended General Brock Elementary School, near East 33rd Avenue and Main Street, in the Little Mountain District of Vancouver. Mike spent one year at a high school where I can't remember its name, but when we moved to our new house on East 44th Avenue in 1957, we then both attended Killarney Secondary School. Not long after graduating from Killarney, Mike married his high school sweetheart and they started a family soon afterward. He eventually fathered two daughters."

"After school, Mike began working as a bag-boy with Safeway, the national grocery-chain at Harold Street and Kingsway Avenue, near our home. Over time, and through his diligence and hard work, he gradually worked his way up the corporate ladder to become the store's manager. Mike then later spent his entire career managing several other Safeway stores throughout the city, and eventually retired at 55 years of age. Sadly, he passed away in 2018."

"Now, Ken is the brother younger than me by six years. Of all the brothers, Ken is considered to be the 'El Guapo'. The handsome one, and the undisputable chick magnet. I mentioned in my last journal, that Ken was living in Australia, but I didn't explain how he got there. After graduating from Killarney High School, Ken attended the University of British Columbia, where he received his degree and teaching certificate in Physical Education. Literally, in the shadow of receiving these distinguished credentials in 1974, Ken noticed a poster tacked to the gymnasium door at UBC. It was from the Australian National School Board, advertising for qualified, English-speaking P.E. teachers to travel to Australia, at the host's expense, to assist in developing a formal, National Physical Education Program for that country. Until then, they had no such program, and P.E. was being taught to Australian students by non-qualified teachers who simply enjoyed sports. The Australian government felt that it was high-time they remedied that deficiency. The only stipulation was that interested parties, must commit to a one-year timeline, in order to receive their paid return tickets home. Ken thought this would be a great opportunity, and off he went, taking a few of his boyhood pals along for the ride. After a whirlwind vacation on the Australian continent, Ken went to work at his teaching job in Brisbane, while his friends returned to Canada."

"Well, Ken immediately fell in love with the country and its people, and when his one-year commitment had been completed, to the sheer delight of the Corinda High School staff and students, he chose to remain there. Ken continued as a Physical Education teacher at the Corinda High School in Queensland, for nearly forty years. Imbedded in the concrete on the school's playground, is a bronze star, depicting Ken's name and his invaluable service and contribution to that institution's successes. Ken was also very active in coaching youth baseball throughout his teaching career, and he went on to coach at the Regional, State, and eventually, the National level. One year, Ken coached the Australian Under-18 National Hardball Team, and they won the championship. To cap his illustrious teaching and coaching career, Ken was installed as a member of the Queensland Sports Hall of Fame. Not bad for a punk kid from East Vancouver, eh? Oh, and I forgot to mention, during the time Ken was making a name for himself, he eventually married an Australian gal, with whom he fathered a son and daughter. Naturally, Ken and his family are still based in Australia, and at 70-years old, he's still actively playing organized league hardball, including one team that his son still plays on."

"Last but definitely not least, is my youngest brother, John. I say that he's the baby of the family, even though he's long since retired from work and also pushing seventy-years old, himself. Now, John is the 'Brain' of the family. I mean to say that we're all smart, but John took his education to the next level. He too graduated high school from Killarney and went on to attend UBC, after first putting in a couple of years at Langara Community College, in South Vancouver. John received his Bachelor of Arts Degree in Psychology from UBC and then later went on the receive his Master's Degree at the University of Guelph, in Ontario. In between obtaining degrees and working, John was a little chip off older brother Bobby's block. He has also traveled extensively, from hitch-hiking across Canada, to traveling throughout Europe, North Africa, Southeastern Asia and Australia, just to name a few places. He also visited many of the same spots I did, just so that he could obtain similar experiences. I always had a soft spot in my heart for John, and I remember that he was one of the few people that would regularly write me letters during my travels. I always enjoyed hearing from him. For approximately one-year, John utilized his psychologist skills in the Lower Mainland, working in the Whalley District of Surrey, counselling drug addicts. Always searching out new experiences, John moved back to Toronto, Ontario where he commenced a fruitful career in sales management with several high-level, National and International technology firms."

"He was now kicking corporate butt, taking names and making 'big bucks'. While back there, he too married a very pretty lady and eventually fathered three children. A daughter and twin boys. John has now come full-circle by moving back to B.C. and living, literally, a few minutes' walk from where we all grew-up."

"You know Kid, if I've said it once, I've said it a hundred times. I'm not much on religion, but if someone asks me about my family and how I felt about them, I'd have to say that I was blessed. My parents and older brother Mike that have already departed, and Ken and John who are still around, were and still are, a wonderful group of human beings. I would like to think that they feel the same way about me."

Although McGuire was completely forthcoming regarding his travels, past experiences and exploits, Jesse was also fully aware that he was the exact opposite when it came to his emotions or personal feelings. The fact that he was now revealing that information to him, meant that he had somewhat let his guard down. Attempting to avert another somber mood developing, Jesse quickly attempted to lighten the atmosphere.

"Hey, Bob. That last travel journal brought us up to the end of 1978. Did you make any other major trips after that time?"

Quickly regaining his composure, McGuire responded. "Nah, no actual extended adventures after that last one to the Orient with Carol. But I still traveled here and there over the years. I visited my brother, Ken in Brisbane, Australia five or six times, Hawaii and a few other American states now and again, a few places within Canada and then finally, the FIFA World Cup Soccer Finals in South Africa in 2010. I guess that was the last big one. However, I never again returned to Europe, North Africa, the Orient, or the Caribbean. They just seemed to drop off my radar." Having made that final commentary, his last few words just trailed off.

Upon receiving those conclusive and anticlimactic comments from McGuire, Jesse had no further questions.

Chapter 11
The Eagle Has Flown

Now wearing a reflective expression, McGuire had already closed the cover of his final journal and gently placed it on top of the coffee table. Once again, they had conversed throughout the entire night and daybreak would soon be upon them. McGuire silently held his gaze out of the living room window for a considerable time-frame, but seemingly, not focusing on anything specific. Jesse remained seated on the couch while intently watching his friend. What was he looking at? Was he expecting someone? Without explanation, McGuire then suddenly rose from his chair and silently walked into his bedroom. Jesse remained seated for several minutes longer. However, when it appeared evident that his host would not be returning anytime soon, he assumed that McGuire had traded his favorite living room chair in favor of his bed. Jesse, now too left the living room and entered his own bedroom. He also needed a bit of a nap, and then afterward, there were still some things he wanted to check on his computer.

Sometime later that day, Jesse awoke to hear rustling in the living room and then the sound of music playing. Surprisingly, he recognized the singer and the song. It was Joni Mitchell and the tune was 'Free Man in Paris'. His mother also owned this album but it was on compact disc rather than vinyl. The album was called 'Court and Spark' and she played it frequently. Jesse could probably recite the words to every song on the album, since it was played at home so often, even when he was a small child. His mother never tired of it. This record was a bit of a departure for McGuire's musical preferences since this was mellow, emotional music with a hint of Jazz. During Jesse's short time on Texada Island, Mr. McGuire's mainstay music seemed to be psychedelia, hard rock and blues. Was he in another one of his pensive moods? Perhaps that accounted for his silence and unexplained exit from the living room after relating his final travel story.

Since he had not been summoned to join McGuire in the other room, Jesse tip-toed down the hallway to the bathroom. Returning to his room minutes later, he then decided to temporarily remain there and browse his computer. He had recently come to a personal conclusion and had made an important decision. He had some vital news to impart to both McGuire and his aunt Pearl, but he wanted to wait for the perfect time to inform them. He had also just forwarded his militant friend, or perhaps former friend, on the mainland an email. He wondered if a response would be forthcoming.

From the sound of the music playing on the classic Marantz stereo, Jesse instinctively knew that the final tune on Side A was nearing its conclusion. When it did, there was a momentary silence and then the sound of a replacement platter beginning to play. Amazingly, he recognized this one as well. It was the Beatles 'Rubber Soul' album. Although they were still just young kids when the Beatles disbanded in 1970, both his parents were huge fans, and owned many of their albums. Rubber Soul was one of them, and he often heard it played.

McGuire had still not enquired as to his whereabouts, so Jesse returned to his computer activities. As the afternoon waned and the natural light within Jesse's room grew dimmer, he had still not communicated with McGuire. However, his presence was still detected by the changes in record albums on the stereo and the occasional clinking of glass upon glass inside the kitchen. Jesse assumed that McGuire had been refilling his drinking glass with the abundant and much enjoyed Golden Gage plum wine from the gallon jug. During the past half hour or so, the landline telephone on the small table located in the entry hallway had rung twice. But each time, it remained unanswered. Did McGuire not hear it ringing, being so near the stereo, or was he just ignoring it? Jesse was aware that McGuire possessed a cell phone but he rarely utilized it. For him, it was more of an inconvenience than a benefit, so it was switched-off more often than not. As a result, he relied on the landline for outside contact. It was strange that he wasn't responding to the calls. Jesse too, had little use for cellular telephones, and in fact, did not own one. This was extremely unusual for a person of his generation, since most others of his age group not only frequently used these devices, they rarely put them down.

Jesse preferred his laptop computer. For him, this was his primary and only electronic method of communication.

It was now sufficiently faded within the house to warrant turning on the lights. For several minutes, the music in the living room had stopped playing and Jesse wondered whether or not that might be his opportunity to rejoin his friend. He rose to turn on the bedroom light. As he did, Jesse heard the landline in the hallway commence ringing for a third time. Now there would be no reason for McGuire not to hear the sound. When the ringing continued unabated, Jesse decided to answer it. As he loped from his bedroom into the hallway, he quickly glanced into the gloomy living room. What he observed was McGuire lying on the floor in front of the stereo, gently snoring. Beside him was his empty, overturned drinking glass. Jesse managed to lift the telephone's receiver at the climax of the fourth ring.

"Hello, Jesse," stated the voice on the other end of the line. It was his aunt Pearl. "I've been calling for a while now. Were you guys out somewhere?" Not wishing to lie to his aunt but also not wanting to inform her that McGuire was passed-out on the living room floor with a snoot full of wine. He responded with a half-truth.

"Sorry, Aunt Pearl. We're in the living room listening to music and we've got it cranked pretty loud. I guess we didn't hear the phone ring before. As a matter of fact, I only heard it ring now because we're in between albums, and Bob is in the bathroom."

Jesse didn't like the subterfuge but at least it was somewhat close to the truth. "Okay, no problem," Pearl acceptingly remarked. "Tomorrow is supposed to be a beautiful day and I'm not working. I was wondering if you and Bobby wanted to join me and the kids for a picnic. I was thinking on driving out to Shelter Point Park at about one o'clock and you two could meet us there. I'm going to make some fried chicken and potato salad as well as bring along a watermelon. I'm even go to make some Bannock since, I know that both you and Bobby really enjoy eating it."

"That sounds great, Aunt Pearl," replied Jesse. "I'm pretty sure that Bob would love an outing to break-up the boredom. I'll ask him when he gets out of the bathroom. Do we need to bring anything? Can I call you back in a few minutes?"

"Yes, that's fine. But please don't leave it too long because I need to start preparing the food. If you guys are coming, I need to make extra. If you want, you can bring something sweet for dessert."

"Alright, I'll call-back within the next few minutes," Jesse assured Pearl. Immediately upon replacing the receiver in its cradle, Jesse returned to the living room to find McGuire in the identical position as before. He was still breathing through his mouth and the atmosphere within the room was alcohol saturated. Jesse was certain that if he lit a match at that moment, they would probably have both been blown to 'kingdom come'. Jesse was aware that McGuire held his liquor well, so, to see him in this condition, he must have consumed a considerable amount. Several times, Jesse shook McGuire by the shoulder, and each time, a little firmer than the last. Eventually, he managed to gain a few grunts from the prone inebriate, and a gradual opening of unfocussed eyes. A couple of minutes later, Jesse managed to bring McGuire to a sitting position and asked him if he was hungry for dinner. The response he received was a slurred and muffled "Nooooo..." plus a slow-motion shaking of his head.

Jesse was not a physically strong person but he had acquired a certain newly found strength these past few weeks laboring for his mentor. Somehow by sheer weight of effort and determination, he managed to raise McGuire to his feet. He then positioned himself under the left armpit of his charge and draped the limp arm over his shoulder. Holding tightly onto McGuire's left arm by the wrist, he then very deliberately, but slowly began guiding him toward the master bedroom, while first turning on the light from the wall switch. Thankfully, McGuire's bedroom was right off the living room and as such, not a great distance to travel. However, it was still taking everything Jesse had within himself in order to breach that distance.

Luckily, McGuire was able to dispel a certain level of his inebriated state and became aware of this colossal effort being undertaken by Jesse. As a result, he was able to muster an added degree of mobility. Although still very foggy and unsteady in his own right, he somehow managed to assist Jesse in bringing him to the foot of his bed. However, once released by Jesse, he lacked the necessary power and coherence in order to keep himself upright. As a result, he fell face first on to the mattress but then instinctively rolled onto his back. Now becoming somewhat more focused from the exertion, he began mumbling. "It's not fair…it's not fair," McGuire repeated several times.

"What's not fair," enquired Jesse. McGuire did not respond but kept mumbling incoherently. Jesse realized that McGuire was about to pass-out, so he quickly posed his question. "Bob. Aunt Pearl called a while ago and asked if we wanted to picnic with her and the kids tomorrow afternoon. She's making all of the food. Are you up for that?"

There was a lengthy pause before McGuire's response. Did he understand the question? Finally, "Yahhh, dat shounds guud wiff me," was McGuire's pie-eyed reply. That was followed with, "I like dat Pearl. Sheee'z a wonnerful person an' a greaaat fren." Now that Jesse felt McGuire had sufficiently understood the question, he added, "Okay, I'll call her back and let her know that we're coming. By the way, what isn't fair?" Unfortunately, by then, Jesse had lost his audience to unconsciousness, and McGuire was once again in a foggy dream-state. His breathing was slow and steady while the timbre of his snoring gradually increased. After placing a blanket over McGuire's prostrate form, Jesse left the bedroom but kept the door open. He called back to Pearl and informed her of their acceptance to her invitation. Afterward, he turned-off the power to the stereo, returned all of the loose record albums to their appropriate sleeves, and carried the overturned but unbroken drinking glass to the kitchen sink.

Jesse then decided to watch T.V. in the living room after first eating. Maybe McGuire drank his appetite into oblivion, but Jesse was famished. He quickly scrambled a few eggs and slipped a couple slices of white bread into the toaster. It was hardly a gourmet meal but it was good enough for tonight. After cleaning-up after himself, he returned to the great room.

He had to admit that he was concerned for his host. Why did he once again become so broody after relating the final portion of his travels? And why did he drink so much this afternoon? As he had previously formulated, there was more to Bob McGuire than just the easy living, happy-go-lucky guy that he projected toward most people.

Throughout that evening, Jesse had chosen a variety of different shows from McGuire's vast library of programs and viewed them all in their entirety. Occasionally, he would peek his head into Bob's bedroom to ensure that all was well. It was. Near midnight, he found that he was beginning to nod-off and decided to call it a night. Rising from the couch, he stopped momentarily and looked toward McGuire's bedroom. Then walking into his own bedroom, he removed the top blanket and a pillow from the bed and returned to the living room. Tonight, he would sleep on the couch. He was probably over-reacting, but he wasn't going to take any chances.

The next morning, Jesse was awoken by the clattering of dishes from within the kitchen. Rousing to investigate, he found McGuire in his BVD's preparing breakfast. He looked a little worse for wear but Jesse was certain that he would never admit to it. "Good morning, Kid!" was his immediate reaction upon seeing Jesse enter the kitchen with a confused look on his face. "Why did you sleep on the couch last night?"

"I was worried about you," replied Jesse. "You were so drunk last night, I barely got you to bed. I wanted to make sure that you weren't going to choke on your own vomit during the night or something crazy like that. What happened?"

McGuire thought for a moment, perhaps reluctant to respond. Eventually, he did. "Thanks for helping me into the sack last night. But you never need to worry about me. I can't count the number of times that I've been 'gunned' on my wine and each time I'm just fine the next morning. I would never do anything so stupid as barf-up that golden nectar. It's like medicine to me. This next bit is kind of a secret so, I expect you to keep it to yourself."

McGuire then stopped speaking and intently looked at Jesse awaiting a reply. When Jesse nodded to the affirmative, he continued. "Just to let you know. Sometimes when I talk about my past travels or about so many of the great times I've had in my life, it makes me feel a bit somber. I want to relive them again but I know that I can't. It's not fair, but that's life."

There was those three words again, thought Jesse. "It's not fair." McGuire probably wouldn't recall uttering them last night in his sodden condition, but at least now Jesse was aware of the context in which they were made. "You may not know it to look at me but I'm actually not as young as many people think," reflected McGuire in a humorous tone, while displaying his near naked body to Jesse, and striking a body builder's pose.

"I'm getting older now and although I'm still in reasonably good health, I don't know how much time I have left. Some of my good friends have already left this world and are upstairs right now partying with Jimi and Janice and the rest of the cool heads that went before them. There's still so much that I want to do, but will there be enough time? That's why I get shit-faced every once in a while. It's not to drown my sorrows, I'm trying to relive the good times."

With this final statement McGuire shook his head again and muttered, "It's not fair."

Attempting to dispel the melancholic cloud that this conversation had now created, Jesse announced. "Hey Bob. Do you remember that Aunt Pearl invited us to a picnic with her this afternoon at Shelter Point Park? McGuire quickly responded that he did indeed remember. However, from the momentary look of confusion on his face before answering, Jesse wouldn't testify to that. McGuire consumed an extra-large breakfast that morning to compensate for last night's missed dinner. Jesse ate sparingly. He was fond of his aunt Pearl's cooking, so he wanted to ensure that he kept a good appetite for what was to come later that afternoon. Afterward, Jesse offered to clean the kitchen so that McGuire could take his time in the bathroom. McGuire didn't argue but merely left the table and walked straight down the hallway toward the lavatory."

Jesse took his time straightening-up the kitchen. When all was completed, he could still hear the shower running in the bathroom and understood that McGuire was still soaking away last night's excess. It would be a while yet before Jesse could claim his place in the bathroom. Entering the living room, he retrieved his blanket and pillow from the couch and returned them to his bedroom.

He then opened his laptop computer and began browsing. As he opened the in-box of his email account, the first message to catch his eye was from his school friend on the mainland. There was no title or remark in the header. Somewhat hesitantly, he opened the message and began reading. Within moments, a thin smile creased his face. This was a good start to the day and would lessen the stress of conversations yet to come.

Several minutes later, Jesse heard McGuire calling from the hallway. "Okay, Kid. The bog is all yours. Please excuse the smell but it felt like I was giving birth to a politician in there. I left the window open for you but you might want to give it a minute to air-out." Jesse called back that he understood and took the suggestion to heart. It was no big deal, he could wait a few more minutes, and then returned to his computer. The two men left the residence at 12:30 P.M. en route to their rendezvous with Pearl and her children.

First stopping at the convenience/grocery section at a nearby service station, the pair picked-up a couple bags of Dad's assorted cookies and a 2-liter bottle of Coca-Cola. That was to be their contribution for this afternoon's meal. It wasn't much, but at least the kids would be happy. They arrived right on time and soon located Pearl. She had secured a large picnic table in a shady portion of the park. The table was already bedecked with a checkered cloth and Pearl was just in the process of removing the food stored within her two large Coleman coolers and relocating it all, on top of the table. She gave them both a cheerful wave as she noticed their approach, while all three children ran to greet them.

Pearl definitely did not disappoint when it came to the lunch. The chicken was fried to perfection while the potato salad was comparable to what you could expect to be served at a Westphalian matriarch's Sunday table. The Bannock was the first to go with both McGuire and Jesse vying for the last piece. The children attempted to bypass the apple slices, raw carrot and celery sticks in favor of Dad's cookies but Pearl wouldn't hear of it. No fruit and veggies…no cookies. Reluctantly, they first ate their healthy food, and then literally attacked the cookies. After their meal, the afternoon evolved pleasantly with both Uncles Jesse and Bobby playing Frisbee and Nerf football with the kids, while Pearl quietly chuckled to herself as both men soon became winded from their exertions. The children were just getting started and still had plenty of steam. When the men exhaustively returned to the picnic table fully spent, the children ran to the water's edge and commenced skipping stones on its surface.

Jesse and McGuire caught their breath while Pearl repacked the remaining food into the Coleman coolers. It would stay fresher there, if anyone requested 'seconds' later. The adult's conversation remained animated and congenial. As the sun eventually crested the tree line, it became apparent that the day's activities were winding down. The children had since returned to the table and were becoming fidgety. Just as Pearl was about to suggest a termination to the outing, Jesse spoke up.

"Aunt Pearl. Do you remember asking me to inform you when I wished to return home?" Pearl alarmingly looked at McGuire and then hesitantly responded in the affirmative. "Well, if it's okay with you, I'd like to head back tomorrow." Now, both McGuire and Pearl searched each other's faces as if to ask, "What's going on? Did we do anything wrong?" Jesse saw the looks and immediately wished to dispel their anxiety.

"No…no. Don't get me wrong. Everything's fine. I just think it's time for me to be going back. You guys have both been great, so don't be upset." Pearl and McGuire were both caught off guard by Jesse's request but then realized that it was going to happen anyway, sooner or later. After all Jesse's life was on the mainland with his friends and family. Not here on Texada. Quickly recovering, Pearl responded, "Yah, sure. No problem. What time do you want to leave? I just need to know so that my neighbor can watch the kids again." Jesse replied that late morning would be fine. The trip wasn't a long one, so there was no need for anyone to rise earlier than necessary. At that moment, McGuire piped up. "If you can't make it Pearl, I can take him back." Pearl declined his kind offer and added that anyway, she wanted to see her sister, Willow again.

That settled it. Jesse would be leaving tomorrow. Knowing that he wouldn't be seeing Pearl's children again before his departure, he gave them all a firm embrace and a kiss to their foreheads. He also promised to return to Texada as soon as he possibly could. The return trip to the Leonard Road address in McGuire's Ford F150 was uneventful and quiet. Jesse noticed that his pilot's driving mannerism had now completely changed and he was actually motoring at the posted speed limit. He was also glad that McGuire was not bombarding him with questions or small talk. Jesse would be speaking his piece tomorrow and he was already internally rehearsing what he was going to say. When they finally arrived home, McGuire was upbeat and acting as though tonight was just another evening. "You want to watch a blockbuster movie tonight, Kid? I was thinking Lawrence of Arabia. It's a long one but a real barn-burner. I've got it recorded on my PVR."

This was now the third Friday in a row that McGuire was missing his weekly 'pub night'. He thought to himself, that the regulars are going to think he'd croaked. With Jesse visiting these past three weeks, he had deviated somewhat from his usual routine. However, that was alright. He could make up for lost time starting next week.

"That's fine with me," responded Jesse. And then added, "It's about time we had a little action around here. Even if it is only on the T.V."

McGuire countered that remark with one of his own. "If it's action you want, I'll take you on my next adventure. I promise that you'll have more action than you can handle." And with that comment, he raised both eyebrows. Jesse acknowledged McGuire's comment and facial gesture with his now patented head shake and laugh. He thought, *This guy will never change...and I'm glad.*

Jesse had to agree, for an older movie, Lawrence of Arabia was indeed very good. No pyrotechnics, no computer-generated images or soapy love interests with the lead characters. Just a well written screenplay, excellent cinematography and great acting. The two men were about to drain the remnants of their refreshments when McGuire suddenly raised his glass to propose a toast. "Here's to you, Kid. All the very best with the rest of your life, and don't be a stranger." Jesse raised his own glass in acknowledgement and merely nodded his appreciation. On that note, McGuire switched off the television set and both rose from their seats. As they placed their empty glasses in the kitchen sink, McGuire enquired. "What do you want for breakfast tomorrow, Kid? Seeing it's your last day and all."

Jesse replied, "Nothing special. Steak and eggs with hash browns and a tall stack of flapjacks will be just fine." McGuire immediately caught Jesse's sarcasm and started laughing, "Yah, right. Who was your slave last week?" Jesse then jumped-in again, this time in a more serious tone. "Nothing special, Bob. A bowl of porridge and a couple slices of toast will do."

"A man after my own heart," was McGuire's acceptance. "You got it." With that, both men separated and walked in opposite directions. McGuire turned-off the lights as he stepped past each switch.

The following morning dawned as had many recent others on Texada. The sky was partially overcast but not unpleasantly so. There was still plenty of blue to be seen and with the constant blowing of a gentle but steady breeze, those sporadically placed clouds would probably soon be gone. Jesse and his host did not rise any earlier than usual and with their unhurried breakfast nearing its end, it was time for one last cup of tea. Pearl would probably be arriving soon.

As the last of the dishes were being dried and returned to their rightful places of storage, there was a knock on the front door. McGuire went to answer, while Jesse completed putting away the clean dishes. He immediately recognized Aunt Pearl's voice as McGuire opened the door. He also called-out a greeting and stated that he would be ready in a few minutes. "No rush," exclaimed both Pearl and McGuire almost simultaneously, "Take your time," they parroted. McGuire then asked Pearl if she wanted a cup of tea while Jesse entered his bedroom. She declined the offer. Jesse had little to pack, and most of that was done last night. All he needed to accomplish now was a quick brushing of his teeth.

Within a few short minutes, Jesse entered the living room carrying his belongings. Pearl was seated on the couch while McGuire occupied his favorite chair. "I'm glad you're sitting down. There's something that I want to say to you both." Jesse took a moment to compose himself and clear his thoughts while McGuire and Pearl looked at one another inquisitively. "I want to thank you both for what you've done for me these past few weeks. First, Aunt Pearl. You and my parents all stuck your noses in my personal life. Although, I resented that at first, I know now that it was done out of love, and not because you wanted to control me. Thank you."

"Bob. Never in my wildest dreams did I ever imagine that I would meet a person like you. If someone would have told me three weeks ago that I would be living with a geriatric hippy on an island in the Georgia Strait, tending marijuana plants and doing manual labor, drinking plum wine every night and listening to his former travel adventures as though he were some modern-day Marco Polo, I wouldn't have believed them."

"You've taught me about living and you've taught me about life. You've shown me that hard work is honorable and that you can never make assumptions about strangers or 'judge a book by its cover'. You've taught me that most people, no matter where they're from, are generally good. That tolerance, kindness and generosity still exist, but so does evil. You've shown me the difference between love and lust, although in your case, lust played the bigger role in your life."

At that point, Jesse chuckled, stopped speaking momentarily and waited for a reaction from McGuire. He didn't have long to wait, as he quickly received a stern look from him, as well as his middle finger. Pearl feigned a look of shock and slapped McGuire's shoulder. He wondered if she could tell that he was blushing.

"To simply say thank you, is like saying that the Pyramids at Giza are just a bunch of stone blocks stacked on top of each other. I'll never forget the time I've spent here and I'd like to think that I would still be welcome to visit some other time in the future."

"'Anytime,' was McGuire's one-word reply. By now all this outpouring of love and thanks was getting to Pearl and she was becoming teary-eyed."

"I've got one last thing I'd like to say," continued Jesse. "This past little while I've been messaging back and forth with my dad on the computer. It's kind of been our little secret and even my mom doesn't know. Anyway, we've been discussing a few options for my future and I told him what would interest me. So, after all that, I've decided to go back to school in September. I completed the enrollment application form for U.B.C. on-line and they accepted me."

Upon hearing this revelation, Pearl sprang out of her chair and literally squealed. She ran up to Jesse and grabbed him in a massive bear hug. Now she was crying in earnest, out of sheer joy. "Oh, Jesse. That's great news," she blubbered. "Let the man finishing speaking!" interjected McGuire.

"I've decided that I want to become a 'Human Rights Lawyer'. There's a lot of injustice in the world and I'd like to play some role in defeating it. Dad has a friend that works in that field at the International Ministry of Justice in the Haag, Netherlands. Apparently, he's some big mucky-muck there. Dad has been speaking with him and he has agreed to take me on there as an intern, as soon as I complete my university studies and pass the Bar. It's still a few years away, but at least now, I've got a plan. Even my best friend on the mainland likes the proposal, because he said now, I can help my own People. He still hasn't changed his radical ideas but maybe in time, I'll be able to soften them a bit."

Now it was McGuire's turn to stand and approach Jesse. He held out his right hand and firmly shook Jesse's as it was proffered. For the next several minutes, the trio excitedly discussed Jesse's plans and the bright future that now lay before him. After the initial ferment relaxed, Jesse informed them that he needed to leave. "Here, let me grab your bag," was McGuire's immediate reaction.

"No thanks, Bob," replied Jesse. "I need to start doing things for myself now." McGuire gave him a firm clap on the back and said, "Good man!"

"All three exited the house and as Pearl opened the hatch to her old blue and white Bronco, Jesse placed his bag inside. Pearl looked at McGuire but was too emotional to say anything. She would thank him properly later, when she was better composed. With a thin smile and a brief nod of her head, she turned and entered the driver's side of the Bronco. Jesse stood before McGuire, but like his aunt, he too was unable to speak. He didn't trust opening his mouth for fear of losing his own composure. Instead, he walked to McGuire and embraced him."

Nothing was spoken between the two men but through this gesture, everything was said. As they momentarily held that posture, a single and distinctive screeching cry was detected overhead. Then, moments later, a second. As the men loosened their hold and looked skyward, they observed two adult bald eagles gracefully circling the property. The majestic raptors rode the air currents with the elegance and agility of ballet dancers in their aerial display. They passed high above several times to the delight of their enchanted audience, before finally soaring away.

Looking at Jesse, McGuire asked if he was familiar with Auguries. When Jesse informed him that he was not, he then proceeded to explain. "Augurs were the ancient Roman equivalent of what we know today are psychics or clairvoyants. In Roman times, the ever-powerful Emperor or other influential citizens would hire an Augur to predict their future. Especially if they were about to do something important like go to war or get married, purchase a new home, or something similar. The Augur would sometimes sacrifice an animal such as a lamb or a bull and then gut it afterward. Taking out the animal's entrails and other organs, the Augur would then look for signs within, good or bad. Other times, the Augur would check around from where he was standing, and look for signs in his vicinity, to predict the future. A murder of crows flying overhead or a black cat crossing his path could mean bad luck. While a flock of doves crossing the heavens or a white stallion trotting by, could mean good luck."

"Right now, I'm going to become an Augur and predict that those two eagles we just witnessed flying overhead, are a good omen. After all, your last name means 'eagle-heart' and you're about to embark upon an important decision that will help solidify your future. In my opinion, this was a sign that everything you decide, or do from now on, is going to work out just fine."

Jesse threw McGuire a gratifying smile and thanked him for his positive prediction. Verbally acknowledging him one more time and vowing to speak again soon, Jesse then spun on his heels and walked toward the passenger side of the Bronco. Without turning around, he boarded the vehicle and it slowly began moving forward down the driveway toward Leonard Road. Pearl honked the horn twice as she reached the end of the driveway and then they disappeared.

Finally, now alone again, McGuire surveyed his property. Aside from some brush and debris having been cleared away by Jesse, not much had really changed during these past three weeks. However, then again, everything had changed. He walked toward the wooden garden bench selectively positioned in the front yard and slowly sat down. The sun was shining and the air was gradually beginning to warm. As he gazed upward, all he could see now was pale blue sky. The clouds had vanished. As to the twin eagles, they had not reappeared. He decided to remain seated on the bench a while longer.

McGuire found it very pleasant relaxing in his front yard on this Saturday morning, and the rejuvenating warmth of the sun was beginning to soothe his aching joints. As he held his gaze forward, his eyelids gradually became heavy, and he struggled to keep them open. Dreamily but unquestionably, the familiar images surrounding him gradually altered, as if props on the stage of a theatrical production were being maneuvered by invisible handlers.

Inexplicably, he was now sitting atop a large, smooth boulder at the AUB Beach, overlooking Zaituna Bay. To his right, where once stood a garden shed, he could now clearly see the S.S. Independence in all her nautical splendor. She was firmly secured to the pier at the Beirut dockyards. Scores of tiny people populated the wharf and were climbing the gangway, as if intent on entering her inner sanctum. Behind him where his house previously stood, was now the unmistakable image of the American University of Beirut campus, on Paris Road. This was the bastion of higher learning he never actually attended, but who's very existence allowed him to enjoy hours of bliss in the sand levelled at her foundation.

Squinting his eyes in order to avoid the blinding glare of sunlight on the shimmering water before him, he suddenly beheld a wondrous sight. Gradually emerging from the surf, and gracefully striding toward him in all her naked magnificence, was Antoinette. The watery sheen enveloping her mocha-shaded skin caused her to glisten, as if she were some highly polished, life-sized gem. As she drew ever nearer, her perfect beauty dazzled him.

Smiling, she raised both arms in welcoming promise. Blood coursed rapidly through his veins, and his heart began to race, as he too now rose from his perch in anticipation of embracing her. Before long, she would once again be smothered in his loving caress and they would be forever joined. This was all he ever wanted. All he ever needed. And with the fulfillment of that realization, he was content.

The End...Or Is It?